AF873663

UNITARY SYMMETRY AND ELEMENTARY PARTICLES

Second Edition

UNITARY SYMMETRY AND ELEMENTARY PARTICLES

Second Edition

D. B. LICHTENBERG

Physics Department
Indiana University
Bloomington, Indiana

1978

ACADEMIC PRESS New York San Francisco London
A Subsidiary of Harcourt Brace Jovanovich, Publishers

COPYRIGHT © 1978, BY ACADEMIC PRESS, INC.
ALL RIGHTS RESERVED.
NO PART OF THIS PUBLICATION MAY BE REPRODUCED OR TRANSMITTED IN ANY FORM OR BY ANY MEANS, ELECTRONIC OR MECHANICAL, INCLUDING PHOTOCOPY, RECORDING, OR ANY INFORMATION STORAGE AND RETRIEVAL SYSTEM, WITHOUT PERMISSION IN WRITING FROM THE PUBLISHER.

ACADEMIC PRESS, INC.
111 Fifth Avenue, New York, New York 10003

United Kingdom Edition published by
ACADEMIC PRESS, INC. (LONDON) LTD.
24/28 Oval Road, London NW1 7DX

Library of Congress Cataloging in Publication Data

Lichtenberg, Don Bernett, Date
Unitary symmetry and elementary particles.
Second edition.
Includes bibliographical references and index.
1. Eightfold way (Nuclear physics) I. Title.
QC793.3.S6L5 1978 539.7'21 77-92306
ISBN 0-12-448460-3

PRINTED IN THE UNITED STATES OF AMERICA

CONTENTS

PREFACE TO SECOND EDITION

In November 1974, two groups of high-energy physicists, working on opposite sides of the United States, announced the discovery of a new elementary particle. The ramifications of this discovery were so great that the leaders of the two groups, Sam Ting and Burt Richter, jointly received the 1976 Nobel Prize in Physics. The existence of the new particle, named the J on the east coast and the ψ on the west coast, provided indirect evidence that a new quantum number was needed in elementary particle physics. Less than two years later, direct evidence for the existence of the new quantum number, called charm, was obtained.

The discovery of charm, about a decade after it was predicted to exist by a number of theoretical physicists, required an enlargement of the $SU(3)$ symmetry scheme of Gell-Mann and Ne'eman. The most plausible enlargement of the symmetry is to $SU(4)$.

These important discoveries have led me to revise my book on unitary symmetry to include charm and $SU(4)$. Because nature may not stop with $SU(4)$, I have stressed methods which work for any $SU(n)$. Of course, the relevant symmetry of nature may turn out not to be $SU(n)$, but something else. If so, many of the techniques illustrated in this book will lose some of their physical interest, although I hope that even then the group $SU(n)$ will be considered amusing to study for its own sake.

Another feature of this new edition is that I have revised and enlarged the chapter on the quark model. I have done this, not only because of charm, but because more and more phenomena appear to have a natural explanation if hadrons (strongly-interacting elementary particles) are indeed composed of quarks. Furthermore, I have considerably expanded the idea, which I discussed only briefly in the 1970 edition, that each kind or "flavor" of quark comes in three "colors." If there are n different quark flavors, the most natural underlying symmetry appears to be $SU(n)$, and the most natural underlying symmetry of the three colors appears to be $SU(3)$. Thus, the unitary symmetry of elementary-particle physics may have as its underlying cause the existence of colorful and flavorful quarks bound into hadrons. At the very least, the quark model provides excellent applications of the ideas and methods which I have treated.

PREFACE TO FIRST EDITION

The role of symmetry has come to be emphasized in the study of the physics of the elementary particles. One reason for this, as pointed out by Wigner, is that we do not have a satisfactory dynamical theory to describe the interactions of the elementary particles. Despite this lack of detailed knowledge of dynamics, much information can be obtained by studying the symmetry properties of elementary particle interactions. However, there is a further reason to study the symmetry properties of interactions: The beauty and simplicity of symmetry properties make their use aesthetically rewarding and valuable, even in cases for which a dynamical theory exists. Many results which can be calculated only with difficulty by a dynamical theory follow quite simply from the exploitation of the symmetry properties of a physical system.

In the years since 1936, when Breit and his collaborators postulated that nuclear forces were charge independent, symmetry under rotations in the space of isospin has been well established as an approximate symmetry of the strongly interacting particles. However, after the discovery of hyperons and heavy mesons, physicists began to look for a larger approximate symmetry.

The first important step in this direction was taken by Gell-Mann and independently by Nishijima in 1952, who postulated the existence of the quantum number of strangeness. Another important contribution was made in 1956 by Sakata and his collaborators, with a model based on primacy of the proton, neutron, and Λ hyperon. Sakata's followers in 1959 incorporated his model into the framework of the group $SU(3)$, or, as it is commonly called, unitary symmetry. Then, in 1961, came the basic work of Gell-Mann and Ne'eman who independently proposed a classification of the strongly interacting particles based on the octet version of unitary symmetry. Subsequently a number of still larger symmetries have been proposed, such as $SU(6)$, but with less success. In 1964 Gell-Mann and, independently, Zweig proposed the quark model. This model has since begun to form the basis of a simple dynamical approach to the interactions of elementary particles.

The origin of the present book was a special series of lectures on unitary symmetry and elementary particles which I gave at the University of Nebraska in 1965. Two of my colleagues there, D. W. Joseph and T. A. Morgan, kindly tape-recorded the lectures and edited them into a set of notes. Subsequently, the notes were considerably expanded into a one-semester course for graduate students at Indiana University. This book is based on the notes of that course, revised and brought up to date. The book as now constituted bears little resemblance to the notes of the original lectures.

In writing the book, I have assumed that the reader has had a course in quantum mechanics and some exposure to the theory of abstract groups and group representations. However, for convenience, I have summarized a number of the most useful concepts in the early chapters. Thus, I have quoted a number of theorems without proof in order to remind the reader of them. Often I have given references to the places where the proofs can be found. Occasionally, I have sketched a proof, usually without rigor, to bring out the flavor of the subject.

Considerable space is devoted to a description of techniques using Young tableaux. Most of these methods I discovered for myself, and therefore, my treatment may not be completely standard. I make no claim that my results are new, but I hope that I have at least given methods which are easy to understand and apply.

In my discussion of experimental matters, I have stressed a few examples of $SU(3)$ multiplets which are reasonably well established. Any attempt to include all elementary particles would become rapidly outdated. Periodically, up-to-date lists of elementary particles are published, and anyone can play the game of trying to classify them into $SU(3)$ multiplets.

Of the many topics on the quark model, I have arbitrarily selected a few of interest to me, especially those I have worked on myself. I have also omitted a wealth of other material for various reasons. I have not discussed attempts to generalize $SU(6)$ to the relativistic case because of the conceptual difficulties and contradictions this program has encountered. I also did not treat in detail the role of $SU(3)$ in the weak interactions, partly because I did not have time to discuss this in one semester, and partly because I did not wish to devote much time to the concept of currents in a course devoted primarily to symmetry.

Neither in my lectures nor in this book have I aimed for completeness, either in content or in references. But I have given references to many of the important works which either provide additional background information for the reader or enable him to pursue the subject more deeply. I particularly hope that the reader is encouraged to undertake the quest for a greater understanding of the nature and scope of unitary symmetry in physics.

ACKNOWLEDGMENTS

I am grateful to many physicists for their valuable discussions, especially to John Challifour, Jerrold Franklin, Gordon Frazer, Archibald Hendry, David Joseph, Tom A. Morgan, Yuval Ne'eman, Roger Newton, Enrico Predazzi, Larry Schulman, and John Wills. I thank Arthur Rosenfeld for permission to reproduce a table from the publication of The Particle Data Group, which appeared in the *Reviews of Modern Physics*. Those portions of my own work which are described in this book were supported in part by Indiana University, the National Science Foundation, and the Department of Energy.

UNITARY SYMMETRY AND ELEMENTARY PARTICLES

Second Edition

CHAPTER 1

INTRODUCTION

"... *if we knew all the laws of nature, or the ultimate Law of nature, the invariance properties of these laws would not furnish us new information.*"

—Wigner

1.1 Uses of Symmetry

We are fortunate that there exist today the theories of electromagnetism and gravitation that can be used to describe the behavior of certain physical systems. It is useful to classify the predictions of such theories into two main types: first, the predictions which follow from the detailed dynamics of the theory, and second, those which follow from the invariance properties or symmetries which the theory satisfies.

To illustrate these two kinds of predictions, we consider an example from the domain of quantum electrodynamics. This theory enables us to calculate, for example, the differential cross section for electron–electron scattering with both beam and target particles polarized in the same direction. If we calculate this cross section both for electrons polarized parallel and antiparallel to the beam direction, we shall find we get the same answer—the two cross-sections are equal. But we can obtain this result immediately by noting that the theory is invariant under rotations and under permutations of the two electrons. Thus, by making use of the symmetry properties of the theory, we can make certain predictions more simply than by going through detailed calculations which incorporate the symmetry at every step.

In this example, the use of symmetry to obtain the equality of the two cross sections is convenient, but not necessary. However, in some other cases, the use of symmetry is essential, either because the system is so complicated that we cannot perform the calculations or because a dynamical theory does not exist. For example, in the case of proton–proton scattering, we do not have a theory which enables us to make an accurate calculation of the differential cross section. However, we can again make use of symmetry to predict that the cross sections will be equal for protons polarized parallel and antiparallel to the beam axis.

Thus, at the present time, when we do not have a good dynamical theory of some of the interactions of elementary particles, it is especially important to understand the consequences of symmetry.

1.2 Symmetries and Conservation Laws

It is a striking fact of nature that among the many properties of physical systems that continuously change with time, a few properties remain constant. These constant properties appear in so many different physical systems that they are among the most fundamental laws of physics, and are known as conservation laws. In addition to those properties which, so far as is known from experiment, are exactly conserved, there are other properties which are only approximately conserved. The oldest known conservation laws are those of linear momentum, angular momentum, and energy. A familiar conservation law which holds only approximately is the conservation of parity.

Another remarkable fact about physical systems is the symmetries or invariances they possess under certain transformations. Again, some of these symmetries are exact,[1] and some are only approximate. We have already mentioned the invariance of physical systems under rotations and permutations of identical particles. Other familiar examples are the translations in space and time. An example of an approximate symmetry is symmetry under space inversion. We shall be concerned primarily with a symmetry which holds only approximately—unitary symmetry. However, for clarity, we shall treat the concept of symmetry more generally, and discuss other symmetries in brief.

[1] Whenever we refer to a symmetry or conservation law as exact, we mean only that no violation has been observed. It is possible that future experiments will show that a symmetry now thought to be exact is really only approximate.

The intimate connection between symmetries and conservation laws, two apparently unrelated properties of physical systems, was first noticed in classical mechanics by Jacobi in 1842.[2] In his paper, Jacobi (1884) showed that for systems describable by a classical Lagrangian, invariance of the Lagrangian under translations implies that linear momentum is conserved, and invariance under rotations implies the same of angular momentum. Still later, Schütz (1897) derived the principle of conservation of energy from the invariance of the Lagrangian under time translations. Herglotz (1911) was the first to give a complete discussion of the ten constants of the motion associated with the invariance of the Lagrangian under the group of inhomogeneous Lorentz transformations. For systems not describable by a Lagrangian, the situation is more complicated (Van Dam and Wigner, 1965), and we shall not discuss it here.

Now let us consider the connection between the symmetry of a system and a conservation law in the Hamiltonian formalism, which is more convenient in quantum mechanics. In both classical and quantum mechanics, the conservation of linear momentum, angular momentum, and energy follow from the symmetries of the Hamiltonian under translations, rotations, and time translations. More generally, in quantum mechanics, whenever a conservation law holds for a physical system, the Hamiltonian of the system is invariant under a corresponding group of transformations. The converse of this statement is not true, as even if the system has a Hamiltonian which is invariant under a group of transformations, there may not be a corresponding conservation law. An example is invariance under time reversal.

Wigner (1959) showed that all symmetry transformations of a quantum mechanical state can be chosen so as to correspond to either unitary or antiunitary operators. It is the unitary transformations which are associated with the conservation laws. However, if a Hamiltonian is invariant under an antiunitary transformation, such as time reversal, this has other consequences which can be tested against experiment.

If a Hamiltonian describing a system of interacting elementary particles is invariant under a unitary transformation, then it can be shown that the scattering matrix or S matrix of a system is also invariant under the transformation. The importance of the S matrix in elementary particle physics is well recognized. In fact, some physicists have speculated that perhaps a Hamiltonian describing elementary particle interactions does not exist, and that the laws of physics are contained in the properties of the S matrix.

[2] The history given in this paragraph is based on a footnote in a paper of Van Dam and Wigner (1965) who attribute the historical study to Eugene Guth.

1.3 Symmetries and Groups

A set of symmetry transformations on a physical system has the mathematical properties that are associated with a group. For example, let us consider some properties of rotations.

(1) If we rotate a physical system and follow this rotation by a second rotation, the resulting transformation is also a rotation.

(2) Suppose we perform three successive rotations. It does not matter whether we perform the first rotation and then a rotation equal to the resultant of the other two or perform the resultant of the first two and then the third. In other words, rotating a system is an associative operation.

(3) We may rotate a system through zero degrees so as to leave its position unchanged. This is an identity operation.

(4) If we rotate a system through a certain angle, we may then rotate it back to its original position. The second operation is an inverse operation to the first.

These four properties of rotations are just the properties of a group. The translations in space and time, and indeed most symmetry transformations on physical systems, also have the group properties. It is for this reason that the study of group theory is so important for understanding the consequences of the symmetries of physical systems.

Let us now look at still another property of rotations. If we perform first one rotation on a system and then a second, the system does not necessarily end up in the same orientation as if we perform the rotations in the reverse order. In other words, the operation of rotating a system is not commutative. On the other hand, rotations about a single axis lead to the same result when performed in any order; similarly, order does not count when performing successive translations. We must therefore study groups whose members commute and those whose members do not.

The rotations and translations in space and time may be made through any angle or any distance. For this reason, there are a continuous infinity of transformations which leave certain physical systems invariant. These transformations correspond to continuous groups. In fact, the continuous transformations in which we are chiefly interested are even more restricted; they form sets of transformations corresponding to Lie groups.

One difference between the rotations and translations is that the rotations vary over a finite angular domain, while translations in space and time (in a flat space) vary over an infinite domain. It turns out that the corresponding groups have many different mathematical properties.

To the rotations and the translations in space and time may be added Lorentz transformations, which are also presently believed to be exact symmetries of elementary particle systems. Together, the translations in space and time, the rotations, and the Lorentz transformations, correspond to the inhomogeneous Lorentz group, or Poincaré group.

In addition to the symmetries which physical systems exhibit under certain continuous transformations, physical systems are invariant (or at least approximately so) under certain discrete transformations. These transformations usually constitute a finite set, and appropriate to the study of these symmetries is the theory of finite groups.

1.4 Eigenstates, Quantum Numbers, and Selection Rules

An invariance or symmetry of an interaction under a transformation does not always lead to useful restrictions on the final states that can be reached from a given initial state. However, the consequences of the symmetry of an interaction are particularly striking for certain special quantum mechanical systems, which are represented by the eigenstates of observable conserved operators.

If a physical system is represented by an eigenstate of several such observable operators, then the system can be characterized by a corresponding set of quantum numbers. In any transition of the system from an initial state to other states, only final states with the same values of the quantum numbers are reached. Therefore, there are *selection rules* against transitions to states with different values of the quantum numbers. These selection rules enable us to make predictions about which reactions are possible and which are not. For example, we know that in any reaction the total charge Q of an isolated system remains a constant; or in other words, we have the selection rule $\Delta Q = 0$ in the reaction. This is only one of a number of selection rules which are known to be satisfied in the interactions of elementary particles.

As we have already mentioned, although some symmetries are exact so far as is known, others are only approximate. As an example of an approximate symmetry we may consider the invariance under rotations in isospin space or conservation of isospin. Isospin is believed to be conserved in strong interactions, but it is known that it is not conserved in electromagnetic interactions. Therefore, if we have a reaction involving strongly interacting particles we have the approximate selection rule $\Delta I = 0$. The fact that the selection rule holds to a good approximation is a reflection of the fact we believe that the symmetry holds to a good approximation: That is, that the extent of the violation is of order of the fine structure constant $\alpha = 1/137$.

However, we can also have a selection rule which holds to a good approximation for a dynamical reason. That is, the fact that a particular selection rule holds well may not reflect an underlying symmetry at all, but may be a dynamical accident. For example, in atomic physics there is the well-known approximate electric dipole selection rule for radiative transitions of excited states of atoms. The selection rule states that the orbital angular momentum L of the atom changes by one unit during the transition. This of course is consistent with conservation of angular momentum because the photon carries off one unit of angular momentum. However, the selection rule does not reflect any basic symmetry but the accident that, at low energy, the photon tends to carry off only one unit of angular momentum. If the energy of the transition were much larger we would have quadrupole and higher multipole transitions. In fact, such transitions are observed in nuclear physics, where the gamma rays are much more energetic. So whenever we are confronted with an approximate selection rule we have the additional question of whether it is the result of true approximate symmetry or of a dynamical accident which holds in a limited energy range. If the latter is true, we call the symmetry an approximate dynamical symmetry.

1.5 A Listing of Symmetries

We now briefly discuss the symmetries of interactions, some of which we shall treat in more detail later. We also discuss their associated conservation laws if they exist.

As in classical mechanics, the symmetry under inhomogeneous Lorentz transformations is connected to ten conserved quantities. These are the momentum-energy four-vector p_μ, the three components of the angular momentum $\mathbf{J}$, and three conserved quantities which are connected with pure Lorentz transformations or *boosts*. By a boost we mean a transformation to a coordinate frame moving with constant velocity with respect to the original system, but which is unrotated and which coincides with the original system at some arbitrary time equal to zero.

The conservation laws connected with boosts do not appear to have as much use in physical problems as conservation of angular momentum. The reason is that most physical systems that we shall consider are in eigenstates of the total angular momentum and its z component. Since the operators corresponding to the components of angular momentum do not commute with the operators connected to boosts, a physical system cannot be in an eigenstate of two of these operators simultaneously. Thus we usually do not get useful selection rules from the conservation laws connected to boosts.

The symmetry under the group of inhomogeneous Lorentz transformations is usually assumed to be an exact symmetry, at least so far as elementary particle physics is concerned. If this symmetry is exact, then space is flat. But according to the theory of general relativity, it is possible for space to be curved. Perhaps then, rather than considering the Poincaré group, we should consider a larger group, of which the Poincaré group is only an asymptotic limit. However, we shall not consider this possibility.

Other important symmetries are invariances under global gauge transformations, called gauge invariance of the first kind. These symmetries are associated with conservation of charge Q, baryon number B, and lepton number L. There is extremely good evidence to support exact conservation of charge and baryon number except possibly on a cosmological scale. Also, no reaction in which lepton number is not conserved has been seen, but the weight of evidence is not nearly so great as for charge and baryon number conservation.

It appears useful at present to write the lepton number as the sum of at least two numbers, the electron number L_e (the sum of electrons plus electron neutrinos) and the muon number L_μ (the sum of muons plus muon neutrinos). Because there is some evidence that more charged leptons exist, it will probably be advantageous to include additional terms in the sum, that is, we write $L = L_e + L_\mu + \cdots$. The electron number and the muon number are themselves conserved, although some physicists have speculated that the separate conservation of L_e and L_μ is only approximate.

In addition to being invariant under a global gauge transformation, the electromagnetic interaction is invariant under a local gauge transformation, or, in other words, a transformation which can be different at each space–time point. This so-called gauge invariance of the second kind has as a consequence that charged particles are coupled to a massless vector field—the field whose quanta are photons. It has been suggested that if a Hamiltonian is invariant under a global gauge transformation, but not under a local one, then the symmetry may not be exact. In fact, it has been speculated that conservation of baryon number and lepton number are only approximate conservation laws.

Next, consider symmetries connected with the permutation of identical particles. All observable quantities of physics must be symmetric (that is, must not change) under the permutation of identical particles. This appears to be obvious, for if it were not so, we could distinguish between two particles by the observable which was not symmetric under their interchange. Although observables must be symmetric under the interchange of identical particles, quantum mechanical states (which are not observables) need not have this symmetry. However, since observables are operators which act on quantum mechanical states, the states themselves must have definite symmetry properties under interchange of identical particles.

As far as is known from experiment, identical particles of integral spin are always in states which are symmetric under the interchange of any two of the particles (Bose statistics), while identical particles of half-odd-integral spin are in states which are antisymmetric under this interchange (Fermi statistics). This property of states of identical particles is the content of a theorem, called the spin-statistics theorem, which can be proved from the postulates of certain field theories. See, for example, Streater and Wightman (1964). Despite the fact that no two identical elementary particles have ever been seen in states with other than the usual symmetry properties, we shall discuss later hypothetical states which possess more complicated symmetry properties under permutations (parastatistics). Of course, whether or not states obeying parastatistics exist, all observables must be symmetric under the interchange of two identical particles. Even if states with complicated permutation symmetry do not exist, consideration of such states is very important. This is because the states may have complicated symmetry properties under the interchange of only some of the coordinates of identical particles.

Most of the invariances under discrete transformations, other than permutation symmetries, seem to be approximate. Invariance under inversion of space coordinates, denoted by the symbol P, is an approximate symmetry. Invariance under space inversion is related to the law of conservation of parity. Also approximate is the symmetry under the transformation of replacing particles by antiparticles, called charge conjugation and given the symbol C. The conservation law related to this symmetry is called charge conjugation parity or C parity.

The symmetry corresponding to inverting space coordinates and performing charge conjugation, denoted by CP, is observed to hold experimentally, in some weak interactions at least, to a greater approximation than either C or P separately. However, the experimental evidence on K-meson decay by Christenson *et al.* (1964) is most simply interpreted as evidence that CP is also an approximate symmetry.

Next consider symmetry under inversion of the time coordinate. This symmetry is called time inversion or time reversal, and is denoted by T. Since, in quantum mechanics, time inversion is an antiunitary transformation, there is no corresponding conservation law. In a Lagrangian or Hamiltonian theory, the effects of time inversion make themselves felt as reality conditions on the coupling constants entering the theory.

The experiments on K-meson decay have been interpreted by Casella (1969) as implying that T is not an exact symmetry in weak interactions. Independently of how good this evidence is for T violation, many physicists expect this symmetry to be approximate. The reason is that in certain field theories, a theorem can be proved in which the product CPT is an exact symmetry. See, for example, Streater and Wightman (1964). If the theorem applies to physical systems, then CP noninvariance implies T noninvariance. Whether or not the postulates of these field theories are true, thus far there is no experimental evidence for any violation of CPT. Since C and P correspond to unitary transformations and T to an antiunitary transformation, CPT symmetry is antiunitary, and, like T, it does not correspond to a conservation law. However, invariance under CPT does have some very important consequences; for example, one consequence is that the mass of a particle is equal to the mass of its antiparticle.

The remaining symmetries we discuss are approximate, and seem to have most to do with the strong interactions. However, some of these symmetries appear to be conserved in electromagnetic interactions as well. The charge Q (in units of the proton charge) of every strongly interacting particle (or hadron) discovered thus far can be written as a sum of three numbers: I_3, called the third component of isospin (for reasons to be discussed shortly); $\frac{1}{2}Y$, where Y is called hypercharge; and C, called the charm. (Sometimes the strangeness S is used as a quantum number instead of Y. The strangeness is defined by the equation $S = Y + C - B$.) The conservation of charge, as we have said before, is exact as far as is known, and corresponds to invariance of interactions under a gauge transformation. However, the strong and electromagnetic interactions, but not the weak interactions, seem to be invariant under three gauge transformations corresponding to the conservation of I_3, Y, and C separately.

The need for the quantum number I_3 arose in 1932 with the discovery of the neutron, and was introduced by Heisenberg (1932). The need for Y (or S) arose in 1947 with the discovery of strange particles, although the quantum number was only introduced several years later by Gell-Mann (1953) and Nishijima and Nakano (1953). The need for the quantum number C arose in 1976 with the discovery of charmed particles, although experiments performed in 1974 already suggested that this quantum number should exist. Actually, charm, unlike strangeness, was introduced into the literature by theoretical physicists (e.g. by Bjorken and Glashow, 1964) before its experimental discovery.

As of today, physicists have not discovered any principle which says that the charge of any hadron can be written as a sum of three numbers. In the absence of such a principle, we should not be surprised if new hadrons are discovered in the future which require additional quantum numbers for their specification.

Although conservation of I_3, Y, and C can be considered to be conservation laws associated with gauge transformations, they also can be considered as certain symmetry operations in a larger group. There are three such larger groups which seem to be of particular importance: the special unitary groups in two, three, and four dimensions, denoted by $SU(2)$, $SU(3)$, and $SU(4)$, respectively. The unitary group $U(n)$ is defined as the group of $n \times n$ unitary matrices; and the special unitary group

$SU(n)$ is the subgroup of unitary matrices which have determinant unity.

The group $SU(2)$ is locally isomorphic to the rotation group in three-dimensions $R(3)$. Because of this, we can define a total isospin[3] I in an abstract three-dimensional space, analogously to the total angular momentum J in ordinary space. (In fact, the group $SU(2)$ is also appropriate to describe ordinary angular momentum.) The vector $\mathbf{I}$, which is conserved if the interactions are invariant under rotations in this space, has components I_1, I_2, and I_3, the last of which we have mentioned previously. Since I_3 is related to the charge of a strongly interacting particle, a rotation about the first or second axis in isospin space is a transformation which changes the charge of a state.

Since two particles which have different values of I_3, but are otherwise identical, have different values of the charge, their electromagnetic interactions are different. Thus, isospin invariance is an approximate symmetry of strongly interacting particles which is broken by the electromagnetic interaction. In the weak interactions, not only I, but I_3 is also an approximate or broken symmetry.

By incorporating Y with I into a still larger group $SU(3)$, we have an even more approximate unitary symmetry, because $SU(3)$ is broken even in strong interactions. Still more badly broken is the group $SU(4)$, which incorporates C in addition to Y and I.

In addition to the unitary symmetries $SU(2)$, $SU(3)$, and $SU(4)$, a number of still larger symmetries have been postulated as being approximate symmetries of the strong interactions. For example, if we combine the $SU(2)$ group of ordinary spin with the isospin $SU(2)$, we obtain the direct product group $SU(2) \times SU(2)$. This group in turn can be considered as a subgroup of a still larger group $SU(4)$. Similarly, from the $SU(3)$ containing I_3 and Y, and from the spin group $SU(2)$, we can construct the groups $SU(3) \times SU(2)$ and $SU(6)$. Likewise, from $SU(4)$ and $SU(2)$, we can construct $SU(4) \times SU(2)$ and $SU(8)$. Even if $SU(4)$ were an exact symmetry, which it is not, $SU(8)$ would be only an approximate dynamical symmetry. The reason is that spin is not a conserved quantity; only the total angular momentum is. However, at low energy, spin can be approximately conserved, and $SU(8)$ might be a useful approximate symmetry.

The known hadrons can be classified into two categories: baryons, which have half-integral spin, and mesons, which have integral spin. As of the present, hundreds of mesons and baryons are known. With the existence of so many different particles, it does not seem useful to regard them as elementary. One possibility is that the hadrons are composites of simpler entities called quarks (Gell-Mann, 1964a; Zweig, 1964, 1965). According to

[3] In addition to the word isospin, the words isotopic spin, isobaric spin, and I spin are often used to describe this space.

the quark model, a baryon is a composite of three quarks, and a meson is a composite of a quark–antiquark pair. In the model, there are n different kinds or *flavors* of quarks, distinguished by $n - 1$ additive quantum numbers. But a collection of n states distinguished by $n - 1$ additive quantum numbers can serve as the basis vectors for a fundamental representation of $SU(n)$. In the $SU(4)$ version of this model, the 3 additive quantum numbers which distinguish the quarks are I_3, Y, and C. Although at present there is no direct evidence for additional additive quantum numbers which would require a theory with more quarks, models have already appeared in the literature with 5, 6, 7, 8, and more different quarks. Although a model with n quarks may correspond to a symmetry group other than $SU(n)$, in this book we do not consider other possibilities.

There is indirect evidence that baryon wave functions are symmetric under the interchange of the space, spin, and unitary spin coordinates of the quarks. But quarks have half-integral spin, and therefore, according to the spin-statistics theorem, they should be fermions. It was suggested by Greenberg (1964) that quarks are not fermions, but obey parastatistics. But the quantum mechanical treatment of a collection of particles which obey parastatistics can be reformulated so that the particles satisfy the spin-statistics theorem provided that they have an extra degree of freedom. In the case of quarks, the new degree of freedom is called color. Because there are three quarks in a baryon, the color degree of freedom takes on three values. It is now most commonly assumed that color symmetry is also a unitary symmetry, the symmetry of $SU(3)$. In some versions of the model, it is an exact symmetry, while in others it is approximate.

In our discussion so far, we have treated the strong, electromagnetic, and weak interactions as separate entities, and we have not really discussed the gravitational interactions. As an alternative possibility, the strong, electromagnetic, and weak interactions, and perhaps even the gravitational interaction, may be different aspects of one interaction. Many attempts have been made in recent years to unify two or more of these interactions in theories incorporating gauge-invariance of the second kind. A full discussion of these gauge–invariant theories is outside the scope of this book, although we briefly touch on the subject in Section 12.4. References to some of the original literature are contained in a book on the subject by Taylor (1976) and in a review by Weinberg (1974).

1.6 Space–Time and Internal Symmetries

Some of the symmetries we have discussed, for example, rotation symmetry, refer to invariances of the Hamiltonian of a system under transformations of the space–time coordinates. On the other hand, other symmetries, for example, isospin symmetry, refer to invariances of the Hamiltonian under transformations which do not involve the space–time coordinates. These latter symmetries we call *internal.*

We feel we have some understanding of space–time symmetries. For example, we "explain" the invariance of a system under translation by saying that space is homogeneous. What we really mean is that in many experiments we can isolate the system from its surroundings to such an extent that we get the same result, within experimental error, no matter where we do the experiment. As a consequence of translation invariance, momentum is conserved. Thus, we say we "understand" the conservation of momentum.

However, the fact that momentum is conserved is by no means a trivial fact, even though we take it for granted. To see this, we need only consider the fact that parity is not conserved. Parity conservation is a consequence of symmetry under space reflection. Before 1957 most physicists took it for granted that parity was conserved because they accepted it as self-evident that space is invariant under reflection, or, in other words, does not distinguish right from left. But evidently, in some experiments we can distinguish between right and left. This fact appears to indicate that whether symmetry under a transformation exists or not is a property of an interacting system rather than of space–time.

More mysterious than the space–time symmetries are the internal symmetries. Why should baryon number be conserved? Why should the neutron and proton have such similar properties that they can be considered as states which differ from each other just by a rotation in an abstract isospin space? More generally, why should approximate unitary symmetry apply to the hadrons?

Perhaps additional understanding will come from study of local gauge transformations, which relate apparently internal quantum numbers (like charge) to space–time dependent transformations. The theory of quantum electrodynamics, and the subsequent work of Yang and Mills (1954), Weinberg (1967), and Salam (1968) may give clues to the direction a future theory must go.

1.7 Limitations of Symmetry

In most of this book, we stress symmetry rather than dynamics. But, many symmetries of nature are only approximate, and in some cases the symmetries are so badly broken that they are hard to recognize. The recognition that $SU(3)$ is a broken symmetry had to wait, first, for the discovery of strange particles in 1947, and then for an accumulation of facts and theoretical ideas which led in 1961 to the work of Gell-Mann and Ne'eman.

In order to understand why some symmetries are exact and some are approximate, we must look to the dynamics. This can be done either on a fundamental or phenomenological level. In the last chapter of this book, we consider the quark model as a phenomenological model to help us understand how the breaking of unitary symmetry can come about. The understanding we get is not deep, but we believe it is helpful. The more we learn about the properties and interactions of elementary particles on a phenomenological level, the better we can hope to appreciate how symmetries are broken. Some good books that discuss the quark model and phenomenology of elementary particles are by Gasiorowicz (1966), Kokkedee (1969), Feld (1969), and Perl (1974).

But we can be more ambitious and try to understand broken symmetry on a fundamental level. One of the ways to do this, which at present seems promising, is within the framework of field theory. As we have already remarked, we do not discuss field theory in this book in any detail. But it is interesting that within field theory there are two basic ways a symmetry can be broken. The first way to break a symmetry is to construct a Lagrangian which contains terms which are not invariant under the relevant transformation. Unfortunately, there are so many ways to do this that little insight is gained. The second way to break a symmetry is spontaneously. This means that we construct a Lagrangian which has the symmetry in question, but arrange that (at least some of) the physical states of the theory do not. An example is a ferromagnet, which singles out a preferred direction in space, although the Lagrangian of electrodynamics is rotationally invariant.

There are restrictions on the ways a symmetry can be spontaneously broken, especially if we add as a requirement that the theory remain renormalizable. For further information in the case of renormalizable gauge field theory, see the review by Weinberg (1974).

We mention one other related topic which we shall not touch on at all in the rest of the book: the possibility of the existence of supersymmetry. A supersymmetry is an invariance property under transformations of bosons into fermions and vice versa. There is at present little or no experimental evidence for such a symmetry, so if it exists as a property of the Lagrangian, it must be spontaneously broken. But the idea of supersymmetry has led to the hope that the strong, electromagnetic, weak, and gravitational interactions may some day be unified in a renormalizable field theory. We give only two references, a review by Ferrara (1976) and a short summary by Lubkin (1977) of a sample of recent work.

CHAPTER 2

SOME PROPERTIES OF GROUPS

2.1 Elementary Notions

We have seen that the symmetry transformations on physical systems have the properties of groups. Therefore, in order to exploit the consequences of symmetries, it is useful to study some of the mathematical properties of groups. These fall into two main categories: the properties of abstract groups and the properties of representations of groups. For the most part, we shall discuss representations in the next chapter.

Here we shall consider only a few of the properties of groups, and we shall usually not prove our assertions. For a fuller treatment, see Wigner (1959) or Hamermesh (1962).

A *group* is a set of elements which obey the following rules:

(1) A law of combination, usually called multiplication, of group elements is defined such that if a and b belong to a group G, then the product element ab is unique and belongs to G. We can write this property of a group as follows:

$$(a \in G \quad \text{and} \quad b \in G) \Rightarrow ab \in G, \tag{2.1}$$

where the symbol $\in$ means "belongs to" or "is a member of" and the symbol $\Rightarrow$ means "implies."

(2) The law of combination is associative, that is, given the elements a, b, and c, we have

$$(ab)c = a(bc). \tag{2.2}$$

The elements do not necessarily commute, however.

(3) An identity element e (or 1 or a_0) exists satisfying the property that for all a in G,

$$ea = ae = a. \tag{2.3}$$

(4) To every element a in G, an inverse element a^{-1} in G exists such that

$$a^{-1}a = aa^{-1} = e. \tag{2.4}$$

Sometimes postulates (3) and (4) are replaced by the weaker postulates that a left identity and a left inverse exist. From these weaker postulates, postulates (3) and (4) can be derived.

A group is said to be *finite* if it has a finite number of elements. Otherwise the group is said to be *infinite*. The number of elements of a finite group is called the *order* of the group.

The simplest group is the trivial group consisting of only one element, the identity e. The identity is always its own inverse. The next simplest group consists of two elements, e and a, with a law of multiplication

$$ee = e, \qquad aa = e, \qquad ea = ae = a. \tag{2.5}$$

The element a is its own inverse. To give a familiar example of this group, we can identify the elements e and a with the numbers 1 and -1, respectively. It is obvious that the multiplication table (2.5) is satisfied by the numbers ± 1, with the identification $1 = e$, $-1 = a$. The group consisting of the elements e and a is important in quantum mechanics because it can be associated with such discrete transformations as space reflection and charge conjugation.

If all the elements of a group commute, the group is said to be *commutative* or *Abelian*. The groups of one and two elements are clearly Abelian. The law of combination for Abelian groups is often denoted by the symbol $+$. If so, the identity element is called zero and the inverse to the element a is called $-a$.

An example of an infinite Abelian group under addition is the set of all integers. For any two integers m and n, $m + n$ is also an integer; the zero element exists such that $0 + m = m$; and for any integer m there exists the integer $-m$ such that $m + (-m) = 0$. The associative law holds, of course, for addition of integers. Thus, the integers satisfy the group properties. However, the set of positive integers is clearly not a group, since an inverse and identity do not exist. It is easy to see that the set of real numbers is an Abelian group under addition.

The smallest non-Abelian group is the group of permutations of three objects. We write the symbol

$$\begin{pmatrix} 123 \\ ijk \end{pmatrix}$$

for the permutation in which the object in position 1 is put into position i, the object in position 2 is put into position j, and the object in position 3 is put into position k. In general, if we use this notation, a permutation can be written in more than one way. For example, the symbols

$$\begin{pmatrix} 123 \\ ijk \end{pmatrix} \quad \text{and} \quad \begin{pmatrix} 312 \\ kij \end{pmatrix}$$

denote the same permutation. The permutation group of three objects has the following six elements

$$\begin{aligned} e &= \begin{pmatrix} 123 \\ 123 \end{pmatrix}, \quad a = \begin{pmatrix} 123 \\ 213 \end{pmatrix}, \quad b = \begin{pmatrix} 123 \\ 132 \end{pmatrix}, \\ c &= \begin{pmatrix} 123 \\ 321 \end{pmatrix}, \quad d = \begin{pmatrix} 123 \\ 312 \end{pmatrix}, \quad f = \begin{pmatrix} 123 \\ 231 \end{pmatrix}. \end{aligned} \tag{2.6}$$

That the permutations on n objects form a group can be readily seen as follows: Two successive permutations (which we call the product) are again a permutation, and the operation of making permutations is associative. The identity is

$$\begin{pmatrix} 123 \cdots n \\ 123 \cdots n \end{pmatrix},$$

and the inverse to the permutation

$$\begin{pmatrix} 12 \cdots n \\ ij \cdots k \end{pmatrix} \quad \text{is} \quad \begin{pmatrix} ij \cdots k \\ 12 \cdots n \end{pmatrix}.$$

The permutation group of n objects or symbols is often called the *symmetric group* of *degree n*, and is written S_n.

Another notation for the symbol

$$\begin{pmatrix} 12 \cdots n \\ ij \cdots k \end{pmatrix} \quad \text{is just} \quad (ij \cdots k).$$

With this shorter notation, it is assumed that this is a permutation on the numbers $123 \cdots n$. In this notation, there is only one symbol for a given permutation. The longer, two-rowed notation is more useful for multiplying permutations. Such a multiplication is done as follows: We write the second permutation using a symbol such that its top row is the same as the bottom row of the first permutation. Then the product permutation has the same top row as the first permutation and the same bottom row as the second. As an example, suppose we wish first to make the permutation

$$\begin{pmatrix}123\\213\end{pmatrix},$$

and then the permutation

$$\begin{pmatrix}123\\312\end{pmatrix}.$$

We proceed by writing the second permutation as

$$\begin{pmatrix}213\\132\end{pmatrix}.$$

Then the product permutation is

$$\begin{pmatrix}213\\132\end{pmatrix}\begin{pmatrix}123\\213\end{pmatrix} = \begin{pmatrix}123\\132\end{pmatrix}.$$

We have seen that S_3 is a group of order six by writing down all possible permutations of three symbols. We can count the number of elements more simply in the following way: The first object can go in any one of three places. Once it is fixed, the second object can go in either of the two remaining places. Once the second object is in its place, the third object must go in the remaining place. Thus, there are a total of $3 \cdot 2 \cdot 1 = 3!$ possible permutations. By similar reasoning, the symmetric group S_n is of order $n!$ Using the two-row notation, there are also $n!$ ways of writing a given permutation of S_n.

We have written down in Eq. (2.5) the complete multiplication table for the group with two elements. This is just the multiplication table for the symmetric group S_2, with elements

$$\begin{pmatrix}12\\12\end{pmatrix} \quad \text{and} \quad \begin{pmatrix}12\\21\end{pmatrix}.$$

Likewise, we can form a multiplication table of the group S_3 by actually carrying out the multiplication of each permutation of S_3 with every other permutation of S_3. This multiplication table has been given in so many places (e.g., Wigner, 1959) that we shall not reproduce it here.

2.2 Homomorphism, Isomorphism, and Subgroups

A group H is said to be *homomorphic* to a group G if there is a mapping of G onto H such that multiplication is preserved. In this mapping, more than one element of G may correspond to one element of H. If a and b are two arbitrary elements of G and $c = ab$, then in H the corresponding elements a', b', and c' must satisfy $c' = a'b'$. In any homomorphic mapping the identity of G is mapped into the identity of H.

In a homomorphic mapping of G onto H, if the correspondence between the elements of G and those of H is one-to-one, then H is said to be *isomorphic* to G. Clearly, if H is isomorphic to G, then G is isomorphic to H. Therefore, we usually say simply that the two groups are isomorphic. An isomorphic mapping of a group onto itself is called an *automorphism*. All isomorphic groups have the same multiplication table. In fact, all the properties of an abstract group are contained in its multiplication table.

Another important concept is that of a *subgroup* of a given group G. This is simply a subset of the group which itself has the group properties under the same multiplication law as G. Every group has at least two subgroups, the group itself and the identity or unit element. These are called improper subgroups. All other subgroups are called proper subgroups. The two-element group has only improper subgroups. The symmetric group S_3 has three proper subgroups of order two and one of order three. We denote the subgroups of order two by H_1, H_2, and H_3 with elements

$$(123), (132) \in H_1,$$
$$(123), (321) \in H_2,$$
$$(123), (213) \in H_3.$$

These groups are isomorphic to S_2. In fact, all groups of order two are isomorphic to S_2. This is because they all have the same multiplication table, as can be seen by the following argument.

Let the two distinct elements of the group of order two be e and a. We must have $ee = e$, $ea = ae = a$ by definition. Then the only other product is aa, but this must be $aa = e$, since if $aa = a$, $a = e$. Thus the multiplication table is unique.

2.3 Infinite Groups

If we have a group with a finite number of elements, we can label the group elements g_a where the subscript is a parameter which takes on a finite number of values. We can generalize this idea to a countably infinite group with an infinite number of discrete values of a or to a group with a continuum of elements in which the parameter a can vary continuously. In the latter case a group element is usually denoted by $g(a)$. We can generalize still further to the case where a stands for more than one continuously varying parameter.

The set of values of the parameter or parameters which characterize a group element can be considered to be points in some kind of space. The number of parameters characterizes the dimension of the space. We shall restrict ourselves to a *topological* space, that is, to a space in which *distance* between any two points is defined. Under certain conditions, we can make a correspondence between elements of a group and the points given by the values of the parameters in a topological space. Alternatively, under certain conditions we can consider the group elements themselves to be points in the space.

Suppose we have a group whose elements are labeled by a letter which can stand for either a single parameter or for several parameters which can take on a continuum of values. Furthermore, suppose we demand that if we form the product of any two group elements to give a third:

$$g(a)g(b) = g(c),$$

where a, b, and c are particular values of the parameter, then c should be a continuous function of the parameters a and b. In other words, we assume that a small change in either a or b will produce only a small change in c. If so, the group is said to be a *continuous* group, and the points in parameter space can be identified with the group elements. A mixed continuous group is a continuous group which also contains discrete parameters. Even more restrictive than a continuous group is a *Lie* group, in which, if $g(a)g(b) = g(c)$, then c is an analytic function of a and b.

In the detailed study of continuous and Lie groups, one can bring to bear the methods of functional analysis. We shall not do so in any generality but shall consider principally the concepts which we shall need later. Further information can be found in Pontriagin (1966). See also Hamermesh (1962); Wigner (1959); and Racah (1965) for discussions of continuous groups.

Let us consider groups of coordinate transformations as examples of Lie groups. Many such transformations have applicability in quantum mechanics. First consider the group of translations in one dimension, namely

$$x' = x + a. \tag{2.7}$$

This is a one-parameter Lie group and is an Abelian group under addition. The group elements can be denoted by $g(a)$ or simply by the parameter a itself. The values of the parameter are the points on a line. The line is an example of a topological space with which we associate or identify the group elements. As another example, consider the transformation

$$x' = ax + b. \tag{2.8}$$

This is a two-parameter Lie group. A continuous group which is characterized by a finite number of continuous parameters is sometimes called a *finite continuous* group.

If a Lie group is defined by a number of parameters, say r parameters, and we cannot find a smaller number of parameters to characterize the group, then the parameters are called *essential* parameters, and the group is called an r-parameter Lie group.

As an example of a group which is characterized by two parameters, only one of which is essential, we give the following:

$$x' = x + a + b.$$

In this case, an infinite number of different combinations of a and b correspond to the same group element. The group characterized by one essential parameter would be

$$x' = x + c, \qquad c = a + b.$$

Then to each c, there is only one group element.

Another example of a Lie group is the general linear group in two dimensions. This is the group of general linear transformations on a two-dimensional space and is denoted by $GL(2)$. If the parameters are real, the group is often denoted by $GL(2R)$, whereas if the parameters are complex, the group is denoted either by $GL(2C)$ or simply by $GL(2)$. The transformation is given by

$$\begin{aligned} x_1' &= a_{11}x_1 + a_{12}x_2, \\ x_1' &= a_{21}x_1 + a_{22}x_2. \end{aligned} \tag{2.9}$$

The group $GL(2R)$ is characterized by four essential parameters and $GL(2C)$ is characterized by eight essential parameters. We can write the transformation in matrix form as follows:

$$x' = Ax, \qquad A = \begin{pmatrix} a_{11} & a_{12} \\ a_{21} & a_{22} \end{pmatrix}, \qquad x = \begin{pmatrix} x_1 \\ x_2 \end{pmatrix}. \tag{2.10}$$

The linear group in n dimensions $GL(n)$

$$x' = Ax, \qquad A = \begin{pmatrix} a_{11} & \cdots & a_{1n} \\ \vdots & & \vdots \\ a_{n1} & \cdots & a_{nn} \end{pmatrix}, \qquad x = \begin{pmatrix} x_1 \\ \vdots \\ x_n \end{pmatrix} \tag{2.11}$$

is characterized by $2n^2$ essential parameters. The matrices A are the group elements. If the matrices are unitary, the group is called $U(n)$. The equalities $A_{ji}^* = (A^{-1})_{ij}$ put n^2 independent conditions on the $2n^2$ variables and reduce the number of essential parameters to n^2. A linear group satisfying the one restriction that its determinant is unity is called a *special* linear group and is denoted by $SL(n)$. The group $SU(n)$ is the special unitary group in n dimensions, i.e., the group of unitary matrices with determinant unity. It is characterized by $n^2 - 1$ parameters. We shall be dealing very much with this group.

Other groups of importance are the orthogonal group in 2, 3, or n dimensions. The group $O(n)$ is composed of those elements of $GL(n, R)$ which leave the product

$$x^2 = \sum_i x_i{}^2 \tag{2.12}$$

invariant. This group has $n(n-1)/2$ essential parameters. The rotation group $SO(n)$ or $R(n)$ consists of the orthogonal matrices with determinant unity. It has the same number of parameters as $O(n)$. However, $R(n)$ in an odd number of dimensions does not include the reflections $x' = -x$. The rotation group in two dimensions is Abelian, and is characterized by one essential parameter θ: The transformation is

$$\begin{aligned} x_1{}' &= x_1 \cos\theta - x_2 \sin\theta, \\ x_2{}' &= x_1 \sin\theta + x_2 \cos\theta. \end{aligned} \tag{2.13}$$

The transformation can be written in matrix form as

$$x' = R(\theta)x, \tag{2.14}$$

where

$$x = \begin{pmatrix} x_1 \\ x_2 \end{pmatrix}, \qquad R(\theta) = \begin{pmatrix} \cos\theta & -\sin\theta \\ \sin\theta & \cos\theta \end{pmatrix}. \tag{2.15}$$

Here, the rotation matrix $R(\theta)$ is the group element.

Still another example of a Lie group is the homogeneous Lorentz group (or simply Lorentz group), which is the group which leaves invariant the quantity

$$x^2 + y^2 + z^2 - c^2 t^2.$$

This is a six-parameter group in four dimensions. The group $O(3)$ is a subgroup of the Lorentz group. If we add to the Lorentz group the group of translations, we obtain the ten-parameter inhomogeneous Lorentz group, or Poincaré group.

Of importance in group theory is the idea of a compact group. We shall not give the general definition of a compact group, but just one that is useful for our purposes. To do this, we need the concepts of a bounded and closed set. A set of numbers is said to be *bounded* if none of the numbers in the set exceeds a given positive number M in absolute value. A set is said to be *closed* if the limit of every convergent sequence of points in the set also lies in the set. As an example, the real line between zero and one is bounded. It is open if either of the end points zero and one is not included in the set, but is closed if zero and one are included.

An r-parameter group is compact if the domain of variation of all the parameters is bounded and closed. The group of rotations is compact, because the angle of rotation about any axis must lie in the closed interval

$0 \leq \theta \leq 2\pi$. However, the translation group is not compact, as the parameter a in the transformation $x' = x + a$ can vary between plus and minus infinity. Here the domain of variation of a is unbounded. (If a varies over a bounded region, the translations do not form a group.)

The group of pure Lorentz transformations is characterized by a parameter, the velocity v, which is bounded. However, the limit point $v = c$ is not a member of the group, as a Lorentz transformation yields infinity when $v = c$. Therefore, the Lorentz group is not compact.

A continuous group is *locally compact* if in the neighborhood of any point in parameter space, a closed domain or interval exists. The point in question can be a boundary point of the domain. A compact group is, of course, locally compact. In addition, some noncompact groups such as the translation group and the Lorentz group are locally compact.

2.4 Cosets, Conjugate Classes, and Invariant Subgroups

Let H be a proper subgroup of the group G, and let g belong to G, but not to H. Then the product elements gh_i, where the h_i are all the elements of H, are said to form a *left coset* of H. The coset is denoted by gH. Similarly, we can define a *right coset* Hg with members $h_i g$. (Some authors define cosets to include sets gH and Hg where g is any member of G, including members of H.)

With our definition of a coset, it can be seen that none of the elements of a coset of H belongs to H. We show this by assuming that gh_i belongs to H and obtaining a contradiction. Since $h_i \in H$, $h_i^{-1} \in H$. If $gh_i \in H$, then $gh_i h_i^{-1} \in H$, since the product of any two elements of a group is itself an element. But

$$gh_i h_i^{-1} = ge = g. \tag{2.16}$$

Thus, if $gh_i h_i^{-1} \in H$, then $g \in H$. But this is in contradiction to the assumption contained in our definition of a coset that g does not belong to H. Therefore, none of the elements of gH belongs to H. From this theorem it can be seen that a coset is not a group. The proof follows from the fact that the identity element e, since it belongs to H, is not in any coset of H.

We state without proof two other properties of cosets. The first is that two cosets have either no elements in common or else have identical elements. The second is that the set S of the elements of all the (left or right) cosets of a subgroup H of G, plus the elements of H, contains all the elements of G. We can obtain all the cosets of H by multiplying the elements of H by all the g_i in G which are not in H. However, using this procedure, we may obtain the cosets more than once.

These properties of cosets can be used to prove Lagrange's theorem, which states that the order of a subgroup of a finite group is a factor of the order of the group. By this we mean that if g is the number of elements of G, and h is the number of elements of a subgroup H, then $g = hn$, where n is an integer. We shall not prove the theorem, which is to be found, for example, in Wigner (1959). Here n is called the *index* of H in G.

We now consider the concept of conjugate elements of a group. An element a belonging to G is said to be *conjugate* to an element b belonging to G if there exists an element u in G such that $a = ubu^{-1}$. The set of all elements of G conjugate to a given element is said to be the *conjugate class* of the element. The *relation* of conjugation between elements is an equivalence relation, since it can be shown that (1) a is conjugate to a, (2) if a is conjugate to b, then b is conjugate to a, and (3) if a is conjugate to b, and b is conjugate to c, then a is conjugate to c.

It is clear that the identity element of any group is in a class by itself. This follows because the identity commutes with all members of the group. Therefore

$$ueu^{-1} = euu^{-1} = e. \tag{2.17}$$

Similarly, every element of an Abelian group is in a class by itself. It should also be obvious that, except for the identity, no class is a group.

By actual multiplication of the elements of S_3, we can verify that this group contains three classes. They are the class with the identity (123), the class with the elements (132), (213), and (312), and the class with the elements (312) and (231).

We now state without proof an important theorem about conjugate elements. Let H be a subgroup of G, and let $h \in H$ and $g \in G$. Form the product elements

$$h' = ghg^{-1}, \tag{2.18}$$

for all h in H. Then the h' form a group H' which is isomorphic to H. The group H' is called the *conjugate subgroup* to H in G. Different choices of the element g, in general, give different groups H'. If, for all g in G, the elements of H and gHg^{-1} are identical, then H is said to be an *invariant* or *self-conjugate* subgroup of G. If H is an invariant subgroup of G and if $h \in H$, then $ghg^{-1} \in H$ for any g in G.

A group is called *simple* if it is not Abelian and contains no proper invariant subgroups. It is called *semisimple* if it contains no Abelian invariant subgroups except the unit element. Rather than use these definitions for Lie groups, we shall define simple and semisimple Lie groups in terms of certain properties of their *Lie algebras*. This is done in Chapter 5.

We shall not prove the following theorem about invariant subgroups:

THEOREM. *A subgroup of index two in any group is invariant.*

The group $SU(2)$ contains the elements

$$e = \begin{pmatrix} 1 & 0 \\ 0 & 1 \end{pmatrix} \quad \text{and} \quad a = \begin{pmatrix} -1 & 0 \\ 0 & -1 \end{pmatrix}. \tag{2.19}$$

These elements comprise an invariant subgroup of order two, sometimes called Z_2.

We quote another theorem without proof:

THEOREM. *The left and right cosets of an invariant subgroup H of a group G, formed with any element of G are identical.*

This theorem enables us to prove an important theorem about an invariant subgroup and its cosets.

Suppose we have a group G containing an invariant subgroup H. Consider a set S whose elements are all the distinct cosets of H plus H itself. This set is a group, as can be seen by the following argument: The elements of S are $b_i H$ and H, and b_i are all the elements of G which are not in H. Since H is invariant we have $b_i H = Hb_i$. We can now perform *coset multiplication*

$$b_i H b_j H = b_i b_j HH = b_i b_j H, \tag{2.20}$$

because, since H is a group, the elements of HH are the elements of H itself. Since $b_i b_j$ belongs to G, $b_i b_j H$ is either H itself or a coset of H, depending on whether or not $b_i b_j$ is an element of H. The identity element of S is just H itself, and the inverse to the element $b_i H$ is $b_i^{-1} H$. Thus S is a group whose members are sets: namely H and the cosets of H. It is called the *factor* or quotient group and is denoted by G/H.

As an example, we note that from $SU(2)$ and Z_2 we can construct the factor group $SU(2)/Z_2$. This group is isomorphic to the three-dimensional rotation group $R(3)$.

We now introduce the notion of the direct product of two groups H and H'. The group G is called the *direct product* of H and H' if the following two conditions are satisfied:

(1) Every h in H commutes with every h' in H'.

(2) Every g in G can be written uniquely as a product of an element of H and an element of H', that is

$$(g \in G) \Rightarrow (g = hh'). \tag{2.21}$$

The direct product is written in the form $G = H \times H'$.

Another notation for an element of a direct product is to denote the element g by means of an ordered pair (h, h'), i.e.,

$$g = (h, h'), \tag{2.22}$$

with the following rule of multiplication

$$(h_1, h_1')(h_2, h_2') = (h_1 h_2, h_1' h_2'). \tag{2.23}$$

It follows from the definition of the direct product G of two groups H and H' that H and H' are isomorphic to two invariant subgroups of G having only the identity element in common. Suppose H has m elements h_i, where $h_1 \equiv e$ is the identity of H, and H' has n elements h_j' with identity $h_1' \equiv e'$. Then H is isomorphic to the invariant subgroup of G with elements $h_i e'$, and H' is isomorphic to the invariant subgroup with elements $e h_j'$. The common identity element of both invariant subgroups is ee'. It is often said for brevity that H and H' themselves are invariant subgroups of G with only the identity in common.

CHAPTER 3

SYMMETRY, GROUP REPRESENTATIONS, AND PARTICLE MULTIPLETS

3.1 Linear and Unitary Vector Spaces

We have seen that the transformations which can be made on physical systems and which leave them invariant have the mathematical properties of groups. Thus, the use of group theory is a valuable tool in studying symmetry. However, in order to apply the ideas of group theory to physical problems, we must study the properties of the quantum mechanical states of a system and how these states transform under symmetry operations. To do this, we need the concepts of unitary vector spaces and of unitary and antiunitary operations on these spaces. We need these concepts because the physical states are represented in quantum mechanics by vectors in a unitary vector space, and the symmetry transformations are represented by unitary and antiunitary operators which act on the space.

We begin by defining a *linear vector space* as a set of elements called vectors with the following properties:

(1) If ϕ, ψ, and χ are vectors in a linear vector space V, then a law of combination, called addition, is defined such that

$$\phi + \psi \in V. \tag{3.1}$$

(2) This operation is commutative and associative, i.e.,

$$\phi + \psi = \psi + \phi, \tag{3.2}$$

$$\phi + (\psi + \chi) = (\phi + \psi) + \chi. \tag{3.3}$$

(3) There exists a null vector 0 such that

$$\phi + 0 = \phi. \tag{3.4}$$

(4) For every vector ϕ, there exists a negative $-\phi$ such that

$$\phi + (-\phi) = 0. \tag{3.5}$$

The foregoing properties show that a vector space is an Abelian group under addition.

(5) There are further properties, namely that there exist scalars $a, b, \ldots$ (in general, complex numbers) which satisfy the requirements

$$\phi \in V \Rightarrow a\phi \in V, \tag{3.6}$$

$$a(\phi + \psi) = a\phi + a\psi, \tag{3.7}$$

$$(a + b)\phi = a\phi + b\phi, \tag{3.8}$$

$$(ab)\phi = a(b\phi). \tag{3.9}$$

(6) There exist an identity scalar 1 and a null scalar 0 such that

$$1\phi = \phi, \tag{3.10}$$

$$0\phi = 0. \tag{3.11}$$

We use the same notation for the scalar zero and the null vector. This should not lead to any confusion.

If the scalars are complex numbers, the space is often called a *complex* vector space; if the scalars are real, the space is called a *real* vector space.

A set of vectors $\phi_1, \phi_2, \ldots, \phi_n$ belonging to a vector space is called *linearly dependent* if there exist scalars $a_1, a_2, \ldots, a_n$, not all zero, such that

$$a_1\phi_1 + a_2\phi_2 + \cdots + a_n\phi_n = 0. \tag{3.12}$$

Otherwise, the vectors are said to be *linearly independent*. The *dimension* of a vector space is the maximum number of linearly independent vectors in the space. If there are a finite number of these, the space is called a finite-dimensional vector space; otherwise, the space is infinite-dimensional.

A *linear manifold* is a subspace M of a vector space V, such that M itself is a vector space. If $M = V$ or M contains only the null vector, M is said to be an improper subspace of V; otherwise M is a proper subspace. A vector space or linear manifold is *spanned* by a set of vectors $\phi_1, \ldots, \phi_n$ if every vector in the space can be written as a linear combination of the $\phi_1, \ldots, \phi_n$ with scalar coefficients. A linearly independent set of vectors which spans a vector space is called a *basis*. Any vector in the set is called a *basis vector*.

Of special importance in physics is the concept of a unitary vector space in which a scalar product is defined. A *unitary vector space* is a linear vector space in which the scalar product of any two vectors ϕ and ψ, denoted by (ϕ, ψ), has the following properties:

$$(\phi, \psi) \quad \text{is a scalar (a complex number)}, \tag{3.13}$$

$$(\phi, \psi)^* = (\psi, \phi). \tag{3.14}$$

where the asterisk denotes the complex conjugate of a scalar. It is clear from this postulate that (ϕ, ϕ) is a real number. Also

$$(\phi, a\psi) = a(\phi, \psi) \tag{3.15}$$

$$(\phi, \psi + \chi) = (\phi, \psi) + (\phi, \chi) \tag{3.16}$$

$$(\phi, \phi) \geq 0, \qquad (\phi, \phi) = 0 \Rightarrow \phi = 0. \tag{3.17}$$

Any linear vector space in which (3.13) through (3.16) hold is said to be a space with metric. If (3.17) also holds, the metric is said to be positive definite.[1] The quantity

$$(\phi, \phi)^{1/2} = \|\phi\| \tag{3.18}$$

is called the norm[2] of the vector ϕ. If $(\phi, \phi) = 1$, then ϕ is said to be normalized to unity, or more simply, normalized. The *distance* between two vectors ϕ and ψ is defined to be $\|\phi - \psi\|$. Two vectors ϕ and ψ are called *orthogonal* if

$$(\phi, \psi) = 0.$$

Clearly, the null vector is the only vector orthogonal to every vector in the space. That 0 is orthogonal to any vector ϕ is seen by writing $0 = \phi - \phi$. Then $(0, \phi) = (\phi - \phi, \phi) = (\phi, \phi) - (\phi, \phi) = 0$. That 0 is the only such vector follows immediately from the postulate $(\phi, \phi) = 0 \Rightarrow \phi = 0$.

We next come to the important concept of an orthonormal set of vectors. The vectors $\phi_1, \phi_2, \ldots, \phi_n$ form an orthonormal set if

$$(\phi_i, \phi_j) = \delta_{ij}, \tag{3.19}$$

where δ_{ij} is the Kronecker-δ defined by $\delta_{ij} = 1$, $i = j$; $\delta_{ij} = 0$, $i \neq j$. If an orthonormal set spans a vector space V, it is called a complete orthonormal

[1] Some people define such a metric as nonnegative.

[2] Some authors call (ϕ, ϕ) the norm of the vector ϕ.

set or an *orthonormal basis*. Then any vector ψ in V can be written

$$\psi = \sum_i a_i \phi_i, \tag{3.20}$$

where the a_i are scalars. By taking the scalar product of ϕ_j with (3.20) and using (3.19), we see that

$$(\phi_j, \psi) = \sum_i a_i(\phi_j, \phi_i) = \sum_i a_i \delta_{ji} = a_j. \tag{3.21}$$

An important kind of linear vector space is a *Hilbert space*. This is a unitary vector space which is also *complete*. By completeness we mean that any convergent sequence of vectors converges to a vector within the space. With this definition, any unitary space of finite dimension is a Hilbert space. An example of a Hilbert space of infinite dimension is the space of square integrable functions. The scalar product of two functions in this case is defined to be

$$(\phi, \psi) = \int \phi^*(x)\psi(x)\, dx. \tag{3.22}$$

In quantum mechanics, it is postulated that a normalized vector in Hilbert space stands for a possible state of a physical system. The quantity (ψ, ϕ) stands for a transition amplitude from the state ϕ to the state ψ. The physical significance of the amplitude (ψ, ϕ) is that the square of the absolute value of a transition amplitude represents the probability of a transition from the state ϕ to the state ψ. Thus, (ψ, ϕ) is not physically observable, but $|(\psi, \phi)|$ is.

It follows that if ψ denotes a physical state, then

$$\psi' = e^{i\alpha}\psi, \qquad \alpha \text{ real},$$

equally well denotes the same physical state. The set of all vectors which differ only by a phase factor (a constant of magnitude unity) is said to be a *ray* in Hilbert space. Thus, a physical state can be denoted by a ray in Hilbert space or by any vector of the ray.

3.2 Operators

An operator on a linear vector space is a mapping of the space into itself. In other words, an operator L acting on a vector ψ in a space V yields a new vector ψ' in the space. We use the notation $\psi' = L\psi$ and write this definition as follows:

$$\psi \in V \Rightarrow L\psi \in V. \tag{3.23}$$

The concept of an operator without further restrictions is too general for our purposes. It is convenient at times to restrict ourselves to operators which satisfy in addition to (3.23) the postulate

$$L(\phi + \psi) = L\phi + L\psi. \tag{3.24}$$

Especially important is a *linear operator* which, *in addition* to (3.23) and (3.24), satisfies

$$L(a\phi) = aL\phi. \tag{3.25}$$

An *antilinear* operator K satisfies the first two of these postulates, but instead of (3.25), it satisfies the following postulate:

$$K(a\phi) = a^*K\phi. \tag{3.26}$$

Suppose we have a set of operators L_i which map a vector space into itself. Consider two operators L_1 and L_2 which map any vector ϕ into corresponding vectors ψ_1 and ψ_2; i.e.,

$$L_1\phi = \psi_1, \qquad L_2\phi = \psi_2.$$

Then we can define the *sum* of L_1 and L_2 as the operator L which maps ϕ into the vector $\psi = \psi_1 + \psi_2$. With this definition, we have

$$(L_1 + L_2)\phi = L_1\phi + L_2\phi.$$

Also, the sum is commutative and associative.

Now if the vector ψ belongs to V, then so does $a\psi$ where a is a scalar. Then, given an operator L, such that $L\phi = \psi$, we can define an operator $L' = aL$ such that

$$L'\phi = aL\phi = a\psi.$$

Thus, we have defined addition of operators and multiplication of operators by scalars. We can further define a null operator which maps every vector onto the null vector. Also, if $L\phi = \psi$, we can define $-L$ as the operator which maps ϕ into $-\psi$. With these definitions, it is easy to show that the set of operators and scalars satisfy Eqs. (3.1) through (3.11) and so, the operators *themselves* constitute a linear vector space.

Let us now define a *product* of two operators belonging to a vector space as a mapping that associates the given two operators with another operator in the space. Furthermore, suppose the products satisfy the following equations:

$$(aL_1 + bL_2)L_3 = aL_1L_3 + bL_2L_3,$$
$$L_1(aL_2 + bL_3) = aL_1L_2 + bL_1L_3.$$

Then we say that the operators form an *algebra*. (Some authors also require associativity.)

Suppose $L_1\phi = \psi$ and $L_2\psi = \chi$, where L_1 and L_2 are operators. Then we can define the product of the two operators as the operator L which satisfies

$$L\phi = \chi = L_2L_1\phi.$$

If the operators are linear, it is easy to see that they constitute an algebra. Furthermore, this algebra is *associative*, but not necessarily commutative. In Chapter 5, we shall consider a *nonassociative* algebra, in which the product of two operators L_1 and L_2 is defined as the *commutator* of the operators:

$$L = L_1 L_2 - L_2 L_1 = [L_1, L_2].$$

Such an algebra is called a *Lie algebra*. In the future, when we write the product simply as $L_1 L_2$ we shall mean an associative algebra.

We can define the *hermitian adjoint* $L^\dagger$ of a linear operator L acting on a unitary vector space as follows:

$$(\psi, L\phi) = (L^\dagger \psi, \phi). \tag{3.27}$$

A linear operator is *hermitian* or *self-adjoint* if

$$(\psi, L\phi) = (L\psi, \phi).$$

If

$$(\psi, L\phi) = -(\psi, L\phi),$$

the operator is said to be *antihermitian*. We can also define a hermitian operator by the equality

$$L^\dagger = L.$$

We are assuming that every vector in the space has finite norm. Then, because for every ϕ in the space, $L\phi$ is also in the space, it follows that $\|L\phi\|$ is finite. An operator for which this is true is called a *bounded* operator.

In quantum mechanics, it is useful to consider unbounded operators and also operators which are not defined for every vector in the space. For such operators, there is a distinction between a hermitian operator and one which is self-adjoint. Although we sometimes discuss unbounded operators, like the momentum operator and the Hamiltonian, we shall not distinguish between hermitian and self-adjoint operators. For a more rigorous discussion, see, for example, van der Waerden (1974).

If we operate with a linear operator L on a vector ψ, it may happen that we get a constant c times the same vector. If so, ψ is said to be an eigenvector of L and c is the eigenvalue. It is easy to prove that the eigenvectors of a hermitian operator are real and that its eigenvectors belonging to different eigenvalues are orthogonal. We shall assume that the eigenvectors of a hermitian operator form a complete set or basis.

In quantum mechanics, a self-adjoint operator is an *observable*. There may be a limitation on the observability of certain self-adjoint operators because of the existence of so-called superselection rules postulated by Wick *et al.* (1952). (See, however, Aharonov and Susskind, 1967.) However, these considerations are outside the scope of our work.

Every linear operator can be written as a sum of a hermitian and an antihermitian operator. This can be seen as follows: Any linear operator L can be written

$$L = \tfrac{1}{2}(L + L^\dagger) + \tfrac{1}{2}(L - L^\dagger).$$

Since $(L^\dagger)^\dagger = L$, it follows that $\frac{1}{2}(L + L^\dagger)$ is hermitian and $\frac{1}{2}(L - L^\dagger)$ is antihermitian.

A linear operator is *unitary* if

$$U^\dagger = U^{-1}. \tag{3.28}$$

For U to be unitary, both the following conditions must hold

$$U^\dagger U = UU^\dagger = 1. \tag{3.29}$$

Transformations of vectors by unitary operators leave the scalar product invariant. Let $\psi' = U\psi$ and $\phi' = U\phi$. Then

$$(\psi', \phi') = (U\psi, U\phi) = (U^\dagger U\psi, \phi) = (\psi, \phi). \tag{3.30}$$

The converse is not necessarily true. If an operator U satisfies

$$(U\psi, U\phi) = (\psi, \phi),$$

U does not have to be unitary. It is true that $U^\dagger U = 1$, but not necessarily that $UU^\dagger = 1$. We shall not consider this possibility further.

Any unitary operator can be written as follows:

$$U = e^{iaL}, \tag{3.31}$$

where L is a hermitian operator and a is a real number. The operator L is said to be the *generator* of U. An *antiunitary* operator K is an antilinear operator satisfying

$$(K\psi, K\phi) = (\phi, \psi) = (\psi, \phi)^*. \tag{3.32}$$

In quantum mechanics, if ϕ and ψ describe a possible state of a physical system, the quantity $|(\psi, \phi)|^2$ is interpreted as a transition probability from the state ϕ to the state ψ. It is therefore a quantity closely related to the results of an experimental measurement.

For this reason, operators which transform ψ and ϕ but leave $|(\psi, \phi)|$ invariant have a special importance in quantum mechanics. Unitary and antiunitary operators are among the operators having this property.

Wigner has proved an important theorem about operators which leave $|(\psi, \phi)|$ invariant. The theorem states that in considering operators which leave $|(\psi, \phi)|$ invariant, we lose no generality if we restrict ourselves to unitary and antiunitary operators.

Unitary and antiunitary operators as we have seen, are special members of the class which leave $|(\psi, \phi)|$ invariant. In other words, if an operator A leaves $|(\psi, \phi)|$ invariant, i.e.,

$$|(A\psi, A\phi)| = |(\psi, \phi)|, \tag{3.33}$$

we can find a unitary or antiunitary operator B which, when operating on any vector χ in the space, gives a result $B\chi$ which differs from $A\chi$ only by a constant of magnitude unity. The constant may, of course, depend on the vector χ. The proof of the theorem is given by Wigner (1959). Since an antiunitary operator can be written as the product of a unitary operator times complex conjugation, we shall usually confine our attention to unitary operators. Antiunitary operators are important in considering such transformations as time reversal.

We can use an orthonormal basis ϕ_k to write an operator in the form of a matrix. If we operate on one of the vectors ϕ_j with an operator L, we must obtain a vector which can be written as a linear combination of the vectors ϕ_k

$$L\phi_j = \sum_k L_{kj}\phi_k, \tag{3.34}$$

where the coefficients L_{kj} are complex numbers. Taking the scalar product of ϕ_i with $L\phi_j$, we obtain

$$(\phi_i, L\phi_j) = \sum_k L_{kj}(\phi_i, \phi_k) = \sum_k L_{kj}\delta_{ki} = L_{ij}. \tag{3.35}$$

The L_{ij} can be written as a matrix array, or simply matrix, and the L_{ij} are called matrix elements. The number of basis vectors ϕ_i is the same as the dimensionality of the matrix. The matrix is square if the dimensionality is finite. If the dimensionality is infinite, the matrix is square in the sense that there is a one-to-one correspondence between its rows and columns.

The matrix corresponding to the product of two operators is that matrix obtained by ordinary matrix multiplication of the matrices corresponding to the two operators. This is seen as follows: Let $C = AB$, where A, and B, and C are operators. Now, we have

$$B\phi_j = \sum_k B_{kj}\phi_k.$$

Then

$$AB\phi_j = \sum_k AB_{kj}\phi_k = \sum_k B_{kj}A\phi_k = \sum_k B_{kj}\sum_l A_{lk}\phi_l. \tag{3.36}$$

But

$$C\phi_j = \sum_l C_{lj}\phi_l. \tag{3.37}$$

Comparing (3.36) and (3.37), we get

$$C_{lj} = \sum_k A_{lk} B_{kj}, \tag{3.38}$$

which is the definition of matrix multiplication.

The *hermitian conjugate* or *hermitian adjoint* $L^\dagger$ of a matrix L has matrix elements given by

$$(L^\dagger)_{ij} = L^*_{ji}. \tag{3.39}$$

The notion of the hermitian conjugate of a matrix applies even if the matrix is not square. If L has m columns and n rows, $L^\dagger$ has n columns and m rows.

In matrix notation, a vector is a column matrix of real or complex numbers, and the scalar product of two vectors ψ and ϕ is given by

$$(\psi, \phi) = \psi^\dagger \phi = \Sigma \psi_i^* \phi_i, \tag{3.40}$$

where

$$\phi = \begin{pmatrix} \phi_1 \\ \vdots \end{pmatrix}, \qquad \psi^\dagger = (\psi_1^*, \psi_2^*, \ldots). \tag{3.41}$$

The numbers $\phi_1, \phi_2, \ldots, \psi_1^*, \psi_2^*, \ldots$ refer to the set of scalar products of ϕ and ψ with a set of given basis vectors $f_1, f_2, \ldots$, i.e.,

$$\phi_i = (\phi, f_i), \qquad \psi_i^* = (f_i, \psi), \qquad i = 1, 2, \ldots.$$

An important transformation on a set of operators and vectors is a *similarity transformation*. This is a transformation by means of a linear operator S which leaves unaltered the algebra in the transformed system. All operators L are transformed by

$$L' = SLS^{-1}, \tag{3.42}$$

and all vectors ϕ by

$$\phi' = S\phi. \tag{3.43}$$

Then, if $\phi + \psi = \chi$, it follows that $\phi' + \psi' = \chi'$ and if $L_1 L_2 \phi = L\phi$, it follows that $L_1' L_2' \phi' = L'\phi'$. Note that S must have an inverse. A unitary transformation is a special case of a similarity transformation.

3.3 Some Properties of Representations

We next come to the important concept of a representation of a group. Physicists deal more with group representations than with the properties of abstract groups.

A *representation* of a group G is a homomorphism of G onto a group of linear operators acting on a linear vector space. If the linear operators are matrices, the representation is called a *matrix representation*. Unless otherwise specified, by a representation we shall mean a matrix representation.

We now consider some properties of representations. If, to every member g_i of a group G, we associate a matrix $D(g_i)$, then, in order for the matrices to form a representation of G, we must have the following:

$$D(g_i)D(g_j) = D(g_i g_j). \tag{3.44}$$

If the group is continuous, there is the additional requirement that if $g_i g_j$ is a neighboring element of g_i, then all the numbers of the matrix $D(g_i g_j)$ must differ by only a small amount from the corresponding numbers of $D(g_i)$. We denote the set of all matrices of a representation of a group G by $D(G)$. If we consider more than one representation, we can distinguish between them by superscript $D^{(i)}(G)$.

As a simple example, we consider four representations of the two-element group e, a.

$$\begin{aligned}
&1.\quad D^{(1)}(e) = 1, && D^{(1)}(a) = -1,\\
&2.\quad D^{(2)}(e) = 1, && D^{(2)}(a) = 1,\\
&3.\quad D^{(3)}(e) = \begin{pmatrix} 1 & 0 \\ 0 & 1 \end{pmatrix}, && D^{(3)}(a) = \begin{pmatrix} 1 & 0 \\ 0 & -1 \end{pmatrix},\\
&4.\quad D^{(4)}(e) = \begin{pmatrix} 1 & 0 \\ 0 & 0 \end{pmatrix}, && D^{(4)}(a) = \begin{pmatrix} -1 & 0 \\ 0 & 0 \end{pmatrix}.
\end{aligned} \tag{3.45}$$

All of these representations are by square matrices (including one-dimensional matrices or numbers). We shall always restrict ourselves to this case, with the understanding that a square infinite matrix is one in which the rows and columns can be put in one-to-one correspondence. The matrices of the representations $D^{(4)}$ have vanishing determinants and therefore do not have inverses. Unless otherwise specified, we shall subsequently restrict ourselves to representations by matrices with nonvanishing determinants. The representations $D^{(1)}$, $D^{(3)}$, and $D^{(4)}$ are isomorphic to the group; $D^{(2)}$ is not. If a representation is isomorphic to the group it is said to be *faithful*. All groups of matrices are faithful representations of themselves.

A representation is often called a *vector* representation to distinguish it from a *ray* representation in which, if a group element g_i is represented by the matrix $D(g_i)$, then it is also represented by all the matrices $e^{i\alpha_i}D(g_i)$, where α_i is real. Then the multiplication law becomes

$$D(g_i)D(g_j) = e^{i\alpha_{ij}}D(g_i g_j), \tag{3.46}$$

where α_{ij} is an arbitrary real number which can depend on g_i and g_j. Sometimes α_{ij} is restricted to take on only a finite number of values, in which case the representation is called multiple-valued. Of special importance are the double-valued representations in which $\alpha_{ij} = 0$ or $\alpha_{ij} = \pi$, so that any group element g_i is represented by two matrices $\pm D(g_i)$. Then

$$D(g_i)D(g_j) = \pm D(g_i g_j). \tag{3.47}$$

Unless otherwise specified, we mean, by representation, a single-valued or vector representation.

Two representations are *equivalent* if one can be transformed into the other by a similarity transformation. By this we mean that all matrices of one of the representations can be transformed into their corresponding matrices by the same similarity transformation. A representation by unitary matrices is called a *unitary* representation.

An important theorem states that a representation of a finite or compact Lie group by matrices with nonvanishing determinants can be transformed into a unitary representation by a similarity transformation. For the proof of the theorem see Wigner (1959). Thus, in considering representations of finite or compact Lie groups, we lose no generality in restricting ourselves to unitary representations.

If a representation can be brought into the following form by a similarity transformation:

$$D(g) = \begin{pmatrix} D_1(g) & X(g) \\ 0 & D_2(g) \end{pmatrix}, \tag{3.48}$$

for all $g \in G$, then the representation is called *reducible*. Otherwise it is called *irreducible*. By the representation being brought into a certain form by a transformation, we mean that every matrix of the representation is brought to that form by the *same* transformation. If the representation can be brought into the following form by a similarity transformation:

$$D(A) = \begin{pmatrix} D_1(g) & 0 \\ 0 & D_2(g) \end{pmatrix}, \tag{3.49}$$

then the representation is called *fully reducible* or *decomposable*.

A theorem states that if a representation by unitary matrices is reducible, then it is fully reducible. From this theorem and the fact that representations of a finite or compact Lie group are equivalent to unitary representations, we get the following:

THEOREM. *If a representation of a finite or compact Lie group is reducible, then it is fully reducible.*

The representation is said to be decomposable into a direct sum

$$D = D^{(1)} + D^{(2)} + \cdots \tag{3.50}$$

of irreducible representations. This sum is unique except for the order in which the irreducible representations are written down. Usually, when referring to such representations, we shall just call them reducible.

An important theorem about group representations is known as

SCHUR'S LEMMA. *Let G be a finite or compact Lie group and let D and D′ be irreducible representations of G of dimension n and n′, respectively. Let M be an arbitrary n′ × n matrix (n′ rows and n columns) which is independent of the elements g of G, and let*

$$MD(g) = D'(g)M,$$

for all g in G. Then if $n' \neq n$ it follows that $M = 0$, while if $n' = n$, either $M = 0$ or M is nonsingular and D and D′ are equivalent.

[Note: A corollary, *also called Schur's lemma*, is: *A matrix which commutes with all matrices of an irreducible representation is a multiple of the unit matrix.*]

The *trace* of a matrix is the sum of its diagonal elements. A matrix with trace equal to zero is often said to be traceless. The trace of a matrix $D(g)$ belonging to a representation $D(G)$ is called the *character* of g in the representation D, and is denoted by $\chi(g)$. The set of characters of all members of the representation D is called the character of the representation D and it is denoted by $\chi(D)$.

The importance of character lies in the following:

THEOREM. *All equivalent representations have the same character.*

This theorem is useful in enabling one to decide whether two representations are equivalent.

3.4 Unitary Representations, Multiplets, and Conservation Laws

We now discuss the reasons for the importance of group representations in quantum mechanics. Suppose a group of transformations $T(a)$ leave a physical system invariant. If this is the case, the quantities $|(\phi, \psi)|$, where ϕ and ψ are any vectors (or *functions* or *states*) of the system, are left invariant by the operators representing the transformation. Let us restrict ourselves to the case in which the transformations are represented by unitary operators $U(a)$. Furthermore, consider only those cases in which the Hamiltonian of the system is left invariant by the transformations.

Now consider an eigenfunction ϕ_n of the Hamiltonian H with energy eigenvalue E_n. The eigenvalue equation is

$$H\phi_n = E_n \phi_n. \tag{3.51}$$

If we operate on this equation with the unitary operator $U(a)$, we obtain (omitting the argument a)

$$UH\phi_n = UHU^{-1}U\phi_n = E_n U\phi_n. \tag{3.52}$$

Letting

$$H' = UHU^{-1} \qquad \text{and} \qquad \phi_n' = U\phi_n, \tag{3.53}$$

we get

$$H'\phi_n' = E_n \phi_n'. \tag{3.54}$$

By assumption, the transformation U leaves the Hamiltonian invariant; that is, we have $H' = H$. Therefore, from (3.54) we see that the transformed state ϕ_n' is an eigenstate of the Hamiltonian with the same energy as the state ϕ_n. If we operate with $U(b)$, where b is another value of the parameter, on the state ϕ_n we will obtain, in general, still another eigenstate of H with the same energy E_n. All the states which can be obtained by operating with all U on a given state can be written as linear combinations of a set of basis vectors which span the subspace of eigenstates of the Hamiltonian with the given energy. These vectors are the basis vectors of the unitary representation U of the group of transformations. Furthermore, in general the vectors are the basis vectors of an *irreducible* representation. This can be seen as follows: Suppose the representation is reducible. Then, since the representation is unitary, it is fully reducible and can be decomposed by a similarity transformation into the form

$$U = \begin{pmatrix} U_1 & 0 \\ 0 & U_2 \end{pmatrix}, \tag{3.55}$$

where U_1 and U_2 are themselves unitary matrices. Likewise the vectors take the form

$$\phi_n = \begin{pmatrix} \phi_{1n} \\ \phi_{2n} \end{pmatrix}, \tag{3.56}$$

where ϕ_{1n} and ϕ_{2n} are themselves column matrices. Operating on ϕ_n with U we obtain

$$\phi_n' = \begin{pmatrix} U_1\phi_{1n} \\ U_2\phi_{2n} \end{pmatrix}. \tag{3.57}$$

Thus the transformed vectors of either subspace never involve a linear combination of vectors from the other subspace. Therefore, the vectors in the two subspaces may have different energies. However, if the vectors are the basis vectors of an *irreducible* representation, any one can be transformed into any of the others (or into linear combinations of them) by the unitary transformations of the group. Therefore, all must have the same energy eigenvalue.

The basis vectors of an irreducible unitary representation of a symmetry transformation denote a set of quantum mechanical states. These states are said to constitute a *multiplet*. Since all of the states of a multiplet are eigenstates of the Hamiltonian with the same energy eigenvalue, the states are said to be *degenerate* in the energy. The degeneracy is the number of states with a given energy. If there is only one state corresponding to a given energy, the state is said to be singly degenerate, or more often, nondegenerate.

As an example of a multiplet, we can consider the different states of a particle with spin J. There exist $2J + 1$ vectors representing different orientations of the spin of the particle with respect to an arbitrary z axis. The unitary operators representing rotations transform these states into linear combinations of one another. Because of the rotational invariance of the Hamiltonian, the $2J + 1$ states all have the same energy and constitute a multiplet of degeneracy $2J + 1$. If $J = 0$, the state is nondegenerate.

As another example, the $2I + 1$ different charge states of a particle with isospin I constitute a multiplet. However, since isospin is not an exact symmetry, the different states are not exactly degenerate in energy. It is often said under such circumstances that different members of the multiplet are different particles, rather than different states of the same particle.

Next we point out the relation between symmetry transformations and conservation laws. If $T(a)$ is a symmetry transformation which can be represented by the operator $U(a)$, then we have (again omitting the argument a):

$$H = UHU^{-1}. \tag{3.58}$$

Operating on this equation from the right with U, we obtain $HU = UH$ or

$$[U, H] = 0, \tag{3.59}$$

where

$$[U, H] = UH - HU, \tag{3.60}$$

is the *commutator* of U and H. But in quantum mechanics, the time dependence of an operator F is given by (see, e.g., Schiff, 1968)

$$\frac{dF}{dt} = \frac{\partial F}{\partial t} + \frac{1}{i\hbar}[F, H]. \tag{3.61}$$

Now since the operator U does not depend explicitly on the time, the fact that it commutes with H shows that it is a conserved quantity (a constant of the motion). But any unitary operator U can be written in terms of a hermitian operator A as

$$U = e^{iaA},$$

where a is real. Since U is conserved, so is the operator A. But since A is hermitian it is an observable. Therefore, if a symmetry transformation can be represented by a unitary operator, there exists a conserved observable quantity. The existence of a conserved observable is said to be a conservation law.

As an example, consider a translation operator $U(a)$ which is defined by

$$U(a)\psi(x) = \psi(x + a) \tag{3.62}$$

We expand $\psi(x + a)$ in a Taylor series about x:

$$\begin{aligned}\psi(x + a) &= \psi(x) + a\frac{d}{dx}\psi(x) + \frac{a^2}{2!}\frac{d^2}{dx^2}\psi(x) + \cdots \\ &= \sum_{n=0}^{\infty}\frac{a^n}{n!}\frac{d^n}{dx^n}\psi(x) \\ &= \exp[ia(-id/dx)]\psi(x).\end{aligned} \tag{3.63}$$

Thus $U(a)$ is given by

$$U(a) = e^{iap_x} \tag{3.64}$$

where p_x is just the ordinary linear momentum operator[3] $-id/dx$. Thus if the Hamiltonian of the system is invariant under translations, linear momentum is conserved.

[3] We let $\hbar = 1$.

CHAPTER 4

THE SYMMETRIC GROUP AND IDENTICAL PARTICLES

4.1 Two- and Three-Particle States

Although we are primarily interested in the unitary group $SU(n)$, especially when $n = 2$, 3, or 4, it is convenient to discuss a few properties of the symmetric or permutation group S_n. We mention two reasons for this. The first is that we shall consider a number of systems of several identical particles, and the symmetric group is the relevant group in this regard. The second reason is that we shall introduce the concept of Young diagrams or Young tableaux, which will prove very useful when we discuss the basis functions of the unitary irreducible representations of $SU(n)$. In this and later chapters, we shall place our primary emphasis on the basis functions or multiplets rather than on the representations of the underlying groups of transformations.

Of all the permutations, those that interchange two objects, called *transpositions* are of special importance. This is because any permutation can be written as a product of transpositions. We denote by P_{ij} the transposition which interchanges the object in the ith and jth box. Note that $P_{ji} = P_{ij}$ and $P_{ij}^2 = 1$. A permutation is even or odd according to whether it is a product of an even or odd number of transpositions.

We now wish to make a correspondence between objects in boxes and quantum mechanical states of identical particles. Suppose we have n nonidentical objects, each of which is in a separate identical box. The boxes are labeled by the numbers 1, 2, ..., n. A permutation

$$\begin{pmatrix} 1, & 2, & 3, & \ldots, n \\ a_1, & a_2, & a_3, & \ldots, a_n \end{pmatrix},$$

consists of putting the objects originally in box number 1 into box number a_1, etc. Now, suppose we have n identical particles labeled by 1, 2, ..., n, and each one is in a distinct quantum-mechanical state. Then the identical particles correspond to the identical boxes, and the nonidentical objects correspond to the nonidentical states or wave functions of the particles. This is sometimes a bit confusing, since the objects are in the boxes, while the particles are in states. Nevertheless, the boxes correspond to the particles.

We can use a number of different notations to denote a particular arrangement of objects in boxes or particles in states. Let us denote by a, b, and c three nonidentical objects or three nonidentical single-particle states or wave functions. Then the notation

$$a(1)b(2)c(3),$$

means that object a is in box 1, object b is in box 2, and object c is in box 3. Alternatively, this notation means that particle 1 is in state a, particle 2 is in state b, and particle 3 is in state c. From now on, we shall use the language of quantum-mechanical states. If we perform the permutation $\binom{123}{231}$ on the state $a(1)b(2)c(3)$, we obtain the state $a(2)b(3)c(1)$:

$$\begin{pmatrix} 123 \\ 231 \end{pmatrix} a(1)b(2)c(3) = a(2)b(3)c(1). \tag{4.1}$$

There is a simpler notation for these states of identical particles, based on the convention of always writing first the state of particle 1, second the state of particle 2, etc. Then (4.1) becomes

$$\begin{pmatrix} 123 \\ 231 \end{pmatrix} abc = cab. \tag{4.2}$$

We observe that the permutation $\binom{123}{231}$ can be written as the product of two transpositions

$$P_{13}P_{12} = \begin{pmatrix} 123 \\ 321 \end{pmatrix}\begin{pmatrix} 123 \\ 213 \end{pmatrix} = \begin{pmatrix} 213 \\ 231 \end{pmatrix}\begin{pmatrix} 123 \\ 213 \end{pmatrix} = \begin{pmatrix} 123 \\ 231 \end{pmatrix}. \tag{4.3}$$

Note also that we always operate first with the permutation furthermost on the right. Still another notation for the state $a(1)b(2)c(3)$ is $\psi(123)$. This notation has the advantage that the state $\psi(123)$ may be more general than a product of three one-particle states. In this notation $\psi(231)$ corresponds to $a(2)b(3)c(1)$ or cab.

We now wish to consider states of two and three identical particles as examples. We begin with S_2. Suppose we have a state of two identical particles and want to consider the symmetry of that state under the interchange of the two particles. Let the vector describing this state be $\psi(12)$ where 1 and 2 stand for all the coordinates of the first and second particles, respectively. If the original state $\psi(12)$ has no particular symmetry under the interchange of the two particles, we can construct a symmetric state ψ_s or and antisymmetric state ψ_a as follows:

$$\begin{aligned} \psi_s &= \psi(12) + \psi(21), \\ \psi_a &= \psi(12) - \psi(21), \end{aligned} \tag{4.4}$$

where

$$\psi(21) = P_{12}\psi(12),$$

and P_{12} is a transposition. The state ψ_s is a basis function for a one-dimensional representation of the permutation group. This follows because ψ_s is an eigenstate of any permutation applied to it, as can be directly verified. Thus, ψ_s is a nondegenerate multiplet or singlet. Similarly, ψ_a is an eigenstate of all permutations of S_2 and is a singlet.

Another notation for the two-particle states ψ_s and ψ_a is by means of two boxes, one for each particle. Two boxes in a row denote a symmetric state; two boxes in a column denote an antisymmetric state.

$\psi_s =$ 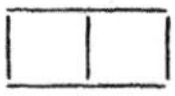 $\psi_a =$ 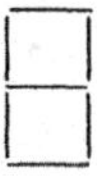 (4.5)

A collection of boxes to represent a state with a particular symmetry property is known as a *Young diagram* or *Young tableau.*[1] For a one-particle state, there is only one diagram, consisting of a single box.

For a three-particle state, we have the following three possible Young tableaux

$\psi_s =$ 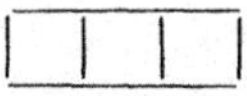 $\psi_a =$  $\psi_n =$ 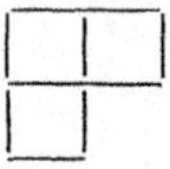 (4.6)

where we adopt the convention that no row is longer than the row above it. Again the row tableau

[1] Some authors distinguish between the terminology *Young diagram* and *Young tableau.* To those authors, a tableau is a diagram with a positive integer in each box.

represents a state which is totally symmetric under the interchange of any two of the particles, and the column tableau

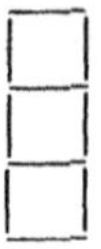

represents a state which is antisymmetric under the interchange. However, there is the possibility of having states of mixed symmetry denoted by the tableau

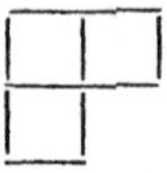

This tableau represents all states which are first made symmetric under the interchange of two of the particles and then made antisymmetric between one of these and the third particle. If we denote the identity operator by 1, then the operator S_{ij} which symmetrizes a state with respect to particles i and j is given by

$$S_{ij} = 1 + P_{ij}. \tag{4.7}$$

Similarly the antisymmetrizing operator A_{ij} is given by

$$A_{ij} = 1 - P_{ij}. \tag{4.8}$$

Then, starting from the state $\psi(123)$, we obtain four independent states ψ_i of mixed symmetry

$$\begin{aligned} \psi_1 &= A_{13} S_{12} \psi(123), \qquad & \psi_2 &= A_{23} S_{12} \psi(123), \\ \psi_3 &= A_{23} S_{13} \psi(123), \qquad & \psi_4 &= A_{12} S_{13} \psi(123). \end{aligned} \tag{4.9}$$

These four mixed states, plus the totally symmetric state ψ_s and the totally antisymmetric state ψ_a, are six linearly independent combinations of the six permutations of $\psi(123)$, including the identity. We could obtain two additional states by using the symmetrizer S_{23}, but these would not be independent of the states (4.9). Alternatively, the mixed tableau

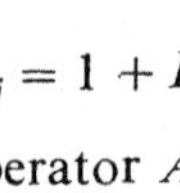

represents the states $\psi_n = S_{ij} A_{ik} \psi(123)$ formed by first antisymmetrizing and then symmetrizing, but again these states are not independent of the states in (4.9). Note, for example, that a state formed by first symmetrizing with respect to particles i and j and then antisymmetrizing with respect to j and k, loses its symmetry under the interchange of i and j. The last-performed symmetrizer or antisymmetrizer is controlling.

The six possible states $\psi(ijk)$ constructed by applying the permutation operators to $\psi(123)$ are the basis functions of a *reducible* representation of S_3. The basis functions for the irreducible representations are those states with definite symmetry properties. Since there is only one symmetric state and one antisymmetric state, the remaining four irreducible states must be of mixed symmetry. However, these states do not form the basis functions of a four-dimensional irreducible representation of S_3, but rather are linear combinations of two two-dimensional irreducible representations. In the next section we shall show how to obtain the dimensionality of the irreducible unitary representations of S_n.

4.2 Standard Arrangements of Young Tableaux

It is true, in general, that the totally symmetric and totally antisymmetric states of S_n are basis functions for one-dimensional irreducible unitary representations. However, all other (inequivalent) irreducible representations of S_n are multidimensional. The dimensionality of these representations can be obtained with the use of Young tableaux. To do this, we must consider so-called standard arrangements of Young tableaux.

Let the number of boxes in the ith row of a tableau be λ_i. We adopt the convention that $\lambda_i \geq \lambda_{i+1}$. Thus a typical tableau looks as follows:

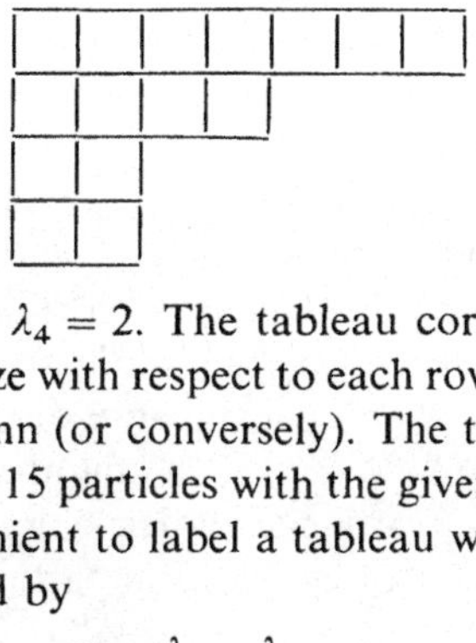

(4.10)

Here $\lambda_1 = 7$, $\lambda_2 = 4$, $\lambda_3 = \lambda_4 = 2$. The tableau corresponds to permutations in which we first symmetrize with respect to each row and then antisymmetrize with respect to each column (or conversely). The tableau (4.10) corresponds to all the possible states of 15 particles with the given symmetry.

It is often more convenient to label a tableau with integers p_i rather than with λ_i. The p_i are defined by

$$p_i = \lambda_i - \lambda_{i+1}. \tag{4.11}$$

For the tableau (4.10) the p_i are given by

$$p_1 = 3, \qquad p_2 = 2, \qquad p_3 = 0, \qquad p_4 = 2.$$

It is easily seen that p_1 is the number of columns with one box, p_2 the number of columns with two boxes, etc. We denote the tableau (4.10) by

$$\lambda = (\lambda_1 \lambda_2 \lambda_3 \lambda_4) = (7422), \qquad \text{or} \qquad p = (p_1 p_2 p_3 p_4) = (3202).$$

Hereafter we shall use only the integers p_i to label a tableau.

We now define a *standard arrangement* of a tableau to be one in which positive integers are placed in the boxes such that the numbers *do not decrease* in going from left to right in a row, and *increase* in going from top to bottom in a column. Each number stands for a possible state in which a single particle can be. If n states are available to a single particle, these states are ordered in some manner, each being assigned a number between 1 and n. Then the number j in each box satisfies $1 \le j \le n$.

As examples of the standard arrangements of tableaux we consider two-particle tableaux with three allowed single-particle states. The symmetric tableau is $\begin{array}{|c|c|}\hline \ & \ \\ \hline\end{array}$ and the standard arrangements are

$$\begin{array}{|c|c|}\hline 1&1\\\hline\end{array}\quad \begin{array}{|c|c|}\hline 1&2\\\hline\end{array}\quad \begin{array}{|c|c|}\hline 1&3\\\hline\end{array}\quad \begin{array}{|c|c|}\hline 2&2\\\hline\end{array}\quad \begin{array}{|c|c|}\hline 2&3\\\hline\end{array}\quad \begin{array}{|c|c|}\hline 3&3\\\hline\end{array} \tag{4.12}$$

The antisymmetric tableau is

$$\begin{array}{|c|}\hline \ \\\hline \ \\\hline\end{array}$$

and the standard arrangements are

$$\begin{array}{|c|}\hline 1\\\hline 2\\\hline\end{array}\quad \begin{array}{|c|}\hline 1\\\hline 3\\\hline\end{array}\quad \begin{array}{|c|}\hline 2\\\hline 3\\\hline\end{array} \tag{4.13}$$

Note that the same number can appear more than once in a given row of a standard arrangement but the numbers in a column must all be different. This is because if two particles are in the same state, this state is symmetric and cannot be antisymmetrized.

We can use the standard arrangements of tableaux to obtain the dimensionality of the irreducible representations of the symmetric group S_n. To do this, we must consider all tableaux with n boxes, each different tableau standing for a different permutation symmetry and hence for a different representation. To obtain the dimensionality of the representation associated with a given tableau of S_n, we count the number of standard arrangements with n states available, and with each particle in a *different* state.

As an example, consider three identical particles, each in a different state, with only three single-particle states available. The standard arrangements of the Young diagrams corresponding to this situation are

$$\begin{array}{|c|c|c|}\hline 1&2&3\\\hline\end{array}\quad \begin{array}{|c|}\hline 1\\\hline 2\\\hline 3\\\hline\end{array}\quad \begin{array}{|c|c|}\hline 1&2\\\hline 3\\\cline{1-1}\end{array}\quad \begin{array}{|c|c|}\hline 1&3\\\hline 2\\\cline{1-1}\end{array} \tag{4.14}$$

We see there is only one way to make a standard arrangement of the tableau

$$\begin{array}{|c|c|c|}\hline \ &\ &\ \\\hline\end{array}$$

and similarly for the tableau

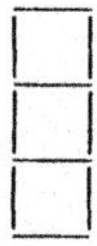

This corresponds to the fact that there is only one symmetric and one antisymmetric combination of three particles. (This is, of course, true for any number of particles.) The nonstandard arrangements correspond to the same states as the standard arrangements after properly symmetrizing, and so must not be counted. For example, the nonstandard arrangement

1	3	2

is the same as the standard arrangement

1	2	3

after symmetrizing with respect to all the particles.

Since there is only one standard arrangement of the symmetric and antisymmetric tableaux, each of them stands for a basis vector of a one-dimensional irreducible unitary representation of S_3. Each is a nondegenerate multiplet or singlet. On the other hand, there are two standard arrangements of the tableau

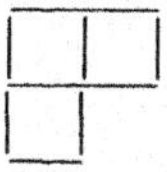

This means that the tableau of mixed symmetry can be used as a shorthand notation for the two functions which are the basis vectors of a two-dimensional irreducible representation of S_3. In other words, the tableau denotes a doublet. So far we have considered only four functions. But we have seen that there are 3! or six functions corresponding to $\psi(123)$ and its permutations. The remaining two functions form the basis vectors of *another* doublet of S_3 represented by the *same* tableau. In general, it can be shown that if an irreducible representation of S_n has N dimensions, there are N such representations of S_n.

As another example, we consider the irreducible representations of S_4. The Young diagrams are

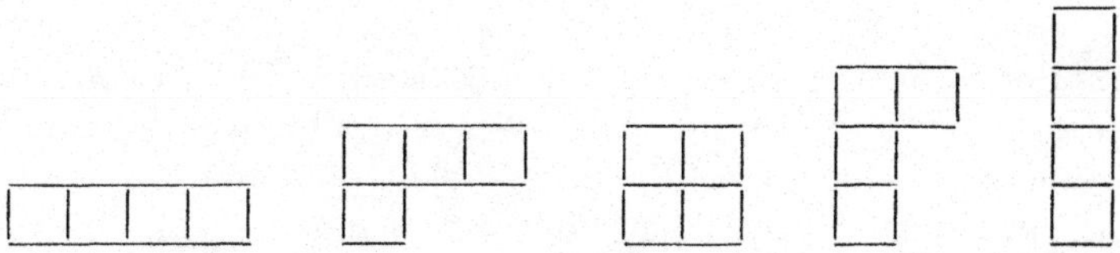

(4.15)

The standard arrangements of these tableaux are

$$
\begin{array}{ccccc}
\begin{array}{|c|c|c|c|}\hline 1&2&3&4\\\hline\end{array} &
\begin{array}{|c|c|c|}\hline 1&2&3\\\hline 4\\\cline{1-1}\end{array} &
\begin{array}{|c|c|}\hline 1&2\\\hline 3&4\\\hline\end{array} &
\begin{array}{|c|c|}\hline 1&2\\\hline 3\\\cline{1-1} 4\\\cline{1-1}\end{array} &
\begin{array}{|c|}\hline 1\\\hline 2\\\hline 3\\\hline 4\\\hline\end{array} \\
 &
\begin{array}{|c|c|c|}\hline 1&2&4\\\hline 3\\\cline{1-1}\end{array} &
\begin{array}{|c|c|}\hline 1&3\\\hline 2&4\\\hline\end{array} &
\begin{array}{|c|c|}\hline 1&3\\\hline 2\\\cline{1-1} 4\\\cline{1-1}\end{array} & \\
 &
\begin{array}{|c|c|c|}\hline 1&3&4\\\hline 2\\\cline{1-1}\end{array} &
 &
\begin{array}{|c|c|}\hline 1&4\\\hline 2\\\cline{1-1} 3\\\cline{1-1}\end{array} &
\end{array}
\tag{4.16}
$$

The tableaux

$$
\begin{array}{|c|c|c|c|}\hline &&&\\\hline\end{array}
\quad \text{and} \quad
\begin{array}{|c|}\hline \\\hline \\\hline \\\hline \\\hline\end{array}
$$

denote the basis vectors of one-dimensional representations. Likewise each of the tableaux

$$
\begin{array}{|c|c|c|}\hline &&\\\hline \\\cline{1-1}\end{array}
\quad \text{and} \quad
\begin{array}{|c|c|}\hline &\\\hline \\\cline{1-1} \\\cline{1-1}\end{array}
$$

denotes the basis vectors of three three-dimensional representations. Finally, the tableau

$$
\begin{array}{|c|c|}\hline &\\\hline &\\\hline\end{array}
$$

denotes the vectors of two two-dimensional representations. This gives a total of

$$1 + 3^2 + 2^2 + 3^2 + 1 = 24$$

basis vectors. But S_4 has $4! = 24$ elements, and therefore 24 functions can be formed from $\psi(1234)$ and its permutations. We have obtained the symmetry properties of the basis functions of the irreducible representations which can be obtained from a reducible 24-dimensional representation of S_4.

In general, a Young diagram with n boxes represents the basis functions for an irreducible representation of S_n. The basis functions are often called *basis tensors*, rather than basis vectors since they depend on n indices.

We now introduce the notion of a Young diagram which is *conjugate* to a given tableau. There are two ways to define conjugation: The first is for the symmetric group and the second is for $SU(n)$. Here we shall define conjugation only for the symmetric group and postpone defining the second kind of conjugation to Section 7.1.

Suppose we consider a given Young tableau with n boxes representing n identical particles, each in a different state. Then the *conjugate tableau* for the symmetric group is that diagram in which each row becomes a column and each column becomes a row. For example, if we consider four-particle states, we have

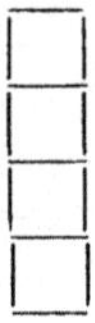 is conjugate to 

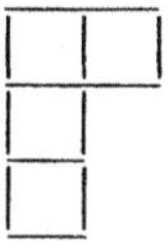 is conjugate to 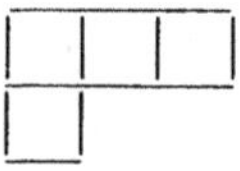

and

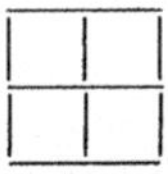 is self conjugate.

Conjugate diagrams belong to different inequivalent representations of the symmetric group with the same dimensionality.

We now reiterate the importance of finding the dimensionality of the irreducible representations. All states which are basis functions of the same irreducible representation of the group (in this case S_n) have the same energy, provided the Hamiltonian of the system is invariant under the group (in this case, permutations of the identical particles). Thus, the dimensionality of an irreducible representation gives the degeneracy of the state, barring any additional accidental degeneracy. The degeneracy associated with the dimensionality of an irreducible representation is often called *essential* degeneracy.

Now suppose the Hamiltonian of a system is invariant under two or more different groups of transformations. These transformations taken together form a larger group, which may or may not be a direct product group, depending on whether or not all the transformations of each group commute with all the transformations of all the others. In any case, the degenerate multiplets are the basis vectors of the irreducible representations of the larger group. If the group is a direct product group, the multiplicities are just the products of multiplicities of the groups comprising the direct product.

All particles observed in nature so far appear to be either bosons or fermions. State vectors describing n identical bosons belong to the symmetric representation of S_n, while the state vectors of n fermions belong to the antisymmetric representation. Thus all particles observed so far appear to belong to one-dimensional representations of the symmetric group, or are singlets.

Why, then, do we study the higher multiplets of S_n? One possible reason is that at some future time particles may be discovered which are neither bosons nor fermions. However, the principal reason for studying higher multiplets of S_n is as follows: Although states of several identical particles are singlets under any permutation of *all* the coordinates of identical particles (i.e., of the particles themselves), nevertheless, the states often exhibit higher multiplicities under permutations of only *some* of the coordinates of the particles. For example, consider three identical spin-1 particles. The overall state of such a system is symmetric. However, the state may have mixed symmetry under the permutations of the spin or space coordinates alone such that the state is symmetric overall.

4.3 Basis Functions of S_3

We construct here a set of orthogonal basis functions for the irreducible unitary representations of S_3 to illustrate how the multiplets are obtained. The state functions of mixed symmetry are not unique, but can be related to others by a similarity transformation. The totally symmetric function is

$$\psi_s = \begin{array}{|c|c|c|}\hline 1 & 2 & 3 \\ \hline \end{array} = S_{123}\, abc$$

where the symmetrizer S_{123} is given by

$$S_{123} = 1 + P_{12} + P_{13} + P_{23} + P_{13}P_{12} + P_{12}P_{13}, \tag{4.17}$$

and the P_{ij} are transpositions. Then

$$\psi_s = abc + bac + cba + acb + cab + bca. \tag{4.18}$$

The totally antisymmetric function is

$$\psi_a = \begin{array}{|c|}\hline 1 \\ \hline 2 \\ \hline 3 \\ \hline \end{array} = A_{123}\, abc$$

where the antisymmetrizer A_{123} is

$$A_{123} = 1 - P_{12} - P_{13} - P_{23} + P_{13}P_{12} + P_{12}P_{13}. \tag{4.19}$$

Then

$$\psi_a = abc - bac - cba - acb + cab + bca. \tag{4.20}$$

Four functions ψ_i of mixed symmetry can be written using the symmetrizers S_{ij} and antisymmetrizers A_{ij} defined in (4.7) and (4.8). We obtain for ψ_1

$$\psi_1 = \begin{array}{|c|c|}\hline 1 & 2 \\ \hline 3 & \multicolumn{1}{c}{} \\ \cline{1-1}\end{array} = A_{13}\, S_{12}\, abc$$

or

$$\psi_1 = abc + bac - cba - cab. \tag{4.21}$$

For ψ_2 we obtain

$$\psi_2 = \begin{array}{|c|c|}\hline 1 & 2 \\ \hline 3 & \multicolumn{1}{c}{} \\ \cline{1-1}\end{array} = A_{23}\, S_{12}\, abc,$$

or

$$\psi_2 = abc + bac - acb - bca. \tag{4.22}$$

For ψ_3 we obtain

$$\psi_3 = \begin{array}{|c|c|}\hline 1 & 3 \\ \hline 2 & \multicolumn{1}{c}{} \\ \cline{1-1}\end{array} = A_{12}\, S_{13}\, abc,$$

or

$$\psi_3 = abc + cba - bac - bca. \tag{4.23}$$

For ψ_4 we obtain

$$\psi_4 = \begin{array}{|c|c|}\hline 1 & 3 \\ \hline 2 & \multicolumn{1}{c}{} \\ \cline{1-1}\end{array} = A_{23}\, S_{13}\, abc,$$

or

$$\psi_4 = abc + cba - acb - cab. \tag{4.24}$$

The functions ψ_1, ψ_2, ψ_3, ψ_4 are not orthogonal to one another. We construct a function ψ which is a linear combination of ψ_1 and ψ_2 and which is orthogonal to ψ_1. It is given by $\psi = \psi_1 - 2\psi_2$, or

$$\psi = 2acb + 2bca - abc - bac - cba - cab. \tag{4.25}$$

It can be seen (by directly operating on ψ_1 and ψ with all six permutations) that any permutation of either of them is a linear combination of the two functions. Thus, ψ_1 and ψ are an orthogonal basis for a two-dimensional representation of S_3. In other words ψ_1 and ψ form a doublet.

We then construct a function ϕ which is a linear combination of ψ_1, ψ, and ψ_3 and which is orthogonal to ψ_1 and ψ. It is given by

$$\phi = 2\psi_3 + \tfrac{1}{2}\psi_1 + \tfrac{1}{2}\psi,$$

or

$$\phi = 2abc - 2bac + cba - cab + acb - bca. \tag{4.26}$$

Finally we construct a linear combination χ of ψ_1, ψ, ϕ, and ψ_4 which is orthogonal to ψ_1, ψ, and ϕ. It is

$$\chi = \tfrac{1}{3}(4\psi_4 - \psi_1 + \psi - \phi),$$

or

$$\chi = -acb + bca - cab + cba. \tag{4.27}$$

It can be directly verified that the functions ϕ and χ also form a doublet of mixed symmetry. All the functions ψ_s, ψ_a, ψ_1, ψ, ϕ, and χ can be easily normalized.

This technique can be generalized to enable us to obtain the basis functions of S_n. To do so, we must make use of symmetrizers and antisymmetrizers of up to n particles. This method is not often used, since it is possible to obtain more useful basis functions which are simultaneously basis functions of $SU(n)$.

For further discussion of the symmetric group, see, for example, Wigner (1959).

CHAPTER 5

LIE GROUPS AND LIE ALGEBRAS

5.1 Some Definitions and Examples

A continuous group with a finite number of parameters is sometimes called a *finite continuous group*. The number of parameters is the *order* of the group. Let us consider a finite continuous group of order r. Denote a group element by $g(a)$, where now a stands for r parameters, namely $a_1, a_2, \ldots, a_r$. The identity element of the group is characterized by the parameter a^0. Without loss of generality we sometimes take $a^0 = 0$. The group elements have the following properties:

$$g(a^0)g(a) = g(a) \qquad \text{or} \qquad g(0)g(a) = g(a), \tag{5.1}$$

$$g(\bar{a})g(a) = g(a^0) \qquad \text{or} \qquad g(\bar{a})g(a) = g(0), \tag{5.2}$$

where we have defined

$$g(\bar{a}) = [g(a)]^{-1} = g^{-1}(a). \tag{5.3}$$

Also if

$$g(c) = g(a)g(b), \tag{5.4}$$

then c is a continuous function of a and b

$$c = \phi(a, b). \tag{5.5}$$

This notation stands for

$$c_i = \phi_i(a_1, \ldots, a_r, b_1, \ldots, b_r), \qquad i = 1, 2, \ldots, r.$$

If further, c is an analytic function of a and b, and $\bar{a}$ is an analytic function of a, then the group is said to be an *r-parameter Lie group*. Instead of calling a group element $g(a)$, we can simply call it a. The domain over which the parameters vary may comprise a number of disconnected regions. If so, the group is said to be a mixed continuous group.

We now consider groups of transformations on an n-dimensional space by means of an r-parameter Lie group. We assume that the r parameters are essential. If the parameters are not essential, we can find a formal procedure for obtaining the essential parameters (for example, see Hamermesh, 1962). Our treatment follows in part that of Racah (1965) and Behrends *et al.* (1962), but is somewhat less rigorous in the interest of simplicity. We also omit the proofs of a number of theorems. Some of these proofs can be found in Racah (1965).

Suppose we have an n-dimensional vector space with basis vectors $x_1, \ldots, x_n$. A transformation on that space can be written

$$x_i' = f_i(x_1, \ldots, x_n, a_1, \ldots, a_r), \qquad i = 1, 2, \ldots, n, \tag{5.6}$$

or in simpler notation:

$$x' = f(x, a). \tag{5.7}$$

The f_i are all analytic functions of their arguments. Suppose we have a set of transformations $x' = f(x, a)$. We assume that an identity transformation exists and that the transformations are associative. Then, in order for the transformations to form a Lie group, the following requirements must hold.

(1) If $x' = f(x, a)$ and $x'' = f(x', b)$, there exists a parameter c satisfying

$$c = \phi(a, b) \tag{5.8}$$

where ϕ is an analytic function of a and b such that

$$x'' = f(x, c) = f(x, \phi(a, b)). \tag{5.9}$$

(2) For every a there exists a unique $\bar{a}$ such that if $x' = f(x, a)$ then

$$x = f(x', \bar{a}). \tag{5.10}$$

As an example of a Lie group we consider the group of transformations

$$x' = a_1 x + a_2, \qquad a_1 \neq 0. \tag{5.11}$$

Identity:	$a_1{}^0 = 1,$	$a_2{}^0 = 0,$
Inverse:	$\bar{a}_1 = 1/a_1,$	$\bar{a}_2 = -a_2/a_1,$
Product:	$c_1 = b_1 a_1,$	$c_2 = b_1 a_2 + b_2.$

Alternatively, a group element can be written as an ordered pair (a_1, a_2). Then the identity is $(1, 0)$, the inverse is $(1/a_1, -a_2/a_1)$ and the product is given by

$$(b_1, b_2)(a_1, a_2) = (b_1 a_1, b_1 a_2 + b_2).$$

If a_1 and a_2 are real, this is a two-parameter group; otherwise it is a four-parameter group. If we restrict the domain of variation of the parameters, the transformations do not necessarily form a group at all. For example, if a_2 is restricted to be between zero and one, the product of two transformations may be outside the interval.

Two Lie groups are isomorphic if they have the same number of parameters, the same domain of variation of the parameters, and the same law of combination. As an example we note that the one-dimensional group

$$x' = ax, \qquad a \neq 0,$$

is isomorphic to the two-dimensional group

$$x_1' = ax_1, \qquad x_2' = ax_2, \qquad a \neq 0.$$

Both of these groups are Abelian Lie groups. It should be clear that we can have a one-dimensional group isomorphic to a group of transformations in any number of dimensions.

We shall enlarge the definition of a Lie group to include a mixed analytic group which has a finite number of discontinuous pieces. For example, the orthogonal group $O(3)$ divides into two pieces, one piece consisting of those elements with determinant 1 and the other piece consisting of those elements with determinant -1. Only the elements of the first piece can be reached continuously from the identity.

5.2 Generators of Lie Groups

The great contribution of Sophus Lie (Lie and Scheffers, 1893) to the theory of continuous analytic groups (Lie groups) was to consider those elements which differ infinitesimally from the identity, and to show that from them one can obtain most of the properties of the group in the large. Specifically, Lie showed that the properties of the elements of a Lie group which can be reached continuously from the identity are determined from the elements lying in neighborhood of the unit element. In view of the restrictive nature of analytic functions, this result is not surprising.

Consider a Lie group of transformations which take x into x':

$$x' = f(x, a). \tag{5.12}$$

The vectors x and x' are n-dimensional and a stands for r parameters. In other words, the transformation $x' = f(x, a)$ is shorthand notation for

$$x_i' = f_i(x_1, \ldots, x_n, a_1, \ldots, a_r), \qquad i = 1, \ldots, n.$$

Let us use the convention that the identity element is given by $a = 0$. Then

$$x = f(x, 0). \tag{5.13}$$

We can make a small change in x by varying the identity element by a small amount da, obtaining

$$x + dx = f(x, da). \tag{5.14}$$

Then to lowest order in da we obtain

$$dx = f(x, da) - f(x, 0) = \frac{\partial f(x, 0)}{\partial a}\, da. \tag{5.15}$$

Define $u(x)$ by

$$u(x) = \frac{\partial f(x, 0)}{\partial a}. \tag{5.16}$$

Then

$$dx = u(x)\ da. \tag{5.17}$$

This equation is shorthand for

$$dx_i = \sum_{\nu}^{r} u_{i\nu}(x)\, da_\nu, \qquad i = 1, 2, \ldots, n, \tag{5.18}$$

where

$$u_{i\nu} = \left.\frac{\partial f_i(x, a)}{\partial a_\nu}\right|_{a=0} \equiv \frac{\partial f_i(x, 0)}{\partial a_\nu}.$$

We are using the convention that Greek indices go from 1 to the number of parameters r, and Latin indices go from 1 to the dimension of the space n.

Now consider the change in a function $F(x)$ under an infinitesimal transformation

$$dF = (\partial F/\partial x)\, dx. \tag{5.19}$$

But $dx = u(x)\, da$. Therefore (5.19) becomes

$$dF = (\partial F/\partial x)u(x)\, da = da\{u(x)(\partial/\partial x)\}F. \tag{5.20}$$

Now define the quantity X by

$$X = u(x)\frac{\partial}{\partial x} = \frac{\partial f(x, 0)}{\partial a}\frac{\partial}{\partial x}. \tag{5.21}$$

The operator X is an r-dimensional vector operator with components

$$X_\nu = \sum_{i=1}^{n} u_{i\nu}(\partial/\partial x_i), \qquad \nu = 1, 2, \ldots, r. \tag{5.22}$$

The component operators X_ν are called the *generators* of the group. There is one generator for each essential parameter.

With a suitable choice of parametrization, these generators can be made hermitian. From a set of hermitian generators we can obtain a representation of the group by means of unitary operators $U(a_\nu)$ which are given by

$$U(a_\nu) = e^{ia_\nu X_\nu} \equiv U_\nu \tag{5.23}$$

where the a_ν are real parameters which vary continuously from the identity. An arbitrary element of the group which can be reached continuously from the identity can be written as a product of operators U_ν. Therefore, to obtain that portion of a Lie group which is connected to the identity element, we need only to study the properties of the generators. All the Lie groups with a given set of generators X_ν are *locally isomorphic*, that is isomorphic in a neighborhood of the identity. Furthermore, there exist homomorphic mappings from each of these groups onto the piece which can be reached continuously from the identity, which constitutes an invariant subgroup.

An important theorem is that the commutators of the generators of a Lie group are linear combinations of the generators. See, for example, Racah (1965) for the proof of the theorem. The statement of the theorem can be written

$$[X_\mu, X_\nu] = \sum_{\lambda=1}^{r} c_{\mu\nu\lambda} X_\lambda \tag{5.24}$$

where the $c_{\mu\nu\lambda}$ are complex numbers called *structure constants*. The algebra of the generators is called a *Lie algebra*. Since commutators are linearly independent and satisfy the relation

$$[X_\nu, X_\mu] = -[X_\mu, X_\nu], \tag{5.25}$$

it follows that the structure constants satisfy

$$c_{\mu\nu\lambda} = -c_{\nu\mu\lambda}. \tag{5.26}$$

Also, since commutators satisfy the Jacobi identity

$$[[X_\nu, X_\mu], X_\lambda] + [[X_\mu, X_\lambda], X_\nu] + [[X_\lambda, X_\nu], X_\mu] = 0, \tag{5.27}$$

we obtain

$$\sum_{\beta=1}^{r} (c_{\nu\mu\beta} c_{\beta\lambda\delta} + c_{\mu\lambda\beta} c_{\beta\nu\delta} + c_{\lambda\nu\beta} c_{\beta\mu\delta}) = 0. \tag{5.28}$$

Since the foregoing is somewhat abstract, we shall give some examples of how to obtain the generators of Lie groups and how to obtain the Lie algebras. As a first example, consider the two-parameter group of transformations on a one-dimensional vector

$$x' = \alpha_1 x + \alpha_2, \qquad \alpha = (\alpha_1, \alpha_2).$$

We write this in another way so that the identity element is zero. This is

$$x' = (1 + a_1)x + a_2, \qquad a = (a_1, a_2)$$

where $a_1 = \alpha_1 - 1$ and $a_2 = \alpha_2$. The infinitesimal transformation now is

$$x + dx = (1 + da_1)x + da_2$$

or

$$dx = x\ da_1 + da_2.$$

Then, comparing this expression for dx with the expression in Eq. (5.18), we obtain

$$u_{11} = x, \qquad u_{12} = 1.$$

Then, using (5.22), we obtain the generators

$$X_1 = x(\partial/\partial x), \qquad X_2 = (\partial/\partial x).$$

We can work out the commutator by operating with it on a function $F(x)$:

$$[X_1, X_2]F = x\frac{\partial^2 F}{\partial x^2} - \frac{\partial}{\partial x}\left(x\frac{\partial F}{\partial x}\right)$$

$$= x\frac{\partial^2 F}{\partial x^2} - \frac{\partial F}{\partial x} - x\frac{\partial^2 F}{\partial x^2} = -\frac{\partial F}{\partial x} = -X_2 F.$$

Thus the commutator is

$$[X_1, X_2] = -X_2,$$

and the structure constants are

$$c_{121} = 0, \qquad c_{122} = -1.$$

It should be clear that the way we parametrize a group is, to some extent, arbitrary. For example the transformation $x' = (1 + a_1)x + a_2$ which we have just considered may be written in terms of new parameters b_1 and b_2 as

$$x' = (1 + b_1 + b_2)x + b_1 - b_2.$$

If we obtain the generators of the group with this parametrization, they will be different from the generators we previously obtained, and the structure constants will be different. In fact, any linear combination of the generators of a Lie group is also a generator. This means that there is a continuum of different sets of structure constants for a given Lie group. However, all the sets of structure constants correspond to only one Lie algebra.

As another example, consider the two-dimensional, two-parameter Abelian group given by the following transformations

$$x_1' = \alpha_1 x_1,$$
$$x_2' = \alpha_2 x_2.$$

We write this so the identity is zero.

$$x_1' = (1 + a_1)x_1,$$
$$x_2' = (1 + a_2)x_2.$$

Then for an infinitesimal transformation

$$x_1 + dx_1 = (1 + da_1)x_1$$
$$x_2 + dx_2 = (1 + da_2)x_2$$

or

$$dx_1 = x_1\, da_1, \qquad dx_2 = x_2\, da_2.$$

Thus

$$X_1 = x_1(\partial/\partial x_1), \qquad X_2 = x_2(\partial/\partial x_2).$$

In this case the commutator is

$$[X_1, X_2] = 0.$$

In fact, all the structure constants are zero for any Abelian Lie group. Therefore, the Lie algebra for an r-parameter Abelian group is trivial, consisting of r operators X_μ, all of which commute. The Lie algebra is then said to be Abelian.

Still another example is given by the two-dimensional, one-parameter group

$$x_1' = (1 + a)x_1$$
$$x_2' = (1 + a)^{-1}x_2.$$

We have written this transformation so that the identity element is zero. The infinitesimal transformation is

$$x_1 + dx_1 = (1 + da)x_1,$$
$$x_2 + dx_2 = (1 + da)^{-1}x_2 = (1 - da)x_2.$$

Then

$$dx_1 = x_1\, da$$
$$dx_2 = -x_2\, da.$$

Then the single generator is

$$X = x_1(\partial/\partial x_1) - x_2(\partial/\partial x_2).$$

Now let us consider the special orthogonal group in three dimensions $SO(3)$ or $R(3)$. We use matrix notation

$$x' = Ax \tag{5.29}$$

where x and x' are three-dimensional vectors and A is a real three-dimensional matrix satisfying $A\tilde{A} = 1$; where $\tilde{A}$ means the transpose of A, that is $\tilde{A}_{ij} = A_{ji}$. Writing the transformation in the form $x' = (1 + a)x$ so that the identity $a^0 = 0$, we obtain

$$x + dx = (1 + da)x, \qquad dx = x\, da.$$

Now

$$A\tilde{A} = 1 \Rightarrow (1 + da)(1 + d\tilde{a}) = 1,$$

or to lowest order in a

$$1 + da + d\tilde{a} = 1, \qquad \text{or} \qquad da = -d\tilde{a}.$$

Thus da must be of the form

$$da = \begin{pmatrix} 0 & da_{12} & -da_{13} \\ -da_{12} & 0 & da_{23} \\ da_{13} & -da_{23} & 0 \end{pmatrix}. \tag{5.30}$$

Thus the group has three essential parameters. Letting $da_{23} = a_1$, $da_{13} = a_2$, $da_{12} = a_3$ we get

$$dx_1 = x_2 a_3 - x_3 a_2,$$
$$dx_2 = -x_1 a_3 + x_3 a_1,$$
$$dx_3 = x_1 a_2 - x_2 a_1.$$

Then the operators are

$$X_1 = x_3 \frac{\partial}{\partial x_2} - x_2 \frac{\partial}{\partial x_3},$$
$$X_2 = -x_3 \frac{\partial}{\partial x_1} + x_1 \frac{\partial}{\partial x_3}, \tag{5.31}$$
$$X_3 = x_2 \frac{\partial}{\partial x_1} - x_1 \frac{\partial}{\partial x_2}.$$

We can obtain the commutation relations in a straightforward way:

$$[X_1, X_2] = X_3, \qquad [X_2, X_3] = X_1, \qquad [X_3, X_1] = X_2. \tag{5.32}$$

In parametrizing a Lie group, we have taken all the essential parameters to be real. Therefore, in general the Lie algebra is an algebra with real scalars. However, we can take certain complex linear combinations of the generators and still preserve the algebra. We shall often have occasion to do this. For example, in the case of $R(3)$, we can let $J_\nu = iX_\nu$. Then the commutation relations of Eq. (5.32) become

$$[J_\mu, J_\nu] = i\varepsilon_{\mu\nu\lambda} J_\lambda, \tag{5.33}$$

where $\varepsilon_{\mu\nu\lambda}$ is antisymmetric under odd permutations and $\varepsilon_{123} = 1$. A comparison of (5.32) and (5.33) again shows that the structure constants multiplying the operators depend on the choice of operators.

We next consider the group $SU(2)$ and show that it has the same Lie algebra as $R(3)$. The importance of this is that if two Lie groups have the same Lie algebra they are locally isomorphic; that is, they are isomorphic in a neighborhood of the identity. Furthermore, if the two groups are not isomorphic in the large, there exists a homomorphic mapping of the larger one on to the smaller one.

The group $SU(2)$ consists of the unitary matrices of determinant unity (denoted by A) acting on a complex two-dimensional vector space. The transformation is given by

$$x' = Ax. \tag{5.34}$$

We write the infinitesimal transformations as

$$x' = (1 + da)x, \qquad dx = da\,x \tag{5.35}$$

Since $AA^\dagger = 1$, we have

$$(1 + da)(1 + da^\dagger) = 1 = 1 + da + da^\dagger$$

or

$$da + da^\dagger = 0. \tag{5.36}$$

Adding the requirement that

$$\det(1 + da) = 1, \tag{5.37}$$

we find we can write da in the form

$$da = \begin{pmatrix} ia_1 & a_2 + ia_3 \\ -a_2 + ia_3 & -ia_1 \end{pmatrix}. \tag{5.38}$$

where the a_i are infinitesimal real numbers.

$$dx_1 = ia_1x_1 + (a_2 + ia_3)x_2,$$
$$dx_2 = (-a_2 + ia_3)x_1 - ia_1x_2.$$

The generators are

$$X_1 = ix_1\frac{\partial}{\partial x_1} - ix_2\frac{\partial}{\partial x_2},$$
$$X_2 = x_2\frac{\partial}{\partial x_1} - x_1\frac{\partial}{\partial x_2}, \tag{5.39}$$
$$X_3 = ix_2\frac{\partial}{\partial x_1} + ix_1\frac{\partial}{\partial x_2}.$$

The commutation relations satisfied by these operators are

$$[X_1, X_2] = -2X_3, \tag{5.40}$$

and cyclic permutations. If we let $X_\mu = -2iJ_\mu$, we find that the J_μ satisfy the commutation relations (5.33). This shows that $SU(2)$ has the same Lie algebra as $R(3)$. The two groups are not isomorphic, but there is a homomorphic mapping of one onto the other. In this case there is a two-to-one mapping of $SU(2)$ onto $R(3)$.

Equations (5.34) through (5.37) hold for the special unitary group in any number of dimensions. For $SU(n)$, however, the infinitesimal element da, instead of being given by (5.38) is given in terms of $n^2 - 1$ parameters. For $SU(3)$, da is given in terms of eight real infinitesimal parameters c_ν.

$$da = \begin{pmatrix} ic_1 & c_2 + ic_3 & c_4 + ic_5 \\ -c_2 + ic_3 & ic_6 & c_7 + ic_8 \\ -c_4 + ic_5 & -c_7 + ic_8 & -ic_1 - ic_6 \end{pmatrix}.$$

With this parametrization, the generators are

$$\begin{aligned}
X_1 &= ix_1 \frac{\partial}{\partial x_1} - ix_3 \frac{\partial}{\partial x_3}, & X_5 &= ix_3 \frac{\partial}{\partial x_1} + ix_1 \frac{\partial}{\partial x_3}, \\
X_2 &= x_2 \frac{\partial}{\partial x_1} - x_1 \frac{\partial}{\partial x_2}, & X_6 &= ix_2 \frac{\partial}{\partial x_2} - ix_3 \frac{\partial}{\partial x_3}, \\
X_3 &= ix_2 \frac{\partial}{\partial x_1} + ix_1 \frac{\partial}{\partial x_2}, & X_7 &= x_3 \frac{\partial}{\partial x_2} - x_2 \frac{\partial}{\partial x_3}, \\
X_4 &= x_3 \frac{\partial}{\partial x_1} - x_1 \frac{\partial}{\partial x_3}, & X_8 &= ix_3 \frac{\partial}{\partial x_2} + ix_2 \frac{\partial}{\partial x_3}.
\end{aligned} \tag{5.41}$$

If a group has r generators, then the number of commutation relations which must be worked out is $\frac{1}{2}(r - 1)r$. This can be seen as follows: Every generator commutes with itself. Thus, there are $r - 1$ commutation relations between the first generator and the others. Likewise, there are $r - 2$ commutation relations between the second and all the others except the first, etc. The total number of commutation relations is then

$$(r - 1) + (r - 2) + \cdots + 1 = \tfrac{1}{2}(r - 1)r. \tag{5.42}$$

Since $SU(n)$ has $r = n^2 - 1$ parameters it has $n^2 - 1$ generators, and there are $\frac{1}{2}(n^2 - 2)(n^2 - 1)$ commutation relations to be worked out. In the next section we shall see how to simplify the problem of obtaining the commutation relations.

5.3 Simple and Semisimple Lie Algebras

Consider a subspace B of a Lie algebra A. If the commutator of any operator in A with any operator in B also lies in B, then B is said to be an *ideal.* A Lie algebra is *simple* if it is not Abelian and its only nonzero ideal is A itself. A Lie algebra is *semisimple* if it has no nonzero Abelian ideals. Since all the generators of an Abelian Lie group commute, the Lie algebra of an Abelian Lie group is not semisimple. Likewise, the Lie algebra of a one-parameter group is not semisimple. We can now give the definitions of a simple and a semisimple Lie group.

If a Lie algebra is simple or semisimple, then so are all the Lie groups corresponding to the algebra.

This definition is different from the definition given in Section 2.4. For example, consider the groups $SU(2)$ and $SO(3)$. As we have seen, both of these groups have the same Lie algebra, and this Lie algebra is simple. Therefore, we call $SU(2)$ simple, even though the group Z_2 defined in Section 2.4 is a proper invariant subgroup of $SU(2)$.

The Lie algebra of a direct product of two groups is just the algebra of the generators of both. The generators of one of the groups in the direct product, of course, all commute with the generators of the other group. The Lie algebra of the direct product of two simple groups is semisimple.

It is often convenient to study Lie algebras rather than Lie groups. The reason is that by considering a single Lie algebra, one is, in effect, considering a whole class of Lie groups, all of which are locally isomorphic. However, the Lie algebra of a group does not tell about the global properties of the group. Thus, the question arises as to the importance of the global properties for physics. As we have remarked previously, we are more interested in the representations of Lie groups than the abstract groups themselves. Now, in general, if two groups differ globally, then one of them has more vector representations than the other. However, in quantum mechanics we should not exclude multivalued representations. When these are included, then all groups with the same Lie algebra have the same representations.

5.4 Standard Form of Lie Algebras

Let us consider further properties of the generators X_μ of a given Lie group. The number of generators which are mutually commuting operators depends on the particular choice of X_μ. For example, suppose we have three operators X_1, X_2, and X_3 such that

$$[X_1, X_2] = 0, \qquad [X_1, X_3] \neq 0, \qquad [X_2, X_3] \neq 0.$$

If we define

$$X_1' = X_1 + X_3, \qquad X_2' = X_2, \qquad X_3' = X_3,$$

then none of the operators X_1', X_2', X_3' commute. Yet this is just another way to describe the same Lie algebra.

An important question is what is the *maximum* number of commuting generators of the algebra. In other words, we want to know how many generators can be made to commute by choosing new generators to be arbitrary linear combinations of the old. If the group is a symmetry group of the system, all commuting generators will also commute with the Hamiltonian. If the generators are chosen to be hermitian, the commuting set of generators will be simultaneously observable constants of the motion.

We now discuss the question of how to find the maximum number of commuting generators of a Lie group, given a particular set of generators X_μ. Let us choose a linear combination A of the X_μ which can be diagonalized. We let

$$A = \sum_\mu a_\mu X_\mu, \tag{5.43}$$

where the sum goes from 1 to r, the number of generators of the group. Consider the "eigenvalue" problem

$$[A, X] = \rho X, \tag{5.44}$$

where the unknown X is again a linear combination of the X_μ:

$$X = \sum_\mu b_\mu X_\mu, \tag{5.45}$$

and ρ is an eigenvalue to be determined. But we know

$$[X_\mu, X_\nu] = \sum_\lambda c_{\mu\nu\lambda} X_\lambda. \tag{5.46}$$

Substituting these expressions for A and X into Eq. (5.44) and using Eq. (5.46), we obtain

$$\left[\sum_\mu a_\mu X_\mu, \sum_\nu b_\nu X_\nu\right] = \sum_{\mu\nu\lambda} a_\mu b_\nu c_{\mu\nu\lambda} X_\lambda = \rho \sum_\lambda b_\lambda X_\lambda,$$

or

$$\sum_\lambda \left[\sum_{\mu\nu} a_\mu b_\nu c_{\mu\nu\lambda} - \rho b_\lambda\right] X_\lambda = 0. \tag{5.47}$$

Since the X_λ are linearly independent, each coefficient must vanish:

$$\sum_{\mu\nu} a_\mu b_\nu c_{\mu\nu\lambda} - \rho b_\lambda = 0,$$

or

$$\sum_\nu \Big(\sum_\mu a_\mu c_{\mu\nu\lambda} - \rho\delta_{\nu\lambda}\Big) b_\nu = 0. \tag{5.48}$$

This is a set of homogeneous linear equations for b. For a solution to exist, the determinant of the coefficients of b_ν must vanish:

$$\det\Big(\sum_\mu a_\mu c_{\mu\nu\lambda} - \rho\delta_{\nu\lambda}\Big) = 0. \tag{5.49}$$

This equation is called the secular equation for the eigenvalues ρ. These eigenvalues are called *roots*.

If we have an r-parameter group, there may be fewer than r linearly independent eigenvectors if the secular equation has degenerate roots. By a degenerate root, we mean the following: The secular equation is an rth order polynomial in ρ, and thus has r roots. However, all these roots need not be different; if two or more are the same, the root is said to be *degenerate*. The *multiplicity* of a root is the number of times it occurs.

For the groups of interest in most physical problems, Eq. (5.44) does have r independent eigenvectors independently of whether any root is degenerate. In particular, this is true for a semisimple Lie group. An important theorem about semisimple Lie groups has been proven by Cartan.

THEOREM. *If A is chosen so that the secular equation has the maximum number of roots, then only the root $\rho = 0$ is degenerate. Furthermore, if l is the multiplicity of this root, there are l linearly independent eigenvectors H_i $(i - 1, 2, \ldots, l)$ which commute with one another*

$$[H_i, H_j] = 0. \tag{5.50}$$

The number l is called the *rank* of the group, and is the maximal number of mutually commuting generators. Since A commutes with itself, the rank of a semisimple Lie group is $l \geq 1$. We shall use Latin letters for indices which can vary from 1 to l, Greek letters α, β, γ for indices which can vary from 1 to $r - l$, and other Greek letters for indices which can vary from 1 to r.

We denote the noncommuting eigenvectors (corresponding to nonzero nondegenerate values of ρ) by the symbols E_α. Thus, for a rank l semisimple Lie group, we have the following eigenvectors.

$$H_i, \quad i = 1, 2, \ldots, -l, \qquad E_\alpha, \quad \alpha = 1, 2, \ldots, r - l. \tag{5.51}$$

The H_i and E_α are, of course, generators of the group.

Let us use $SU(2)$ as an example: We take for the original X_μ the J_μ which satisfy the commutation relations of Eq. (5.33). Let $A = J_3$. Then if we solve the equation

$$[A, X] = \rho X,$$

we obtain for the three values of X

$$\begin{aligned} X &= J_3 \equiv H_1 = A, \qquad &\rho &= 0,\\ X &= (J_1 + iJ_2)/\sqrt{2} \equiv E_1, \qquad &\rho &= 1,\\ X &= (J_1 - iJ_2)/\sqrt{2} \equiv E_2, \qquad &\rho &= -1, \end{aligned} \tag{5.52}$$

or

$$[H_1, H_1] = 0, \qquad [H_1, E_1] = E_1, \qquad [H_1, E_2] = -E_2. \tag{5.53}$$

The operators E_1 and E_2 are just the usual raising and lowering operators $J_\pm$ of angular momentum except for normalization

$$\sqrt{2}\,E_1 = J_+, \qquad \sqrt{2}\,E_2 = J_-. \tag{5.54}$$

Note that the nonvanishing values of ρ are $\rho = \pm 1$. It is true for any semisimple compact group that the nonvanishing roots come in pairs equal in magnitude and opposite in sign.

Let the roots, denoted by $\rho_i(\alpha)$, be solutions to the equations

$$[H_i, E_\alpha] = \rho_i(\alpha) E_\alpha, \qquad i = 1, 2, \ldots, l. \quad \alpha = 1, 2 \ldots, r - l. \tag{5.55}$$

Then the l values of ρ_i for a given α can be considered to be the components of a *root vector* (or just *root*) of dimensionality equal to the rank of the group. The l operators H_i can also be considered as the components of a vector $\mathbf{H}$, and we can write

$$[\mathbf{H}, E_\alpha] = \boldsymbol{\rho}(\alpha) E_\alpha. \tag{5.56}$$

We now quote without proof two theorems for semisimple compact groups.

THEOREM. *If $\boldsymbol{\rho}(\alpha)$ is a root, then $-\boldsymbol{\rho}(\alpha)$ is also a root.*

We shall now use the notation $\boldsymbol{\rho}(-\alpha) \equiv -\boldsymbol{\rho}(\alpha)$ and let α take on $\frac{1}{2}(r - l)$ values rather than $r - l$ values.

THEOREM. *The E_α satisfy the following commutation relations*

$$[E_\alpha, E_{-\alpha}] = \sum_i c_{\alpha, -\alpha, i} H_i, \qquad \alpha = 1, 2, \ldots, \pm\tfrac{1}{2}(r - l), \tag{5.57}$$

$$[E_\alpha, E_\beta] = \begin{cases} c_{\alpha\beta\gamma} E_\gamma & \textit{if}\ \ \boldsymbol{\rho}(\gamma) = \boldsymbol{\rho}(\alpha) + \boldsymbol{\rho}(\beta) \neq 0 \\ 0 & \textit{otherwise}. \end{cases} \tag{5.58}$$

We can normalize the H_i such that

$$\sum_{\alpha=1}^{(r-l)/2} \rho_i(\alpha)\rho_j(\alpha) = \delta_{ij}. \tag{5.59}$$

Then it can be shown that we can normalize the E_α such that

$$[E_\alpha, E_{-\alpha}] = \sum_i \rho_i(\alpha) H_i, \tag{5.60}$$

or in vector notation

$$[E_\alpha, E_{-\alpha}] = \boldsymbol{\rho}(\alpha) \cdot \mathbf{H}. \tag{5.61}$$

We introduce the notation $N_{\alpha\beta} = c_{\alpha\beta, \alpha+\beta}$ and collect Eqs. (5.50), (5.56), (5.58), and (5.61). Then we have the Lie algebra in standard form:

$$\begin{aligned} [H_i, H_j] &= 0, \\ [\mathbf{H}, E_\alpha] &= \boldsymbol{\rho}(\alpha) E_\alpha, \\ [E_\alpha, E_{-\alpha}] &= \boldsymbol{\rho}(\alpha) \cdot \mathbf{H}, \end{aligned} \tag{5.62}$$

$$[E_\alpha, E_\beta] = \begin{cases} N_{\alpha\beta} E_{\alpha+\beta} & \text{if } \boldsymbol{\rho}(\alpha) + \boldsymbol{\rho}(\beta) \text{ is a nonvanishing root} \\ 0 & \text{otherwise.} \end{cases}$$

The $N_{\alpha\beta}$ can be directly computed once **H** and **E** are known. These equations are called the *standard form* of the commutation relations. For $SU(2)$, the standard form of the commutation relations is given by

$$\begin{aligned} [H_1, E_1] &= E_1, \\ [H_1, E_{-1}] &= -E_{-1}, \\ [E_1, E_{-1}] &= H_1, \end{aligned} \tag{5.63}$$

where

$$H_1 = J_3, \qquad \sqrt{2}E_1 = J_+ = J_1 + iJ_2, \qquad \sqrt{2}E_{-1} = J_- = J_1 - iJ_2. \tag{5.64}$$

Since there is only one positive value of α, there are no $N_{\alpha\beta}$ to be computed in this case.

Next we shall quote two important theorems about Lie groups, and then discuss the roots in some more detail.

THEOREM. *From the members of a Lie algebra of a semisimple Lie group of rank l, one can construct l nonlinear invariant operators, called Casimir operators, which commute with every member of the algebra.*

For example, the group $SU(2)$ (rank $l = 1$) has one invariant operator $J^2 = J_1{}^2 + J_2{}^2 + J_3{}^2$.

THEOREM. *There are only a finite number of simple Lie algebras of any rank l.*

The importance of this theorem for applications in elementary particle physics is that we know of only a finite number of additive conserved quantum numbers which serve to distinguish the elementary particles from one another. It is true that this number appears to keep growing with time, as new particles are discovered which do not fit into old schemes. Nevertheless, the number of known additive quantum numbers has so far grown at a much smaller rate than the number of known elementary particles. If this trend continues, the group-theory approach to the classification of elementary particles should continue to be useful.

At any time, we can attempt to classify the known elementary particles according to the representations of a simple Lie group. This group should be large enough to encompass as many of the known additive quantum numbers as possible. Thus far, this program has been quite successful, enabling us to perceive some order in the large number of elementary particles.

Let us focus on the additive quantum numbers I_3, Y, and C. We first consider I_3 by itself. Then, in the symmetry approach to hadron physics, we look to see whether a simple Lie algebra of rank 1 is relevant. There is only one such algebra, the algebra of $SU(2)$ and $O(3)$. We look at the predictions of this algebra, and find that they are in approximate agreement with experiment. Next we consider I_3 and Y together, assuming that they are the commuting operators of a simple Lie algebra of rank 2. There are only three such algebras. By comparing their predictions with experiment, we can pick out the one which agrees best with the facts. This happens to be the algebra of $SU(3)$. It is of course tempting to generalize from $SU(2)$ and $SU(3)$ to $SU(4)$, assuming that the algebra of $SU(4)$ is relevant to the three additive quantum numbers I_3, Y, and C.

We now return to a discussion of the roots of a semisimple compact Lie group. We consider a *root diagram*, which is the graphical representation of the root vectors in an l-dimensional space. Since it is hard to draw root diagrams of three dimensions on a two-dimensional surface and even harder to visualize a space of more than three dimensions, we shall restrict much of our discussion to groups of rank $l \leq 2$, although the theorems we quote apply to groups of any rank.

THEOREM. *If* $\boldsymbol{\rho}(\alpha)$ *and* $\boldsymbol{\rho}(\beta)$ *are two roots then*

$$\frac{2\boldsymbol{\rho}(\alpha)\cdot\boldsymbol{\rho}(\beta)}{\rho^2(\alpha)} \quad \textit{and} \quad \frac{2\boldsymbol{\rho}(\alpha)\cdot\boldsymbol{\rho}(\beta)}{\rho^2(\beta)}, \tag{5.65}$$

are integers, and

$$\boldsymbol{\rho}(\beta) - 2\boldsymbol{\rho}(\alpha)\frac{\boldsymbol{\rho}(\alpha)\cdot\boldsymbol{\rho}(\beta)}{\rho^2(\alpha)} = \boldsymbol{\rho}(\gamma) \tag{5.66}$$

is a root.

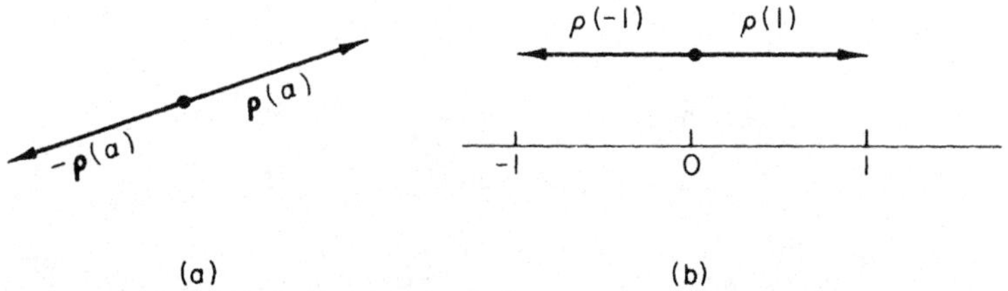

Fig. 5.1. (a) Root diagram obtained from the theorem that states if $\boldsymbol{\rho}(\alpha)$ is a root, then $-\boldsymbol{\rho}(\alpha)$ is a root. (b) Normalized root diagram for all rank 1 simple Lie groups. All these groups [e.g., $SU(2)$, $O(3)$] have the same Lie algebra.

The geometrical interpretation of the theorem is that the root $\boldsymbol{\rho}(\gamma)$ is the reflection of an l dimensional root $\boldsymbol{\rho}(\beta)$ with respect to a hyperplane (of $l-1$ dimensions) perpendicular to $\boldsymbol{\rho}(\alpha)$.

For a rank 1 group, the root diagram is one-dimensional, and all the roots lie on the same line. The hyperplane degenerates to a point in this case, and we obtain no new information from the theorem. However, a previous theorem, that if $\boldsymbol{\rho}(\alpha)$ is a root, then $\boldsymbol{\rho}(-\alpha) = -\boldsymbol{\rho}(\alpha)$ is also a root, is useful. Using this theorem, we obtain the root diagram of Fig. 5.1. In fact this root diagram is the only root diagram for a rank one simple group. The Lie algebra of this group is the Lie algebra of $SU(2)$.

We now consider simple rank 2 groups. The hyperplane in this case is a line. If we have two roots $\boldsymbol{\rho}(\alpha)$ and $\boldsymbol{\rho}(\beta)$ making an angle not equal to 90° or 180°, we obtain a third distinct root [not obtainable from the knowledge that $-\boldsymbol{\rho}(\alpha)$ and $-\boldsymbol{\rho}(\beta)$ are roots]. Figure 5.2 shows the construction. The theorem also states that

$$\frac{\boldsymbol{\rho}(\alpha)\ \boldsymbol{\rho}(\beta)}{\rho^2(\alpha)} = \frac{1}{2}m, \qquad \frac{\boldsymbol{\rho}(\alpha)\cdot\boldsymbol{\rho}(\beta)}{\rho^2(\beta)} = \frac{1}{2}n,$$

where m and n are integers. Letting the angle between $\boldsymbol{\rho}(\alpha)$ and $\boldsymbol{\rho}(\beta)$ be ϕ

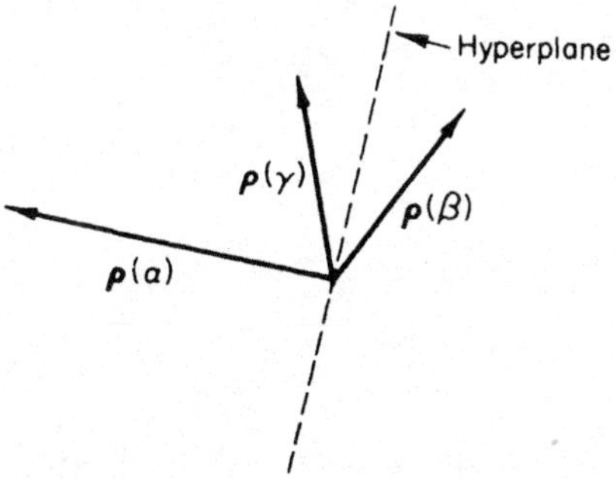

Fig. 5.2. Given two roots $\boldsymbol{\rho}(\alpha)$ and $\boldsymbol{\rho}(\beta)$, we construct a third root $\boldsymbol{\rho}(\gamma)$ by drawing a hyperplane (in two dimensions a line) perpendicular to $\boldsymbol{\rho}(\alpha)$. Then the reflection of $\boldsymbol{\rho}(\beta)$ with respect to this hyperplane is the third root.

and multiplying the two equations together, we obtain

$$\cos^2 \phi = \tfrac{1}{4}mn.$$

The solutions to this equation are

$$\phi = 0°, \quad 30°, \quad 45°, \quad 60°, \quad 90°.$$

This makes plausible the theorem that there are only a finite number of simple compact groups of a given rank. There is another relevant theorem due to Cartan (1933):

THEOREM. *Every semisimple compact group can be written as a direct product of simple compact groups.*

This theorem implies that there are only a finite number of semisimple compact algebras of a given rank. Also we need consider only the simple algebras to obtain all the relevant information about the semisimple ones. All the Lie algebras of simple compact groups have been classified by Cartan (1933). We shall give here only the root diagrams for the ones of rank 2. The first is $SU(3)$ (or A_2 in Cartan's classification). The groups A_l of Cartan are of order $(l+1)^2 - 1$ and of rank l. They are just the groups $SU(l+1)$. The root diagram of $SU(3)$ is shown in Fig. 5.3. With the aid of the root diagram of Fig. (5.3) and Eqs. (5.62), we can write down the commutation relations of $SU(3)$ in standard form. However, the normalization constants $N_{\alpha\beta}$ are not given in terms of the roots but must be computed separately. In Fig. 5.4 are given the only other root diagrams of simple rank 2 groups. See Cartan (1933) or Racah (1965) for further information about these groups.

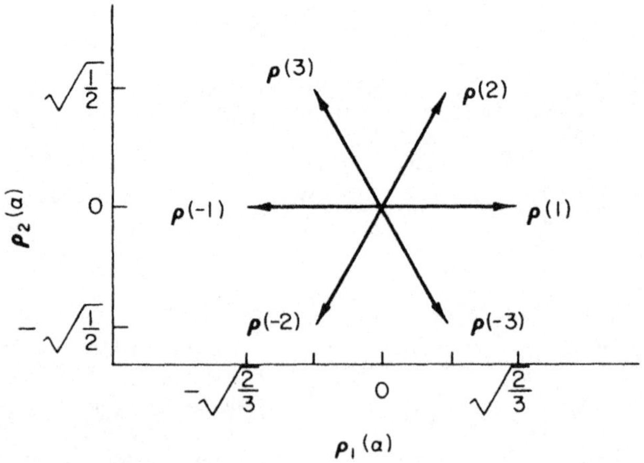

Fig. 5.3. Normalized root diagram for $SU(3)$, or A_2 according to Cartan's classification.

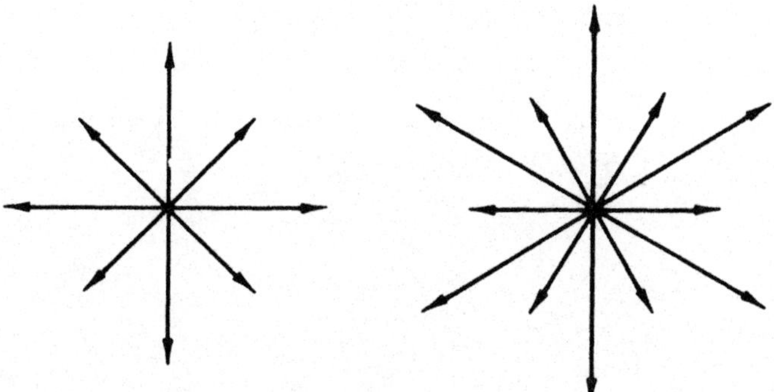

C_2 G_2

Fig. 5.4. Root diagrams of the rank 2 groups C_2 and G_2 according to the classification of Cartan (1933). See also Racah (1965). (The group B_2 of Cartan's classification has the same diagram as C_2 rotated by 45°.)

The root diagram of $SU(4)$ is three-dimensional. Rather than draw the diagram, we give the roots as ordered triples of numbers. For $\alpha = 1, 2, \ldots, 6$, these triples are

$$\boldsymbol{\rho}(1) = \left(\frac{1}{\sqrt{2}}, 0, 0\right), \qquad \boldsymbol{\rho}(2) = \left(\frac{1}{\sqrt{8}}, \sqrt{\frac{3}{8}}, 0\right),$$

$$\boldsymbol{\rho}(3) = \left(-\frac{1}{\sqrt{8}}, \sqrt{\frac{3}{8}}, 0\right), \qquad \boldsymbol{\rho}(4) = \left(\frac{1}{\sqrt{8}}, \frac{1}{2\sqrt{6}}, \frac{1}{\sqrt{3}}\right), \tag{5.67}$$

$$\boldsymbol{\rho}(5) = \left(-\frac{1}{\sqrt{8}}, \frac{1}{2\sqrt{6}}, \frac{1}{\sqrt{3}}\right), \qquad \boldsymbol{\rho}(6) = \left(0, -\frac{1}{\sqrt{6}}, \frac{1}{\sqrt{3}}\right).$$

The roots for negative $\alpha = -1, -2, \ldots, -6$ are given by the relation $\boldsymbol{\rho}(-\alpha) = -\boldsymbol{\rho}(\alpha)$.

For $SU(n)$, there are $n(n-1)$ roots and each root is a vector of $n-1$ dimensions.

CHAPTER 6

MULTIPLETS

6.1 Diagonal Generators and Weights

We shall now discuss the states which are the basis vectors of irreducible unitary representations of Lie groups. The basis vectors of such a representation constitute a multiplet.

As an example, we first consider the multiplets of the rotation group in three dimensions $R(3)$. The basis vectors for the representations of this group are the ordinary spherical harmonics[1] $Y_j^m(\theta\phi)$. These functions are eigenstates of the generator J_3 and of the invariant operator

$$J^2 = J_1{}^2 + J_2{}^2 + J_3{}^2, \tag{6.1}$$

which commutes with all the generators. The quantum number j labels the representation and the quantum number m labels the different vectors belonging to the representation, i.e., the different members of the multiplet. Since m takes on all integral values from $-j$ to j, the dimensionality of the representation is $2j + 1$.

If we are interested merely in how the Y_j^m transform among themselves when operated on by the generators, we can represent the Y_j^m by column vectors and J_i $(i = 1, 2, 3)$ by square matrices in a space of $2j + 1$ dimensions.

[1] It is customary to use the notation Y_l^m for the spherical harmonics, but we reserve the letter l for the rank of a group.

If double-valued representations are included, j can take on both integral and half-integral values. The multiplicity N of a state belonging to the representation specified by j is

$$N = 2j + 1. \tag{6.2}$$

We shall adopt the notation that the quantum numbers labeling a representation are written as superscripts, and the quantum numbers distinguishing the different vectors belonging to the representation by subscripts. We sometimes omit the superscripts when this can be done without confusion. Thus we write $\psi_m^{(j)}$ for the basis vectors of the representations of $R(3)$. When the double-valued representations are included, the multiplets of $R(3)$ are the same as the multiplets of $SU(2)$.

Now it is well known that the J_i can be represented in terms of the Pauli spin matrices σ_i $(i = 1, 2, 3)$:

$$J_1 = \tfrac{1}{2}\sigma_1, \qquad J_2 = \tfrac{1}{2}\sigma_2, \qquad J_3 = \tfrac{1}{2}\sigma_3, \tag{6.3}$$

where

$$\sigma_1 = \begin{pmatrix} 0 & 1 \\ 1 & 0 \end{pmatrix}, \qquad \sigma_2 = \begin{pmatrix} 0 & -i \\ i & 0 \end{pmatrix}, \qquad \sigma_3 = \begin{pmatrix} 1 & 0 \\ 0 & -1 \end{pmatrix}. \tag{6.4}$$

It can be directly verified by matrix multiplication that the J_i of Eq. (6.3) satisfy the commutation relations of Eq. (5.33).

The Pauli matrices are sometimes said to be a *representation* of the Lie algebra for $SU(2)$ [or $R(3)$]. The term "representation" may be confusing in this connection, for in different representations of a Lie algebra, the structure constants may be different.

The representation of the Lie algebra of $SU(2)$ by the Pauli matrices is the one lowest possible dimensionality. This is true because one-dimensional matrices (numbers) always commute. This fact that the lowest-dimensional representation of $SU(2)$ is by 2×2 matrices is a special case of a general theorem which we quote without proof. The theorem states that the lowest-dimensional representation of the Lie algebra of $SU(n)$ is by n-dimensional matrices. The matrices are traceless, and there are $n^2 - 1$ independent ones. These $n \times n$ matrices are called the *matrix generators* or more often the *generators* of the group. The generators of $SU(n)$ may also be represented by matrices of larger dimension. The n-dimensional representation of $SU(n)$ is called a fundamental representation. In Section 6.3, we shall give a more general definition of a fundamental representation of a simple Lie group. It will turn out that a rank l simple Lie group has l fundamental representations. Therefore $SU(n)$ has $n - 1$ fundamental representations, of which we have discussed only one.

If we write the Lie algebra of $SU(2)$ in standard form, the diagonal operator J_3 is denoted by H_1. The vector $\psi_m^{(j)}$ is an eigenstate of H_1:

$$H_1\psi_m^{(j)} = m\psi_m^{(j)}.$$

The eigenvalue m is called a *weight.*

Now consider the general case of a compact semisimple Lie group of rank l in which there are l diagonal operators H_i, $i = 1, 2, \ldots, l$. Then we can write

$$H_i\psi_m^{(j)} = m_i\psi_m^{(j)}, \tag{6.5}$$

where here j denotes all the quantities necessary to specify a particular representation. The m_i can be considered as the components of an l-component vector called the *weight vector* or just the *weight* of the eigenstate $\psi_m^{(j)}$. Thus we can write

$$\mathbf{H}\psi_m^{(j)} = \mathbf{m}\psi_m^{(j)}. \tag{6.6}$$

The l-dimensional vector space spanned by the weights $\mathbf{m}$ is called the weight space. A drawing of all the weights of a representation is called the *weight diagram* of the representation.

The number of different eigenvectors with the same value of a weight is called the *multiplicity* of the weight. The multiplicity of a weight should not be confused with the multiplicity of a representation, which is the number of different eigenvectors of the representation. The term *multiplet* will be reserved for all the eigenstates of an irreducible unitary representation.

If a weight belongs to only one eigenvector of a representation, it is called *simple.* This definition should not be confused with the term *simple* as applied to a Lie algebra.

6.2 Generators of $SU(2)$ and $U(2)$

Let us return to $SU(2)$ and consider the weights of the fundamental or two-dimensional representation. In this representation, we can express the Lie algebra in standard form in terms of the Pauli matrices. We obtain

$$H_1 = \tfrac{1}{2}\sigma_3, \qquad E_1 = \tfrac{1}{2}(\sigma_1 + i\sigma_2)/\sqrt{2}, \qquad E_{-1} = \tfrac{1}{2}(\sigma_1 - i\sigma_2)/\sqrt{2}.$$

or

$$H_1 = \frac{1}{2}\begin{pmatrix}1 & 0\\ 0 & -1\end{pmatrix}, \quad E_1 = \frac{1}{\sqrt{2}}\begin{pmatrix}0 & 1\\ 0 & 0\end{pmatrix}, \quad E_{-1} = \frac{1}{\sqrt{2}}\begin{pmatrix}0 & 0\\ 1 & 0\end{pmatrix}. \tag{6.7}$$

The basis vectors of this representation are

$$\psi_{1/2}^{(1/2)} = \begin{pmatrix}1\\0\end{pmatrix} = u_1, \qquad \psi_{-1/2}^{(1/2)} = \begin{pmatrix}0\\1\end{pmatrix} = u_2. \tag{6.8}$$

Operating on these basis vectors with H_1, we obtain

$$H_1 u_1 = \tfrac{1}{2} u_1, \qquad H_1 u_2 = -\tfrac{1}{2} u_2. \tag{6.9}$$

Thus, for $SU(2)$ there are two one-dimensional weights in the fundamental representation given by

$$m(1) = \tfrac{1}{2}, \qquad m(2) = -\tfrac{1}{2}. \tag{6.10}$$

where the argument of $m(i)$, $i = 1, 2$, refers to the eigenvector to which it belongs. (We reserve the use of subscripts to the different components of a single weight.)

The states u_1 and u_2 can have a variety of physical interpretations. For example, they can represent the two possible spin eigenstates of a spin-$\frac{1}{2}$ particle (a doublet) such as the electron or proton. The weights or eigenvalues $m = \pm\frac{1}{2}$ then are the values of the z component of the spin (in units of $\hbar$) with respect to an arbitrary z axis. Alternatively, u_1 and u_2 can represent the two possible isospin eigenstates of an isospin-$\frac{1}{2}$ particle: for example, the nucleon. In this case the proton and neutron are regarded as the two possible orientations of the nucleon in isospin space, the proton having third component (or z component) of the isospin $m = I_3 = \frac{1}{2}$ and the neutron having third component $m = I_3 = -\frac{1}{2}$. The fact that the proton and neutron are not exactly degenerate in energy shows that isospin symmetry is approximate.

The situation is analogous to the case in which spin-$\frac{1}{2}$ particles are in a magnetic field pointed in the z direction. Then the degeneracy of the states with $m = \frac{1}{2}$ and $m = -\frac{1}{2}$ is broken. The crucial difference between the case of spin and isospin is of course that in the former case the symmetry-breaking magnetic field can be turned off, while in the latter case the symmetry-breaking interaction is always present. In the case of isospin, the proton and neutron differ in their electric charge, and the symmetry-breaking a relation between the charge of an uncharmed hadron (in units of the proton charge) and the third component of the isospin I_3. This is the relation of Gell-Mann (1953) and Nishijima and Nakano (1953) which states that for all members of a given isospin multiplet, Q and I_3 are related by

$$Q = I_3 + Y/2, \tag{6.11}$$

where Y is a constant for all members of the isospin multiplet. The constant Y is known as the *hypercharge*. The Gell-Mann–Nishijima formula is sometimes written in terms of the *strangeness* S, defined for uncharmed particles by

$$S = Y - B,$$

where B is the *baryon number*. (The baryon number equals 1 for baryons, -1 for antibaryons, and 0 for mesons.)

Since I_3 and Y are additive quantum numbers, they can be interpreted as the diagonal generators of a group of rank 2. The group $SU(2)$ [or $R(3)$] must be a subgroup of this rank 2 group because of the known properties of isospin. Therefore we incorporate the nondiagonal generators of $SU(2)$, I_1 and I_2, into the algebra.

The smallest group of rank two which has the generators I_1, I_2, I_3, and Y is the group which has no other generators. The Lie algebra of the operators I_1, I_2, I_3, and Y is not semisimple, since Y commutes with all the operators of the set. The group $U(1) \times SU(2)$ has the algebra of the operators I_μ and Y. This group is not semisimple, but is the direct product of the Abelian group $U(1)$ and the simple group $SU(2)$. The group $U(2)$ also has the same algebra, and therefore $U(2)$ and $U(1) \times SU(2)$ are locally isomorphic.

There is a two-to-one homomorphism of $U(1) \times SU(2)$ onto $U(2)$, the elements

$$e^{i\phi}\begin{pmatrix} 1 & 0 \\ 0 & 1 \end{pmatrix} \quad \text{and} \quad e^{i(\phi+\pi)}\begin{pmatrix} -1 & 0 \\ 0 & -1 \end{pmatrix},$$

of $U(1) \times SU(2)$ both corresponding to the element

$$\begin{pmatrix} e^{i\phi} & 0 \\ 0 & e^{i\phi} \end{pmatrix},$$

of $U(2)$. This is a trivial correspondence, and in the future we shall not distinguish between the groups $U(2)$ and $U(1) \times SU(2)$. Since analogous statements hold in n dimensions, we shall not distinguish between $U(n)$ and $U(1) \times SU(n)$.

To obtain the matrix generators of $U(2)$ we can add the unit matrix σ_0 given by

$$\sigma_0 = \begin{pmatrix} 1 & 0 \\ 0 & 1 \end{pmatrix},$$

to the Pauli matrices σ_ν, $\nu = 1, 2, 3$.

The physical interpretation of $U(2)$ for the strongly interacting particles is that the multiplets of $SU(2)$ are the isospin multiplets, such as proton and neutron, and the generator of $U(1)$ is the hypercharge, which is a constant for a given isospin multiplet.

6.3 Generators of $SU(3)$ and $U(3)$

If the symmetry group of the strongly interacting particles, or hadrons, is taken to be $U(2)$ rather than a larger group of rank 2, then no relation is predicted to exist between hypercharge and isospin. In other words, any isospin multiplet can have any hypercharge. However, if Y and I_3 are assumed to be the commuting generators of a *simple* rank 2 group, then there will be a relation between Y and I_3 for each member of a multiplet. In other words, more can be predicted. The Lie algebras of all the rank 2 simple groups contain the Lie algebra of $U(1) \times SU(2)$ as a subalgebra. This means that if any of the simple rank 2 groups is a symmetry of the hadrons, then isospin and hypercharge will be conserved. Thus, all of the algebras of the simple rank 2 groups should be examined to see how well their predictions agree with the experimental facts. This has been done by Ne'eman (1964), who found that the predictions of $SU(3)$ are in closest agreement with experiment.

We now consider the rank 2 simple group $SU(3)$. The hermitian generators [analogous to the Pauli spin matrices of $SU(2)$] are

$$\lambda_1 = \begin{pmatrix} 0 & 1 & 0 \\ 1 & 0 & 0 \\ 0 & 0 & 0 \end{pmatrix}, \qquad \lambda_2 = \begin{pmatrix} 0 & -i & 0 \\ i & 0 & 0 \\ 0 & 0 & 0 \end{pmatrix}, \qquad \lambda_3 = \begin{pmatrix} 1 & 0 & 0 \\ 0 & -1 & 0 \\ 0 & 0 & 0 \end{pmatrix},$$

$$\lambda_4 = \begin{pmatrix} 0 & 0 & 1 \\ 0 & 0 & 0 \\ 1 & 0 & 0 \end{pmatrix}, \qquad \lambda_5 = \begin{pmatrix} 0 & 0 & -i \\ 0 & 0 & 0 \\ i & 0 & 0 \end{pmatrix}, \qquad \lambda_6 = \begin{pmatrix} 0 & 0 & 0 \\ 0 & 0 & 1 \\ 0 & 1 & 0 \end{pmatrix}, \tag{6.12}$$

$$\lambda_7 = \begin{pmatrix} 0 & 0 & 0 \\ 0 & 0 & -i \\ 0 & i & 0 \end{pmatrix}, \qquad \lambda_8 = \frac{1}{\sqrt{3}} \begin{pmatrix} 1 & 0 & 0 \\ 0 & 1 & 0 \\ 0 & 0 & -2 \end{pmatrix}.$$

These matrices are given with the notation and normalization used by Gell-Mann (1961). It is seen that λ_1, λ_4, and λ_6 are like σ_1 except that each has an extra row and column of zeros. Similarly, except for extra zeros, λ_2, λ_5, and λ_7 are like σ_2, and λ_3 is like σ_3. The matrix λ_8 has no analogy in $SU(2)$, as it is a matrix which commutes with λ_3. If we wish to consider the group $U(3)$ we can add the generator λ_0 given by

$$\lambda_0 = \begin{pmatrix} 1 & 0 & 0 \\ 0 & 1 & 0 \\ 0 & 0 & 1 \end{pmatrix}. \tag{6.13}$$

The group $U(3)$ is a rank 3 group and is not semisimple. The generator λ_0, which commutes with all the λ_ν, can be interpreted physically as associated with the baryon number B. In this case, I_3 and Y can vary within a given multiplet of $U(3)$ but all members of a multiplet have the same value of B.

Let us return to the special unitary groups. The generators λ_ν can be readily extended to $SU(n)$. In any number of dimensions, the λ_ν satisfy the commutation relations

$$[\lambda_\kappa, \lambda_\mu] = 2i \sum_\nu f_{\kappa\mu\nu} \lambda_\nu, \qquad \kappa, \mu, \nu = 1, 2, \ldots, n^2 - 1, \tag{6.14}$$

TABLE 6.1

NONZERO STRUCTURE CONSTANTS $f_{\kappa\mu\nu}$ OF $SU(3)$[a]

$\kappa\mu\nu$	$f_{\kappa\mu\nu}$	$\kappa\mu\nu$	$f_{\kappa\mu\nu}$
123	1	345	$\frac{1}{2}$
147	$\frac{1}{2}$	367	$-\frac{1}{2}$
156	$-\frac{1}{2}$	458	$\sqrt{3}/2$
246	$\frac{1}{2}$	678	$\sqrt{3}/2$
257	$\frac{1}{2}$		

[a] The $f_{\kappa\mu\nu}$ are antisymmetric under the permutation of any two indices.

where $f_{\kappa\mu\nu} = c_{\kappa\mu\nu}/(2i)$, and the $c_{\kappa\mu\nu}$ are the structure constants defined in Eq. (5.24). We shall apply the terminology "structure constants" to the $f_{\kappa\mu\nu}$ as well as to the $c_{\kappa\mu\nu}$. For the $SU(n)$ groups, the $f_{\kappa\mu\nu}$ are antisymmetric under the permutation of any two indices. In the remainder of this section, we restrict ourselves to $SU(3)$. For this group, the nonzero structure constants $f_{\kappa\mu\nu}$ satisfying $\kappa < \mu < \nu$ are given in Table 6.1.

The following properties of the λ_ν are also useful:

$$\begin{aligned} &\mathrm{Tr}(\lambda_\mu \lambda_\nu) = 2\delta_{\mu\nu}, \qquad \{\lambda_\kappa, \lambda_\mu\} = 2 \sum_\nu d_{\kappa\mu\nu} \lambda_\nu + \tfrac{4}{3}\delta_{\kappa\mu}, \\ &\mathrm{Tr}(\lambda_\kappa[\lambda_\mu, \lambda_\nu]) = 4if_{\kappa\mu\nu}, \qquad \mathrm{Tr}(\lambda_\kappa\{\lambda_\mu, \lambda_\nu\}) = 4\, d_{\kappa\mu\nu}. \end{aligned} \tag{6.15}$$

In Eqs. (6.15), the symbol $\{a, b\}$ means the anticommutator of a and b, defined by $\{a, b\} = ab + ba$. The $d_{\kappa\mu\nu}$ are given in Table 6.2. Relations (6.15) are not properties of the algebra but just of the three-dimensional λ-matrices.

TABLE 6.2

NONZERO INDEPENDENT ELEMENTS OF THE TENSOR $d_{\kappa\mu\nu}$[a]

$\kappa\mu\nu$	$d_{\kappa\mu\nu}$	$\kappa\mu\nu$	$d_{\kappa\mu\nu}$
118	$1/\sqrt{3}$	355	$\frac{1}{2}$
146	$\frac{1}{2}$	366	$-\frac{1}{2}$
157	$\frac{1}{2}$	377	$-\frac{1}{2}$
228	$1/\sqrt{3}$	448	$-1/(2\sqrt{3})$
247	$-\frac{1}{2}$	558	$-1/(2\sqrt{3})$
256	$\frac{1}{2}$	668	$-1/(2\sqrt{3})$
338	$1/\sqrt{3}$	778	$-1/(2\sqrt{3})$
344	$\frac{1}{2}$	888	$-1\sqrt{3}$

[a] The $d_{\kappa\mu\nu}$ are symmetric under permutations of any two indices.

The λ-matrices operate on the basis vectors of the first fundamental representation of $SU(3)$. But $SU(3)$, being a simple rank-2 group, has a second fundamental representation. The generators of this second set λ_μ' are given by

$$\lambda_\mu' = -\lambda_\mu^* = -\tilde{\lambda}_\mu, \tag{6.16}$$

where the tilde over a symbol for a matrix denotes the transpose of the matrix. If we apply a similar transformation to the Pauli matrices of $SU(2)$

$$\sigma_\mu' = -\sigma_\mu^* = -\tilde{\sigma}_\mu,$$

we obtain the generators of an equivalent two-dimensional representation. In $SU(3)$, the new representation generated by the λ_μ' of Eq. (6.19) is inequivalent to the representation generated by the λ_μ.

When the Pauli matrices refer to isospin rather than spin, it is conventional to label them by τ_μ rather than σ_μ. The isospin generators I_μ are given by $I_\mu = \frac{1}{2}\tau_\mu$. Similarly, we can define $SU(3)$ generators F_μ of F-spin given by

$$F_\mu = \tfrac{1}{2}\lambda_\mu, \qquad \mu = 1, 2, \ldots, 8. \tag{6.17}$$

From Eqs. (6.14) and (6.15), we immediately obtain that the F_μ satisfy

$$[F_\kappa, F_\mu] = i \sum_\nu f_{\kappa\mu\nu} F_\nu, \qquad \mathrm{Tr}(F_\mu F_\nu) = \tfrac{1}{2}\delta_{\mu\nu},$$

$$\{F_\kappa, F_\mu\} = \sum_\nu d_{\kappa\mu\nu} F_\nu + \tfrac{1}{3}\delta_{\kappa\mu},$$

$$\mathrm{Tr}(F_\kappa[F_\mu, F_\nu]) = \tfrac{1}{2} i f_{\kappa\mu\nu}, \qquad \mathrm{Tr}(F_\kappa\{F_\mu, F_\nu\}) = \tfrac{1}{2} d_{\kappa\mu\nu}. \tag{6.18}$$

We can also define a new set of operators D_μ, which are the following bilinear combinations of the F_μ:

$$D_\mu = \tfrac{2}{3} \sum_{\alpha\beta} d_{\mu\alpha\beta} F_\alpha F_\beta. \tag{6.19}$$

We shall now show that the D_μ satisfy the commutation relations

$$[D_\kappa, F_\mu] = i \sum_\nu f_{\kappa\mu\nu} D_\nu. \tag{6.20}$$

The proof makes use of the identity

$$[F_\mu, \{F_\gamma, F_\alpha\}] = \{[F_\mu, F_\gamma], F_\alpha\} + \{[F_\mu, F_\alpha], F_\gamma\},$$

which holds for any three linear operators. Making use of the relations of Eq. (6.18), this identity becomes

$$\sum_\beta d_{\gamma\alpha\beta}[F_\mu, F_\beta] = \sum_\beta (i f_{\mu\gamma\beta}\{F_\beta, F_\alpha\} + i f_{\mu\alpha\beta}\{F_\beta, F_\gamma\}).$$

We now multiply this expression by F_κ and take the trace, obtaining

$$\sum_\beta d_{\gamma\alpha\beta} f_{\kappa\mu\beta} = \sum_\beta (f_{\mu\gamma\beta} d_{\kappa\beta\alpha} + f_{\mu\alpha\beta} d_{\kappa\beta\gamma}). \tag{6.21}$$

We now substitute Eq. (6.19) into the left-hand side of Eq. (6.20), obtaining

$$[D_\kappa, F_\mu] = \tfrac{2}{3} \sum_{\alpha\beta} d_{\kappa\alpha\beta}[F_\alpha F_\beta, F_\mu].$$

Expanding the commutator, we obtain

$$[D_\kappa, F_\mu] = \tfrac{2}{3} \sum_{\alpha\beta} d_{\kappa\alpha\beta}([F_\alpha[F_\beta, F_\mu] + [F_\alpha, F_\mu]F_\beta).$$

Using (6.18), we get

$$[D_\kappa, F_\mu] = \tfrac{2}{3}i \sum_{\alpha\beta\gamma} d_{\kappa\alpha\beta}(f_{\beta\mu\gamma} F_\alpha F_\gamma + f_{\alpha\mu\gamma} F_\gamma F_\beta).$$

Relabeling the indices in the second term of the sum, we obtain

$$[D_\kappa, F_\mu] = \tfrac{2}{3}i \sum_{\alpha\beta\gamma} (d_{\kappa\alpha\beta} f_{\beta\mu\gamma} + d_{\kappa\beta\gamma} f_{\beta\mu\alpha}) F_\alpha F_\gamma .$$

Then, using the fact that the $d_{\kappa\mu\nu}$ are symmetric under all permutations of the indices and the $f_{\kappa\mu\nu}$ are antisymmetric under odd permutations, we can write this expression as

$$[D_\kappa, F_\mu] = \tfrac{2}{3}i \sum_{\alpha\beta\gamma} (d_{\kappa\beta\alpha} f_{\mu\gamma\beta} + d_{\kappa\beta\gamma} f_{\mu\alpha\beta}) F_\alpha F_\gamma .$$

Now, using Eq. (6.21), we obtain

$$[D_\kappa, F_\mu] = \tfrac{2}{3}i f_{\kappa\mu\beta} \sum_{\alpha\beta\gamma} d_{\beta\alpha\gamma} F_\alpha F_\gamma ,$$

Finally, using the definition of D_β from Eq. (6.19), we obtain Eq. (6.20).

Note that we have proved Eq. (6.20) only for 3×3 matrices. However, we can easily verify that these commutation relations are a property of the algebra which holds in any number of dimensions. To do this, we let

$$D_\kappa = D_\kappa^{(1)} + D_\kappa^{(2)}, \qquad F_\mu = F_\mu^{(1)} + F_\mu^{(2)}.$$

Then we have

$$[D_\kappa, F_\mu] = [D_\kappa^{(1)}, F_\mu^{(1)}] + [D_\kappa^{(2)}, F_\mu^{(2)}],$$

since generators with different superscripts commute, e.g.,

$$[F_\kappa^{(1)}, F_\mu^{(2)}] = 0.$$

Then, if Eq. (6.20) holds for $[D_\kappa^{(1)}, F_\mu^{(1)}]$ and $[D_\kappa^{(2)}, F_\mu^{(2)}]$ separately, it holds for $[D_\kappa, F_\mu]$. Since any D_κ and F_μ can be written as a sum of the corresponding three-dimensional generators, Eq. (6.20) holds in any number of dimensions.

From the λ_μ, we can obtain the generators in standard form. These are given by

$$H_1 = \frac{1}{\sqrt{6}}\begin{pmatrix}1 & 0 & 0\\ 0 & -1 & 0\\ 0 & 0 & 0\end{pmatrix}, \quad H_2 = \frac{1}{3\sqrt{2}}\begin{pmatrix}1 & 0 & 0\\ 0 & 1 & 0\\ 0 & 0 & -2\end{pmatrix},$$

$$E_1 = \frac{1}{\sqrt{3}}\begin{pmatrix}0 & 1 & 0\\ 0 & 0 & 0\\ 0 & 0 & 0\end{pmatrix}, \quad E_2 = \frac{1}{\sqrt{3}}\begin{pmatrix}0 & 0 & 1\\ 0 & 0 & 0\\ 0 & 0 & 0\end{pmatrix}, \quad E_3 = \frac{1}{\sqrt{3}}\begin{pmatrix}0 & 0 & 0\\ 0 & 0 & 1\\ 0 & 0 & 0\end{pmatrix},$$

$$E_{-1} = \frac{1}{\sqrt{3}}\begin{pmatrix}0 & 0 & 0\\ 1 & 0 & 0\\ 0 & 0 & 0\end{pmatrix}, \quad E_{-2} = \frac{1}{\sqrt{3}}\begin{pmatrix}0 & 0 & 0\\ 0 & 0 & 0\\ 1 & 0 & 0\end{pmatrix}, \quad E_{-3} = \frac{1}{\sqrt{3}}\begin{pmatrix}0 & 0 & 0\\ 0 & 0 & 0\\ 0 & 1 & 0\end{pmatrix}. \tag{6.22}$$

We see that

$$E_{-\alpha} = E_\alpha^{\dagger}. \tag{6.23}$$

If we operate on the three-dimensional basis vectors

$$u_1 = \begin{pmatrix}1\\0\\0\end{pmatrix}, \quad u_2 = \begin{pmatrix}0\\1\\0\end{pmatrix}, \quad u_3 = \begin{pmatrix}0\\0\\1\end{pmatrix}, \tag{6.24}$$

with the diagonal vector operator $\mathbf{H} = (H_1, H_2)$ we obtain

$$\begin{aligned}\mathbf{H}u_1 &= \left(\frac{1}{\sqrt{6}}, \frac{1}{3\sqrt{2}}\right)u_1,\\ \mathbf{H}u_2 &= \left(-\frac{1}{\sqrt{6}}, \frac{1}{3\sqrt{2}}\right)u_2,\\ \mathbf{H}u_3 &= \left(0, -\frac{\sqrt{2}}{3}\right)u_3.\end{aligned} \tag{6.25}$$

Thus, the three weights of the first fundamental representation of $SU(3)$ are

$$\mathbf{m}(1) = \left(\frac{1}{\sqrt{6}}, \frac{1}{3\sqrt{2}}\right), \quad \mathbf{m}(2) = \left(-\frac{1}{\sqrt{6}}, \frac{1}{3\sqrt{2}}\right), \quad \mathbf{m}(3) = \left(0, -\frac{\sqrt{2}}{3}\right). \tag{6.26}$$

In this case we have three two-dimensional weights. For the basis vectors of any irreducible unitary representation of $SU(3)$ we have

$$\mathbf{H}\psi_m = \mathbf{m}\psi_m,$$

where $\mathbf{m} = (m_1, m_2)$.

From Eq. (6.16) we see that $-\mathbf{H}$ is the diagonal vector operator for the second fundamental representation. It follows that the weights of the second fundamental representation are just the negatives of the weights of the first.

The interpretation of the weights of $SU(3)$, as applied to the hadrons is that, except for normalization, m_1 is the third component of isospin I_3 and m_2 is the hypercharge Y. The exact relations are

$$I_3 = \tfrac{1}{2}\sqrt{6}\,m_1, \qquad Y = \sqrt{2}\,m_2 .$$

Later we shall normalize the weights differently for convenience.

We have given a physical example of a multiplet belonging to the fundamental representation of $SU(2)$—the nucleon doublet. At the present time there is no firm evidence for the existence of any fundamental triplet of $SU(3)$ among the hadrons. However, some charmed particles have been discovered. If the relevant enlarged group is $SU(4)$, as seems likely, there should exist definite $SU(3)$ triplets of charmed hadrons. As another example where the triplet representation of $SU(3)$ is useful, we cite the three quarks of the original model of Gell-Mann and Zweig. These particles are members of the first fundamental triplet of $SU(3)$. In the present quark model, there are at least four quarks, and in some versions at least, the four quarks are members of the first fundamental quartet of $SU(4)$. It is to $SU(4)$ that we now turn.

6.4 Generators of $SU(4)$ and Beyond

The hermitian matrix generators of $SU(4)$, analogous to the Pauli matrices of $SU(2)$ and the Gell-Mann matrices of $SU(3)$, are

$$\lambda_1 = \begin{pmatrix} 0 & 1 & 0 & 0 \\ 1 & 0 & 0 & 0 \\ 0 & 0 & 0 & 0 \\ 0 & 0 & 0 & 0 \end{pmatrix}, \qquad \lambda_2 = \begin{pmatrix} 0 & -i & 0 & 0 \\ i & 0 & 0 & 0 \\ 0 & 0 & 0 & 0 \\ 0 & 0 & 0 & 0 \end{pmatrix},$$

$$\lambda_3 = \begin{pmatrix} 1 & 0 & 0 & 0 \\ 0 & -1 & 0 & 0 \\ 0 & 0 & 0 & 0 \\ 0 & 0 & 0 & 0 \end{pmatrix}, \qquad \lambda_4 = \begin{pmatrix} 0 & 0 & 1 & 0 \\ 0 & 0 & 0 & 0 \\ 1 & 0 & 0 & 0 \\ 0 & 0 & 0 & 0 \end{pmatrix},$$

$$\lambda_5 = \begin{pmatrix} 0 & 0 & -i & 0 \\ 0 & 0 & 0 & 0 \\ i & 0 & 0 & 0 \\ 0 & 0 & 0 & 0 \end{pmatrix}, \qquad \lambda_6 = \begin{pmatrix} 0 & 0 & 0 & 0 \\ 0 & 0 & 1 & 0 \\ 0 & 1 & 0 & 0 \\ 0 & 0 & 0 & 0 \end{pmatrix},$$

$$\lambda_7 = \begin{pmatrix} 0 & 0 & 0 & 0 \\ 0 & 0 & -i & 0 \\ 0 & i & 0 & 0 \\ 0 & 0 & 0 & 0 \end{pmatrix}, \qquad \lambda_8 = \frac{1}{\sqrt{3}} \begin{pmatrix} 1 & 0 & 0 & 0 \\ 0 & 1 & 0 & 0 \\ 0 & 0 & -2 & 0 \\ 0 & 0 & 0 & 0 \end{pmatrix},$$

$$\lambda_9 = \begin{pmatrix} 0 & 0 & 0 & 1 \\ 0 & 0 & 0 & 0 \\ 0 & 0 & 0 & 0 \\ 1 & 0 & 0 & 0 \end{pmatrix}, \qquad \lambda_{10} = \begin{pmatrix} 0 & 0 & 0 & -i \\ 0 & 0 & 0 & 0 \\ 0 & 0 & 0 & 0 \\ i & 0 & 0 & 0 \end{pmatrix}$$

$$\lambda_{11} = \begin{pmatrix} 0 & 0 & 0 & 0 \\ 0 & 0 & 0 & 1 \\ 0 & 0 & 0 & 0 \\ 0 & 1 & 0 & 0 \end{pmatrix}, \qquad \lambda_{12} = \begin{pmatrix} 0 & 0 & 0 & 0 \\ 0 & 0 & 0 & -i \\ 0 & 0 & 0 & 0 \\ 0 & i & 0 & 0 \end{pmatrix},$$

$$\lambda_{13} = \begin{pmatrix} 0 & 0 & 0 & 0 \\ 0 & 0 & 0 & 0 \\ 0 & 0 & 0 & 1 \\ 0 & 0 & 1 & 0 \end{pmatrix}, \qquad \lambda_{14} = \begin{pmatrix} 0 & 0 & 0 & 0 \\ 0 & 0 & 0 & 0 \\ 0 & 0 & 0 & -i \\ 0 & 0 & i & 0 \end{pmatrix},$$

$$\lambda_{15} = \frac{1}{\sqrt{6}} \begin{pmatrix} 1 & 0 & 0 & 0 \\ 0 & 1 & 0 & 0 \\ 0 & 0 & 1 & 0 \\ 0 & 0 & 0 & -3 \end{pmatrix}. \tag{6.27}$$

In addition we define the 4×4 unit matrix λ_0 by

$$\lambda_0 = \begin{pmatrix} 1 & 0 & 0 & 0 \\ 0 & 1 & 0 & 0 \\ 0 & 0 & 1 & 0 \\ 0 & 0 & 0 & 1 \end{pmatrix}. \tag{6.28}$$

The matrices λ_1 through λ_{15} satisfy the Lie algebra given by the commutation relations of Eq. (6.14). The nonzero structure constants $f_{\kappa\mu\nu}$ of $SU(4)$ satisfying $\kappa < \mu < \nu$ are given in Table 6.3. With our normalization of the λ_μ, if κ, μ, and ν are all less than or equal to 8, the $f_{\kappa\mu\nu}$ are identical to the $SU(3)$ structure constants $f_{\kappa\mu\nu}$ of Table 6.1. In fact, the generators of $SU(n)$ can be so normalized that if the indices of $f_{\kappa\mu\nu}$ satisfy the inequality

$$\kappa, \mu, \nu \leq (n - m)^2 - 1,$$

where n and m are positive integers satisfying $n > m$, then the $f_{\kappa\mu\nu}$ are the structure constants of $SU(n - m)$.

It can be seen by generalizing from Eqs. (6.4), (6.12), and (6.27), how to write down the hermitian generators of any $SU(n)$. We leave this as an

TABLE 6.3

NONZERO STRUCTURE CONSTANTS $f_{\kappa\mu\nu}$ OF $SU(4)$[a]

κ	μ	ν	$f_{\kappa\mu\nu}$	κ	μ	ν	$f_{\kappa\mu\nu}$
1	2	3	1	4	9	14	$\frac{1}{2}$
1	4	7	$\frac{1}{2}$	4	10	13	$-\frac{1}{2}$
1	5	6	$-\frac{1}{2}$	5	9	13	$\frac{1}{2}$
2	4	6	$\frac{1}{2}$	5	10	14	$\frac{1}{2}$
2	5	7	$\frac{1}{2}$	6	11	14	$\frac{1}{2}$
3	4	5	$\frac{1}{2}$	6	12	13	$-\frac{1}{2}$
3	6	7	$-\frac{1}{2}$	7	11	13	$\frac{1}{2}$
4	5	8	$\sqrt{3}/2$	7	12	14	$\frac{1}{2}$
6	7	8	$\sqrt{3}/2$	8	9	10	$1/(2\sqrt{3})$
1	9	12	$\frac{1}{2}$	8	11	12	$1/(2\sqrt{3})$
1	10	11	$-\frac{1}{2}$	8	13	14	$-1/\sqrt{3}$
2	9	11	$\frac{1}{2}$	9	10	15	$\sqrt{2/3}$
2	10	12	$\frac{1}{2}$	11	12	15	$\sqrt{2/3}$
3	9	10	$\frac{1}{2}$	13	14	15	$\sqrt{2/3}$
3	11	12	$-\frac{1}{2}$				

[a] The $f_{\kappa\mu\nu}$ are antisymmetric under the permutation of any two indices.

exercise for the reader. Note that the normalization factor for the diagonal generator $\lambda_{i(i+2)}$ ($i = 1, \ldots, n-1$) is $[2/(i^2+i)]^{1/2}$. In this representation of the λ_μ, $SU(5)$ has 66 nonvanishing $f_{\kappa\mu\nu}$ satisfying $\kappa < \mu < \nu$, and $SU(6)$ has 125. Although these structure constants are known, we omit tables of them as they are too bulky.

We next write the matrix generators of $SU(n)$ in standard form. For convenience we make a change in our notation for the nondiagonal generators from E_α and $E_{-\alpha}$ ($\alpha = 1, 2, \ldots, n(n-1)/2$) to E_{ij} ($i, j = 1, 2, \ldots, n$, $i \neq j$). Then the matrix elements of the H_i ($i = 1, 2, \ldots, n-1$) and the E_{ij} are

$$(H_i)_{\mu\nu} = \begin{cases} [i(i+1)n]^{-1/2}\,\delta_{\mu\nu}, & \mu < i+1 \\ -i[i(i+1)n]^{-1/2}\,\delta_{\mu\nu}, & \mu = i+1 \\ 0, & \mu > i+1 \end{cases} \tag{6.29}$$

$$(E_{ij})_{\mu\nu} = n^{-1/2}\,\delta_{i\mu}\,\delta_{j\nu}, \qquad i \neq j. \tag{6.30}$$

From Eq. (6.29), it is easy to obtain the n weights (each of $n-1$ dimensions), of the first fundamental representation of $SU(n)$. For example, we obtain for $SU(4)$

$$\mathbf{m}(1) = \left(\frac{1}{2\sqrt{2}}, \frac{1}{2\sqrt{6}}, \frac{1}{4\sqrt{3}}\right), \qquad \mathbf{m}(2) = \left(-\frac{1}{2\sqrt{2}}, \frac{1}{2\sqrt{6}}, \frac{1}{4\sqrt{3}}\right),$$

$$\mathbf{m}(3) = \left(0, -\frac{1}{\sqrt{6}}, \frac{1}{4\sqrt{3}}\right), \qquad \mathbf{m}(4) = \left(0, 0, -\frac{\sqrt{3}}{4}\right).$$

6.5 Properties of the Weights

We shall now give some of the properties of weights associated with the multiplets of compact simple Lie groups. For references, see Racah (1965) and Behrends *et al.* (1962).

First we note that there is only one simple Lie algebra of rank one, namely the algebra of $SU(2)$. For any representation j of this group, there are $2j + 1$ eigenvectors. Furthermore, there are $2j + 1$ different weights, each weight belonging to a different eigenvector. The weights are

$$m = -j, -j + 1, \ldots, j, \tag{6.31}$$

as is known from any book on quantum mechanics. Thus, all weights of simple rank-1 groups are simple. For groups of rank $l \geq 2$, the weights are not always simple.

Suppose we know the value of a weight $\mathbf{m}$ belonging to a particular representation of a simple rank l group. We can use this knowledge to obtain other weights of the representation if we also know the root diagram of the group. We obtain other weights of the representation by making use of the following:

THEOREM. *For any weight* $\mathbf{m}$ *and root* $\boldsymbol{\rho}(\alpha)$, *the quantity* k *defined by*

$$k = 2\mathbf{m} \cdot \boldsymbol{\rho}/\rho^2, \tag{6.32}$$

is an integer and

$$\mathbf{m}' = \mathbf{m} - \boldsymbol{\rho}(2\mathbf{m} \cdot \boldsymbol{\rho})/\rho^2, \tag{6.33}$$

is a weight with the same multiplicity as $\mathbf{m}$.

According to Eq. (6.33), if we have a weight $\mathbf{m}$, then by reflecting in the hyperplane perpendicular to $\boldsymbol{\rho}$, we obtain the new weight $\mathbf{m}'$. Weights related by reflections in hyperplanes are said to be *equivalent*. Equivalent weights should not be confused with equivalent representations, which are representations connected by a similarity transformation.

It is useful to be able to order the weights. The usual convention is that a weight $\mathbf{m}$ is higher than a weight $\mathbf{m}'$ if $\mathbf{m} - \mathbf{m}'$ has a positive number for its first nonvanishing component. However, with this ordering, the state vectors u_1, u_2, u_3 of Eq. (6.24) have weights which are not conveniently ordered. The vector u_1 has the highest weight, u_3 the next, and u_2 the lowest weight.

A more convenient convention for $SU(n)$ is that $\mathbf{m}$ is *higher* than $\mathbf{m}'$ if the *last* component of $\mathbf{m} - \mathbf{m}'$ is positive; if it is zero, we apply the criterion to the next to last component, etc. This is the convention we shall adopt for $SU(n)$. Then

$$\mathbf{H}u_i = \mathbf{m}(i)u_i .$$

The weights of Eq. (6.26) conform to this ordering. The highest weight of a set of equivalent weights is said to be *dominant*.

As we have remarked, the weights of the conjugate representation of $SU(3)$ (the second fundamental representation) are negatives of the weights of the first. With our convention they are ordered as follows:

$$\mathbf{m}(1) = \left(0, \frac{\sqrt{2}}{3}\right), \qquad \mathbf{m}(2) = \left(\frac{1}{\sqrt{6}}, -\frac{1}{3\sqrt{2}}\right), \qquad \mathbf{m}(3) = \left(-\frac{1}{\sqrt{6}}, -\frac{1}{3\sqrt{2}}\right) \tag{6.34}$$

An important property of compact semisimple Lie groups is that there exists a highest weight for any irreducible representation. Furthermore this weight is simple. Another property is that two equivalent representations have the same highest weight, and two representations which have the same highest weight are equivalent.

Cartan (1933) has proved the following:

THEOREM. *For every simple group of rank l, there are l dominant weights called fundamental dominant weights denoted by $\mathbf{M}^{(i)}$, $i = 1, 2, \ldots, l$, such that any other dominant weight $\mathbf{M}$ is a linear combination of the fundamental dominant weights*

$$\mathbf{M} = \sum_{i=1}^{l} p_i \mathbf{M}^{(i)} , \tag{6.35}$$

where p_i are nonnegative integers. Furthermore, there exist l so-called fundamental irreducible representations which have the l fundamental dominant weights as their highest weights.

It can also be proved that there exists a representation with $\mathbf{M}$ as its highest weight. This representation is characterized by the l integers p_i.

In general, to obtain the dimensionality of the fundamental representations of simple Lie groups, we need to discuss the characters of the representations. We shall not do this because the results happen not to be needed for the groups $SU(n)$. To find out how one uses the character to obtain the dimensionalities of the fundamental representations, see Racah (1965) and Behrends *et al.* (1962).

We now give some examples of how to obtain the weight diagrams of fundamental representations of simple Lie groups. First consider $SU(2)$. The root diagram consists of the roots $\rho(1) = 1$ and $\rho(-1) = -1$. Using Eq. (6.32) with $k = 1$, we find $m = \frac{1}{2}$ is a nonvanishing weight. Using Eq. (6.33), we obtain an equivalent weight $m' = -\frac{1}{2}$. These are the two weights of the fundamental representation of $SU(2)$, the weight $m = \frac{1}{2} = M^{(1)}$ being the fundamental dominant weight. The weight diagram of this representation is given in Fig. 6.1.

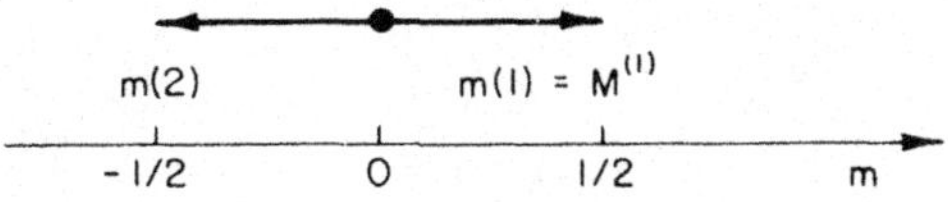

Fig. 6.1. Weight diagram of the fundamental representation of $SU(2)$.

Using Eq. (6.35), we see that the dominant weight M of any representation of $SU(2)$ is given by

$$M = pM^{(1)} = \tfrac{1}{2}p.$$

Thus, the dominant weight of a representation characterized by the integer p is either integral or half-integral according to whether p is even or odd.

Next we consider $SU(3)$. The root diagram is given in Fig. 5.3. From this figure we see that

$$\boldsymbol{\rho}(1) = (\sqrt{\tfrac{2}{3}}, 0).$$

Using this value of $\boldsymbol{\rho}(1)$ in Eq. (6.32) and letting $\mathbf{m} = (m_1 m_2)$, we obtain

$$k = \frac{2[m_1\rho_1(1) + m_2\rho_2(1)]}{\rho^2(1)} = \frac{2m_1\sqrt{\frac{2}{3}}}{\frac{2}{3}},$$

or

$$m_1 = k/\sqrt{6}.$$

Letting $k = 1$, we obtain the smallest nontrivial value of m_1:

$$m_1 = 1/\sqrt{6}.$$

Now consider another root from Fig. 5.3;

$$\boldsymbol{\rho}(2) = (1/\sqrt{6}, 1/\sqrt{2}),$$

and again take the scalar product with a weight **m** having the value $m_1 = 1\sqrt{6}$. We get

$$k = \frac{2[m_1\rho_1(2) + m_2\rho_2(2)]}{\rho^2(2)} = \frac{1}{2} + \frac{3}{\sqrt{2}} m_2 .$$

Letting $k = 1$, we obtain $m_2 = 1/(3\sqrt{2})$, and letting $k = 0$, we obtain $m_2 = -1/(3\sqrt{2})$. Thus, we have obtained two weights

$$\mathbf{m} = (1/\sqrt{6}, 1/3\sqrt{2}) = \mathbf{M}^{(1)}, \tag{6.36}$$

and

$$\mathbf{m} = (1/\sqrt{6}, -1/3\sqrt{2}). \tag{6.37}$$

The weight of Eq. (6.36) is the fundamental dominant weight of the first fundamental representation of $SU(3)$, as can be seen by comparing with Eq. (6.26). Using this fundamental dominant weight and reflecting with respect to the lines (hyperplanes) perpendicular to the roots, we obtain the weight diagram of the first fundamental representation. This is shown in Fig. 6.2. The weights of Fig. 6.2 are just the ones we have obtained previously in Eq. (6.26).

The weight of Eq. (6.37) belongs to the second fundamental representation, as can be seen from Eq. (6.34). Reflecting with respect to the lines perpendicular to the roots, we obtain the weight diagram of Fig. 6.3. The fundamental dominant weight of this conjugate representation is (with our ordering convention)

$$\mathbf{M}^{(2)} = (0, \sqrt{2}/3).$$

A dominant weight M of any irreducible representation can be written

$$\mathbf{M} = p_1\mathbf{M}^{(1)} + p_2\mathbf{M}^{(2)}, \tag{6.38}$$

where p_1 and p_2 are nonnegative integers. The representation with **M** as its highest weight is fully characterized by (p_1p_2). Alternatively, the representation may be denoted by its dimensionality N, but this single number does not fully characterize the representation. The first fundamental representation is characterized by (10) or **3**, and the second by (01) or $\overline{\mathbf{3}}$. The bar on the number 3 is used to denote the conjugate of the representation **3**. Another notation for a representation of dimension N characterized by integers (p_1p_2) is $D^{(N)}(p_1p_2)$.

We have obtained the weight diagrams of the two fundamental representations of $SU(3)$ from Eqs. (6.32) and (6.33) without a knowledge of the

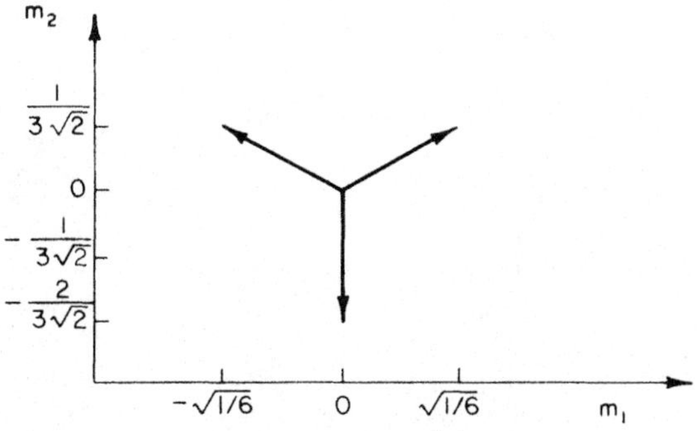

Fig. 6.2. Weight diagram for the first fundamental representation of $SU(3)$. This representation is denoted by (10) or **3**.

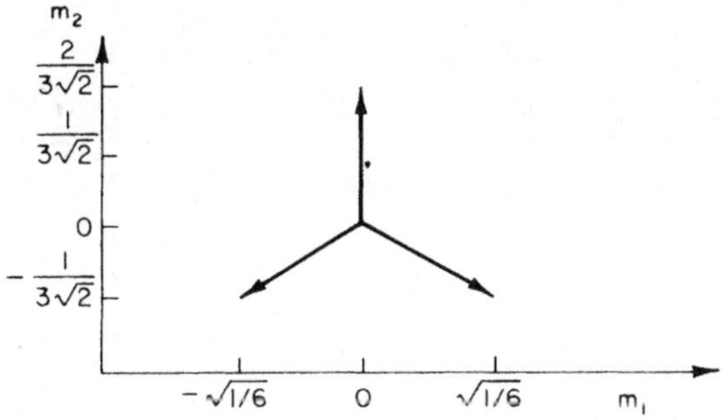

Fig. 6.3. Weight diagram of the second fundamental representation of $SU(3)$, denoted by (01) or $\bar{\mathbf{3}}$.

characters of the representations. This is because all weights of a fundamental representation of $SU(3)$ are simple and equivalent. However, the other rank two simple groups C_2 and G_2 do not have this property. Using Eqs. (6.32) and (6.33), we can obtain sets of equivalent weights, but cannot obtain the complete weight diagrams. Since we do not wish to discuss these groups in any detail, we shall just show the weight diagrams of the fundamental representations without illustrating here the procedure for obtaining them. The fundamental weight diagrams for C_2 are given in Fig. 6.4 and those for G_2 in Fig. 6.5. A method by which these weight diagrams (or the weight diagram of any simple compact group) can be obtained is described in Section 8.2. For a fuller treatment of C_2 and G_2, see, for example, Behrends *et al.* (1962).

The weight diagrams of C_2 and G_2 correspond to multiplets which do not have the properties of the known strongly interacting particles, as has been discussed by Ne'eman (1964). Therefore, if the hypercharge Y and third

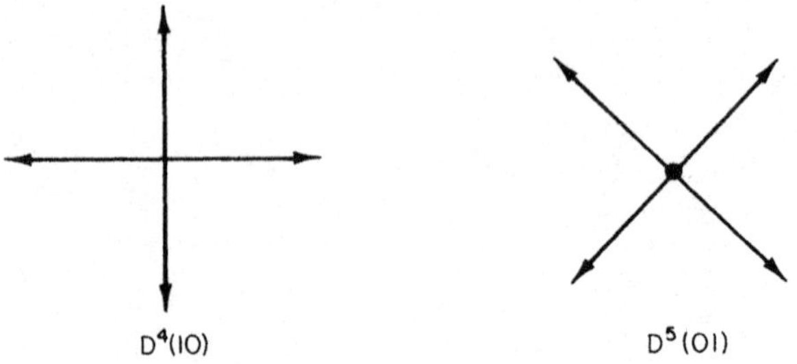

Fig. 6.4. Weight diagrams of the two fundamental representations of C_2. A solid small circle indicates a weight of zero length.

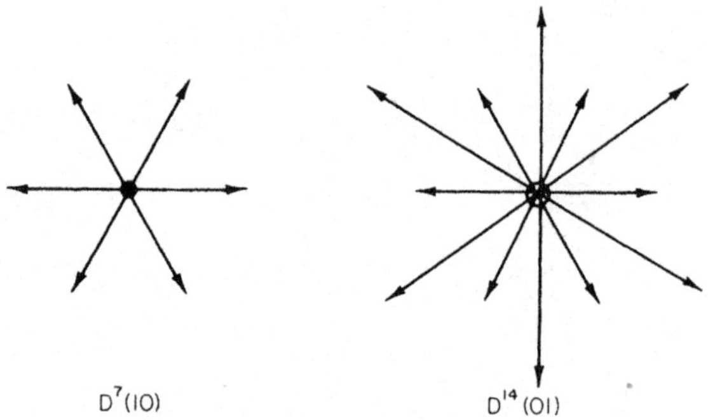

Fig. 6.5. Weight diagrams of the two fundamental representations of G_2. A solid circle indicates a simple weight of zero length and an open circle a double weight.

component of the isospin I_3 are to correspond to the diagonal operators of a compact simple Lie algebra, the only candidate is the Lie algebra of $SU(3)$. If we compare the predictions of $SU(3)$ with experiment, we find that the symmetry must hold only approximately. However, the evidence points very strongly toward $SU(3)$ being an underlying broken symmetry of the strong interactions of hadrons. We shall discuss this question further in Chapter 9.

It should be clear that we can multiply all weights by a common factor without losing any essential information. If we do this we merely change the normalization of the weights. The normalization we used so far enabled us to use Eqs. (6.32) and (6.33) to obtain the weights from the roots, as we have done in this section. This normalization has an important disadvantage, however. This is that the first component of a weight of $SU(3)$ is not a weight of $SU(2)$ but the constant $\sqrt{\frac{2}{3}}$ multiplied by the $SU(2)$ weight. It is convenient to change the normalization of $SU(3)$ weights so that the constant is unity. This change conforms to common usage. We can adopt this same procedure for any $SU(n)$ so that the first $k-1$ components of a weight of $n-1$ dimensions is a weight of $SU(k)$. We shall use this new normalization in the remainder of this chapter. In Chapter 8, when we discuss Clebsch–Gordan coefficients, it will prove convenient to change the normalization of the weights once again.

6.6 Weight Diagrams of $SU(3)$

In Figs. 6.1, 6.2, and 6.3 are given the weight diagrams of the fundamental representations of $SU(2)$ and $SU(3)$. We wish to give a procedure for obtaining the weight diagram of any irreducible representation of $SU(3)$.

First, we make some preliminary remarks about $SU(2)$. The dimensionality N of a representation of $SU(2)$ associated with an integer p is

$$N = p + 1. \tag{6.39}$$

Since the dimensionality is also given in terms of the quantum number j as

$$N = 2j + 1,$$

we obtain a relation between p and j:

$$p = 2j.$$

In Eq. (6.31) are given the values of all the weights associated with the representation j. These values can be plotted on a one-dimensional weight diagram. A knowledge of the weights of $SU(2)$ is useful in obtaining the weight diagrams of $SU(3)$ because $SU(2)$ is a subgroup of $SU(3)$.

In $SU(3)$, there is also a relation between the dimensionality of a representation and the integers $(p_1 p_2)$ which denote it. It is given by

$$N = \tfrac{1}{2}(p_1 + 1)(p_1 + p_2 + 2)(p_2 + 1). \tag{6.40}$$

In Chapter 7 we shall use Young tableaux to prove Eq. (6.39) and to see how Eq. (6.40) arises.

It is convenient to change our notation slightly from the previous section. First, instead of drawing a vector from the origin to indicate a weight, we shall place a dot at the place where the end of the vector should go. If the weight has multiplicity n greater than one, we shall place $n-1$ small concentric circles around the dot. These changes are to avoid cluttering up the weight diagram. We shall also draw dashed lines around the perimeter of a diagram to emphasize its shape. Another change is that we shall multiply all the weights of $SU(3)$ by a common factor $\sqrt{\frac{3}{2}}$, as we have remarked at the end of the last section. If we relate the first component of a weight to I_3 and the second component to Y, we have the relations

$$I_3 = m_1, \qquad Y = 2m_2/\sqrt{3}. \tag{6.41}$$

This normalization supersedes our previous one. With this normalization, m_1 is the eigenvalue of the generator F_3, and m_2 is the eigenvalue of F_8.

With these changes, the weight diagrams of the two fundamental representations of $SU(3)$ are given in Fig. 6.6. The dots in a horizontal row

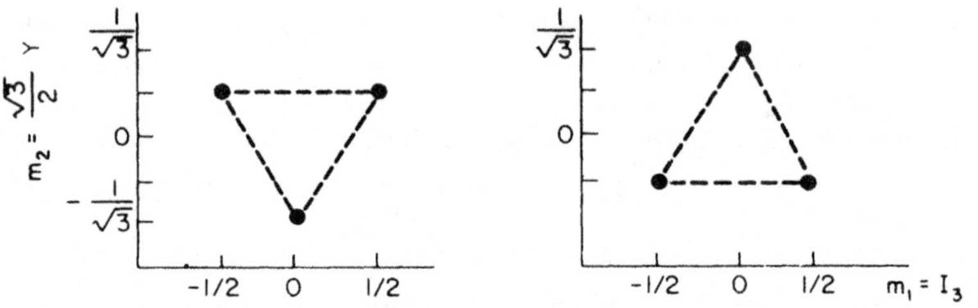

Fig. 6.6. Weight diagrams of the fundamental representations of $SU(3)$ normalized so that the values of m_1 are also weights of $SU(2)$.

of an $SU(3)$ weight diagram are the weights of $SU(2)$ multiplets. If the weights of any horizontal row are all simple, they constitute a single multiplet of $SU(2)$. Thus, the $SU(3)$ weight diagram of the representation (10) contains an $SU(2)$ doublet and an $SU(2)$ singlet, and likewise for the weight diagram of (01).

Note that the two fundamental weight diagrams of $SU(3)$ exhibit symmetry under rotations of 120° about the origin. This occurs because of the symmetry of the root diagram, and is therefore true for all representations. By reflecting a single nonvanishing weight with respect to lines perpendicular to the six roots, we obtain a set of three or six equivalent weights. If the original weight is vertical or makes an angle of 30° with the horizontal, then reflections produce a set of only three equivalent weights with a perimeter in the form of an equilateral triangle. This is the case for the two fundamental representations as well as many others. If the original weight is at another angle, reflections produce a set of six equivalent weights with a perimeter in the shape of a hexagon. This hexagon has the shape of an equilateral triangle with its corners removed by a horizontal line and lines making 60° with the horizontal. Note that all diagrams have an additional symmetry of reflections with respect to the vertical axis going through the origin.

Some $SU(3)$ diagrams have still greater symmetry. This can be seen as follows. If we rotate any weight diagram by 180°, we obtain the conjugate weight diagram. But a self-conjugate diagram must be unchanged by a rotation of 180°. Therefore such a diagram has both 120° and 180° symmetry, or 60° symmetry. Its perimeter is a regular hexagon.

Since all weight diagrams of $SU(3)$ have (at least) 120° symmetry, we must have $SU(2)$ multiplets, not only in horizontal rows, but in rows making angles of 120° with the horizontal. For example, *any* two weights of a three-dimensional representation of $SU(3)$ can be considered as a doublet of $SU(2)$ and the third weight can be considered an $SU(2)$ singlet. Thus, there are three different $SU(2)$ subgroups of $SU(3)$. We shall discuss these subgroups in more detail in Chapter 9.

With these preliminaries, we are ready to construct the weight diagram of any representation of $SU(3)$ specified by the integers $(p_1 p_2)$.

From Eq. (6.38), we can construct the highest weight

$$\mathbf{M} = p_1\mathbf{M}^{(1)} + p_2\mathbf{M}^{(2)},$$

where

$$\mathbf{M}^{(1)} = (1/2,\ 1/2\sqrt{3}), \qquad \mathbf{M}^{(2)} = (0,\ 1/\sqrt{3}), \tag{6.42}$$

with our new normalization. Making use of 120° symmetry and reflections through the vertical axis, we obtain a set of three or six equivalent weights. We may alternatively get a set of equivalent weights by reflecting with respect to the lines perpendicular to the roots. If the representation is $(p_1 0)$ or $(0 p_2)$, this procedure gives a triangular perimeter with three equivalent weights; otherwise it gives a set of six equivalent weights.

Next we may be able to obtain a number of other weights by making use of three different $SU(2)$ subgroups of $SU(3)$. We know the weights of $SU(2)$ decrease in integral steps along the horizontal and 120° lines. These are the only other weights of the diagram except for the problem of multiple weights.

All weights around the periphery of a diagram are simple by a theorem we have previously quoted; the others may be multiple. The maximum multiplicity v of any weight of a diagram is given by the formula

$$v = \tfrac{1}{2}(p_1 + p_2) - \tfrac{1}{2}\,|p_1 - p_2| + 1. \tag{6.43}$$

In the next chapter, we will indicate (but not prove) how the use of Young tableaux leads to this formula. Furthermore the multiplicities must not increase in going away from the center of a diagram, as otherwise we could not have complete multiplets of $SU(2)$. The actual rule of variation of multiplicities is somewhat stricter, as the multiplicity of two weights must either be the same or differ by unity. This can also be justified with Young tableaux.

These rules are sufficient to enable us to obtain any weight diagram of $SU(3)$. We shall illustrate the procedure with a few examples. First we shall obtain the weight diagram of the representation (30). The highest weight $\mathbf{M}$ is

$$\mathbf{M} = 3\mathbf{M}^{(1)} = (3/2,\ \sqrt{3}/2).$$

Using 120° symmetry we obtain the weights $(-3/2, \sqrt{3}/2)$ and $(0, -\sqrt{3})$. These weights are shown in Fig. 6.7a. Since the weights of the $SU(2)$ subgroups must differ by unity, we can add seven more weights as shown in

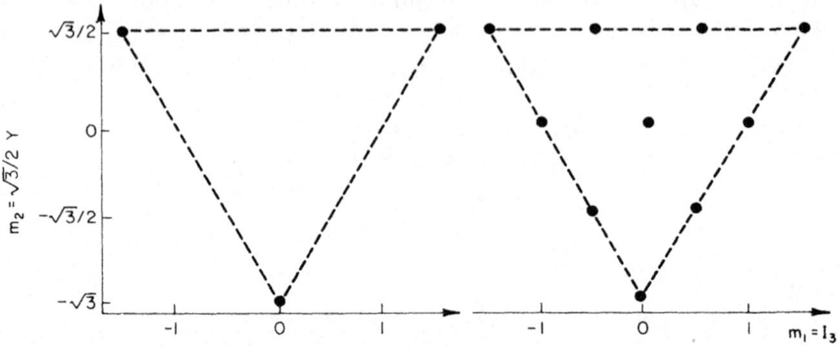

Fig. 6.7 (a) (left) A set of weights equivalent to the dominant weight of the representation (30) or **10**. (b) (right) The complete weight diagram of the representation (30).

Fig. 6.7b. From Eq. (6.43) we see that all the weights of any diagram $(p_1 0)$ are simple, so the diagram is complete. Furthermore, as a check we see from Eq. (6.40) that the dimensionality is $N = 10$.

As a second example consider the representation $p = (1, 1)$. The highest weight of this representation is

$$\mathbf{M} = (1/2, 1/2\sqrt{3}) + (0, 1/\sqrt{3}) = (1/2, \sqrt{3}/2).$$

The 60° symmetry of this self-conjugate representation gives a set of six equivalent weights which are shown in Fig. 6.8a. Since the weights of the $SU(2)$ subgroups must differ by unity, there must be a weight in the center. The maximum multiplicity of any weight of this diagram is 2 by Eq. (6.43).

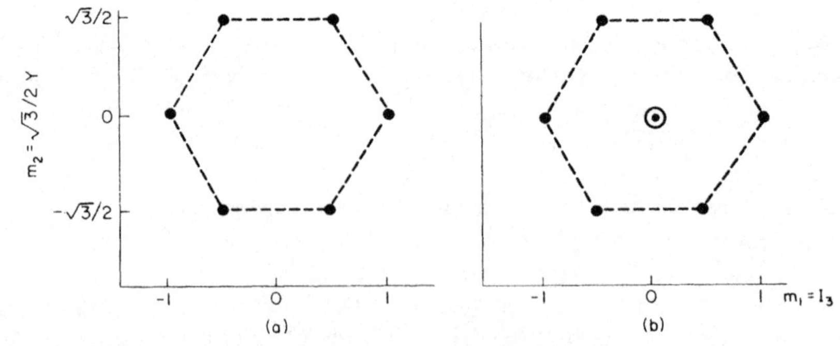

Fig. 6.8 (a) A set of weights equivalent to the dominant weight of (11) or **8**. (b) The complete weight diagram of (11).

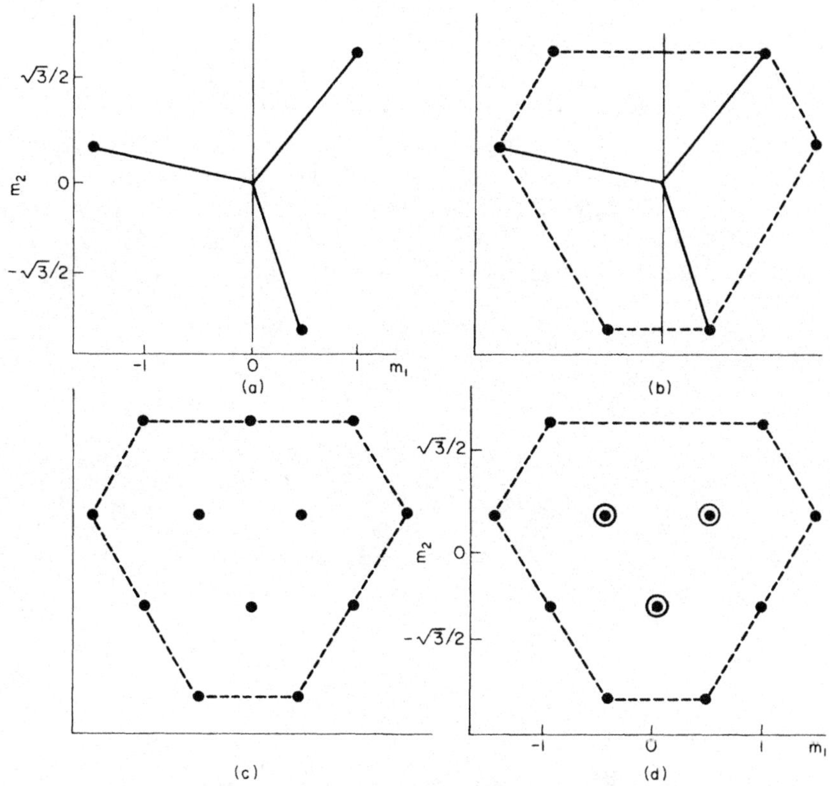

Fig. 6.9 (a) An incomplete set of weights equivalent to the dominant weight of (21) or **15**. (b) A complete set of equivalent weights formed by including the reflections of the weights of (a) with respect to the vertical axis. (c) Weights including those required by $SU(2)$ symmetry. (d) Complete weight diagram.

Furthermore, since the weights on the periphery are simple, the double weight must be in the center. The complete weight diagram is shown in Fig. 6.8b. As a check, we see from Eq. (6.40) that the dimensionality is $N = 8$.

As a third example, consider the representation $p = (2, 1)$. The highest weight of this representation is

$$\mathbf{M} = 2(1/2, 1/2\sqrt{3}) + (0, 1/\sqrt{3}) = (1, 2/\sqrt{3}).$$

Making use of 120° symmetry with respect to the origin, we obtain an incomplete set of three equivalent weights as shown in Fig. 6.9a. Reflecting these weights through the vertical axis, we obtain three additional equivalent weights, as shown in Fig. 6.9b. Making use of the fact that the weights of the $SU(2)$ subgroups differ in unit steps, we obtain Fig. 6.9c. Counting, we see the diagram has 12 weights thus far. From Eq. (6.40) we see that $N = 15$ for this weight diagram. By symmetry the three weights closest to the center must be double weights. This checks with the maximum multiplicity given in Eq. (6.43). The complete diagram is shown in Fig. 6.9d.

As a last example, consider the representation (2, 2) or **27**. The highest weight is $\mathbf{M} = (1, \sqrt{3})$. Making use of 60° symmetry, we obtain six equivalent weights in a regular hexagon. Making use of the $SU(2)$ subgroups, we obtain 13 additional weights. From Eq. (6.43) the maximum multiplicity of a weight for this diagram is three. Also since multiplicities cannot increase on out from the center, the central weight must be a triple weight. This gives two

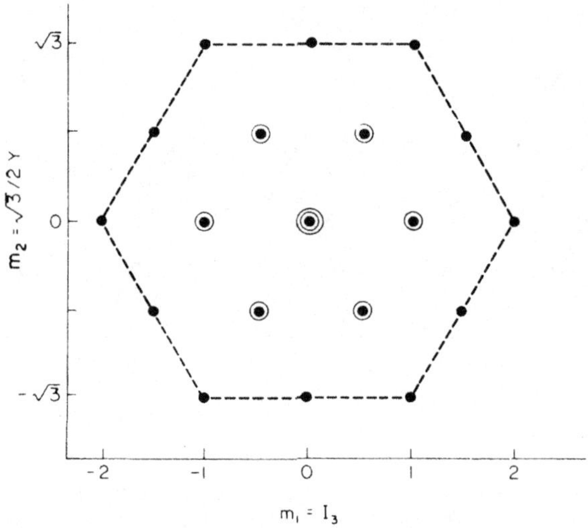

Fig. 6.10. Weight diagram of (22) or **27**.

more weights or 21 so far. But by Eq. (6.40) the dimensionality of the diagram is 27. By symmetry the six weights surrounding the center must be double weights. The complete weight diagram is shown in Fig. 6.10.

As an additional check on any weight diagram, we must see that we have complete multiplets of $SU(2)$. For example, in the diagram of Fig. 6.10, we have nine weights on the horizontal line with $m_2 = 0$. These weights have just the pattern to form a quintet, a triplet, and a singlet of $SU(2)$.

6.7 Weights of $SU(n)$

Weight diagrams are most convenient in two dimensions. In one dimension the diagrams are trivial, in three dimensions they are somewhat difficult to visualize when drawn on a two-dimensional surface, and in more than three dimensions they cannot be visualized at all.

In this chapter we shall not display any weight diagrams of $SU(4)$. However, we shall obtain the weights of the first fundamental representation of $SU(n)$ in our new normalization. With this normalization, the kth weight ($k = 1, 2, \ldots, n$) is the set of the kth eigenvalues of the $n \times n$ matrices $\frac{1}{2}\lambda_{i(i+2)}$, $i = 1, 2, \ldots, n-1$. Then, with the normalization we have chosen for $\lambda_{i(i+2)}$, the ith component of the kth weight $m_i(k)$ is given by

$$m_i(k) = \begin{cases} [2i(i+1)]^{-1/2}, & k < i+1 \\ -i[2i(i+1)]^{-1/2}, & k = i+1 \\ 0, & k > i+1. \end{cases}$$

For example, the $SU(4)$ weights are

$$\begin{aligned} \mathbf{m}(1) &= \left(\frac{1}{2}, \frac{1}{2\sqrt{3}}, \frac{1}{2\sqrt{6}}\right), & \mathbf{m}(2) &= \left(-\frac{1}{2}, \frac{1}{2\sqrt{3}}, \frac{1}{2\sqrt{6}}\right), \\ \mathbf{m}(3) &= \left(0, -\frac{1}{\sqrt{3}}, \frac{1}{2\sqrt{6}}\right), & \mathbf{m}(4) &= \left(0, 0, -\frac{1}{2}\sqrt{\frac{3}{2}}\right). \end{aligned} \tag{6.44}$$

As another example, the first weight of $SU(6)$ is

$$\mathbf{m}(1) = \frac{1}{2}\left(1, \frac{1}{\sqrt{3}}, \frac{1}{\sqrt{6}}, \frac{1}{\sqrt{10}}, \frac{1}{\sqrt{15}}\right).$$

In the next chapter, we shall see how, with the aid of Young Tableaux, we can obtain the weights of any representation of $SU(n)$ from the weights of the first fundamental representation.

6.8 Casimir Operators and the Labeling of States

We shall now briefly discuss the invariant operators or so-called *Casimir operators* of simple compact groups, especially $SU(n)$. Let us begin with the example of $SU(2)$. This group has one nonlinear invariant operator which can be constructed from the members of the Lie algebra of the group, namely the square of the total angular momentum (or isospin)

$$J^2 = J_1^{\,2} + J_2^{\,2} + J_3^{\,2}.$$

This invariant operator commutes with all the generators and is a constant for all multiplets of a given representation. The quantum number j associated with the invariant operator J can be used to label a representation of $SU(2)$ instead of the integer p.

An important theorem states that a simple compact group of rank l has l Casimir or invariant operators which are nonlinear functions of the generators. From this theorem we see that there are two Casimir operators of $SU(3)$. We can label all the states of a given multiplet with the values of these Casimir operators, or alternatively we can label all members of the multiplet by the two numbers p_1 and p_2. Since $SU(n)$ has rank $n-1$, we can label all members of a multiplet with the $n-1$ p_i or Casimir operators C_i. A theorem states that the Casimir operators of $SU(n)$ can be written in the form

$$C_i = \sum_{\mu} a_{\mu i} X_{\mu}^{i+1}, \qquad i = 1, 2, \ldots, n-1, \tag{6.45}$$

where the X_μ are the generators of the group. By demanding that the C_i commute with all the generators, we can, in principle, determine the coefficients $a_{\mu i}$. However, it may be tedious to find the Casimir operators for $SU(n)$ if n is large. Furthermore, in an internal symmetry group like $SU(3)$ there does not seem to be any obvious physical significance to the Casimir operators.

Of course, if one wants to write down scalar or invariant interactions which depend on the internal symmetry coordinates, knowledge of the Casimir operators becomes quite useful. However, if one just wants to label the states of a multiplet, it is much more convenient to use the p_i.

We are now ready to consider the problem of how many numbers are necessary to label a state of $SU(n)$. First, we need to specify $n-1$ Casimir operators or p_i to distinguish which representation the state belongs to. Second, we need to specify the weight $\mathbf{m}$, which has $n-1$ components. For a multiplet in which all weights are simple, this is sufficient. However, for $n > 2$, we need more labels to distinguish the different members of a multiplet with the same weight.

For $SU(3)$, in which the weight signifies the value of the third component of isospin I_3 and the hypercharge Y, we solve this problem by also giving the value of the isospin I. Thus, for example, there is a double weight at the center of the octet weight diagram with $I_3 = Y = 0$. One of the states with this weight is an isospin singlet ($I = 0$) and the other is a member of an isospin triplet ($I = 1$). Thus, we complete the specification of the state by giving the value of the Casimir operator of one of the three $SU(2)$ subgroups. Alternatively, we could specify the value of p for the appropriate $SU(2)$ subgroup.

This solution to the labeling problem for $SU(3)$ can be generalized to $SU(n)$. In this case we can specify in addition to the p_i (or Casimir operators) and weights of $SU(n)$, the values of p_i (or Casimir operators) of $SU(n-1)$, $SU(n-2), \ldots, SU(2)$. This may not be a convenient way to label the states, but it is complete. Thus, a state of $SU(n)$ can be labeled by

$$2(n-1) + (n-2) + (n-3) + \cdots + 1 = \tfrac{1}{2}n(n+1) - 1$$

numbers. Two numbers are necessary to label a state of $SU(2)$ and five are necessary for a state of $SU(3)$.

The first of the two Casimir operators of $SU(3)$, denoted by F^2, can be given in terms of the F-spin generators defined in Eq. (6.17). It is

$$F^2 = \sum_{\mu=1}^{8} F_\mu{}^2. \tag{6.46}$$

The second Casimir operator G^3 is more conveniently given in terms of nine traceless operators A_{ij} defined by

$$(A_{ij})_{kl} = \delta_{ik}\,\delta_{jl} - \tfrac{1}{3}\delta_{ij}\,\delta_{kl} \qquad (i, j, k, l = 1, 2, 3).$$

Only eight of these operators are independent, since

$$A_{11} + A_{22} + A_{33} = 0.$$

Then G^3 is given by

$$G^3 = \sum_{ijk} A_{ij}\,A_{jk}\,A_{ki}. \tag{6.47}$$

See de Swart (1963) for further properties of the A_{ij} operators.

We can verify directly that F^2 and G^3 commute with every generator. It follows that matrix representations of F^2 and G^3 commute with all matrices of an irreducible representation, since all the matrices of an irreducible representation are functions of the generators. Therefore, by Schur's lemma (Section 3.3), F^2 and G^3 are constant multiples of the unit matrix. This proves that they are invariant operators. Thus, a given irreducible representation of $SU(3)$ can be labeled either by the two numbers p_1 and p_2 or by the quantum numbers f and g associated with the Casimir operators F^2 and G^3. Either pair of numbers completely specifies a representation. We shall use the numbers p_1 and p_2. The operator I^2 defined by

$$I^2 = \sum_{i=1}^{3} F_i{}^2, \tag{6.48}$$

is not an invariant operator of $SU(3)$, but it is an invariant operator of the isospin $SU(2)$ subgroup of $SU(3)$. A multiplet of $SU(3)$ can be chosen in a basis such that all members are eigenstates of I^2. The eigenvalues are $I(I + 1)$, where I is integral or half integral.

6.9 Tensor Operators

Let us consider the matrix element of a self-adjoint operator T (an observable) between two states ϕ, ψ which are basis functions of a representation of a symmetry group which we take to be a compact Lie group. This matrix element M is given by

$$M = (\phi, T\psi). \tag{6.49}$$

We now ask how the matrix element M compares to the matrix element of the same observable between transformed states ϕ', ψ' given by

$$\phi' = U\phi, \qquad \psi' = U\psi \tag{6.50}$$

where U is a unitary operator of the symmetry group. The matrix element M' of T with respect to the transformed states is given by

$$M' = (\phi', T\psi') = (U\phi, T\psi), \tag{6.51}$$

or

$$M' = (\phi, U^{-1}TU\psi), \tag{6.52}$$

since U is unitary.

From Eq. (6.52), we see that the matrix element of T between the transformed states ϕ', ψ' is the same as the matrix element of a transformed observable T' given by

$$T' = U^{-1}TU, \tag{6.53}$$

between the original states. Thus, we are led to consider the transformation properties of operators. It is clear from Eq. (6.53) that if T commutes with U, then the transformed operator T' is equal to the original operator T. Such an operator is called a *scalar* under the transformation. However, in general, the operator will have more complicated transformation properties. We shall restrict ourselves to consideration of so-called tensor operators which have definite properties under the transformation of Eq. (6.53). Let us first consider the special case in which the operator T is X_μ, where X_μ is one of the generators of the symmetry group. We transform T by means of a unitary operator

$$U_\nu = \exp(ia_\nu X_\nu).$$

Then

$$T' = X_\mu' = \exp(-ia_\nu X_\nu)X_\mu \exp(ia_\nu X_\nu). \tag{6.54}$$

If the transformation is an infinitesimal one, then to lowest order

$$U_\nu = 1 + ia_\nu X_\nu. \tag{6.55}$$

Then Eq. (6.54) becomes

$$X_\mu' = (1 - ia_\nu X_\nu)X_\mu(1 + ia_\nu X_\nu) = X_\mu + ia_\nu[X_\mu, X_\nu]. \tag{6.56}$$

But the commutators of the generators are given as linear combinations of the generators

$$[X_\mu, X_\nu] = \sum_\lambda c_{\mu\nu\lambda} X_\lambda .$$

Then Eq. (6.56) becomes

$$X_\mu' - X_\mu = dX_\mu = i a_\nu \sum_\lambda c_{\mu\nu\lambda} X_\lambda . \tag{6.57}$$

If an operator T_μ has the same transformation properties as the generators, i.e.,

$$dT_\mu = i a_\nu \sum_\lambda c_{\mu\nu\lambda} T_\lambda , \tag{6.58}$$

the operator is said to be a vector operator. If we have an operator with n indices $T_{\mu_1, \mu_2, \ldots, \mu_n}$ which transforms like $X_{\mu_1} X_{\mu_2} \cdots X_{\mu_n}$, then $T_{\mu_1, \mu_2, \ldots, \mu_n}$ is called a tensor operator of rank n.

Of more interest to us are the so-called irreducible tensor operators. Let us consider the basis vectors $\psi_m^{(j)}$ of an irreducible representation of a Lie group. The index j identifies the irreducible representation and m distinguishes different vectors of the representation by some ordering process. If we operate on $\psi_m^{(j)}$ with a unitary operator U_ν of the symmetry group, we obtain

$$\psi_m'^{(j)} = U_\nu \psi_m^{(j)}$$

or, letting the matrix elements of U_ν be $D_{m'm}^{(j)}(a_\nu)$, we have

$$\psi_m'^{(j)} = \sum_{m'} D_{m'm}^{(j)}(a_\nu) \psi_{m'}^{(j)} . \tag{6.59}$$

We now define an irreducible tensor operator $T_m^{(j)}$ as an operator which transforms the same way as $\psi_m^{(j)}$ in Eq. (6.59). Thus we have

$$T_m'^{(j)} = \sum_{m'} D_{m'm}^{(j)}(a_\nu) T_{m'}^{(j)} = U_\nu^{-1} T_m^{(j)} U_\nu . \tag{6.60}$$

Other authors define a tensor operator by using the inverse or transpose of $D_{m'm}^{(k)}(a_\nu)$.

The *rank* of an irreducible tensor operator may also be defined. For $SU(2)$ if the number of operators $T_m^{(j)}$ is $2j + 1$, then j is the rank. For a more general simple compact Lie group, we may order the irreducible unitary representations in some manner and define the rank according to this ordering. Alternatively, we may regard the rank as a quantity specified by more than one number, for example by $\mathbf{p}$ for $SU(n)$.

We see from Eqs. (6.18) and (6.20) that for $SU(3)$ there are two sets of irreducible vector operators, the F_μ and D_μ. This is different from the case of $SU(2)$, for which there is only one set of irreducible vector operators associated with the generators.

CHAPTER 7

YOUNG TABLEAUX AND UNITARY SYMMETRY

7.1 Dimensionality of Multiplets of $SU(n)$

In the previous chapter, we represented the fundamental doublet of $SU(2)$ by the states

$$u_1 = \begin{pmatrix} 1 \\ 0 \end{pmatrix} \quad \text{and} \quad u_2 = \begin{pmatrix} 0 \\ 1 \end{pmatrix}. \tag{7.1}$$

Let us assume that these states denote states of a single particle of spin $\frac{1}{2}$. Another notation for these one-particle states is by means of a single-box Young tableau $\boxed{}$. We make the identification

$$u_1 = \boxed{1}, \qquad u_2 = \boxed{2}. \tag{7.2}$$

The single-box tableau without a number stands for both members of the doublet. Now suppose we have a two-particle state. We know that in order to be a multiplet of the symmetric group S_2, the state must be either symmetric, corresponding to the Young tableau $\boxed{}\!\boxed{}$, or antisymmetric, corresponding to the tableau

$$\begin{array}{|c|}\hline \\ \hline \\ \hline \end{array}$$

These same tableaux also represent multiplets of $SU(2)$. Consider first the symmetric state. If both particles are in the state u_1, the corresponding tableau is $\begin{array}{|c|c|}\hline 1 & 1 \\ \hline\end{array}$ whereas if both particles are in the state u_2, the tableau is $\begin{array}{|c|c|}\hline 2 & 2 \\ \hline\end{array}$. The only other symmetric possibility is for one of the particles to be in the state u_1 and the other in the state u_2. This is represented by the tableau $\begin{array}{|c|c|}\hline 1 & 2 \\ \hline\end{array}$. Thus, the symmetric two-particle state of $SU(2)$ is a triplet. The multiplicity is given by all the standard arrangements of the Young tableau $\begin{array}{|c|c|}\hline \ & \ \\ \hline\end{array}$ with integers restricted to be 1 or 2. The nonstandard arrangement $\begin{array}{|c|c|}\hline 2 & 1 \\ \hline\end{array}$ is obviously the same as the standard arrangement $\begin{array}{|c|c|}\hline 1 & 2 \\ \hline\end{array}$, since the state is symmetric. Therefore the nonstandard arrangement must not be counted. The only antisymmetric two-particle state is given by the tableau

$$\begin{array}{|c|}\hline 1 \\ \hline 2 \\ \hline\end{array}$$

This is a singlet state.

Next, consider multiplets composed of three spin-$\frac{1}{2}$ particles. The two possible tableaux are

$$\begin{array}{|c|c|c|}\hline \ & \ & \ \\ \hline\end{array} \quad \text{and} \quad \begin{array}{|c|c|}\hline \ & \ \\ \hline \ \\ \cline{1-1}\end{array}$$

This is because one cannot construct a totally antisymmetric state of three particles when only two states are available. The standard tableaux of the symmetric multiplet are

$$\begin{array}{|c|c|c|}\hline 1 & 1 & 1 \\ \hline\end{array} \quad \begin{array}{|c|c|c|}\hline 1 & 1 & 2 \\ \hline\end{array} \quad \begin{array}{|c|c|c|}\hline 1 & 2 & 2 \\ \hline\end{array} \quad \begin{array}{|c|c|c|}\hline 2 & 2 & 2 \\ \hline\end{array}$$

Thus, this state is a quartet. The standard arrangements of the tableau of mixed symmetry are

$$\begin{array}{|c|c|}\hline 1 & 1 \\ \hline 2 \\ \cline{1-1}\end{array} \quad \text{and} \quad \begin{array}{|c|c|}\hline 1 & 2 \\ \hline 2 \\ \cline{1-1}\end{array}$$

Thus, this tableau

$$\begin{array}{|c|c|}\hline \ & \ \\ \hline \ \\ \cline{1-1}\end{array}$$

is a doublet, just like the single-box diagram $\begin{array}{|c|}\hline \ \\ \hline\end{array}$. The reason that the tableaux

$$\begin{array}{|c|c|}\hline \ & \ \\ \hline \ \\ \cline{1-1}\end{array} \quad \text{and} \quad \begin{array}{|c|}\hline \ \\ \hline\end{array}$$

have the same multiplicity is that there is only one way to make an antisymmetric state of two particles, if only two different single-particle states are available. Thus, if we are interested in the multiplicity, but not in the number of particles, we can omit all columns with two boxes.

These examples illustrate the fact that a Young tableau can be used to denote any multiplet of $SU(2)$. The individual members of the multiplet are denoted by the standard arrangements and the multiplicity by the total number of standard arrangements.

An analogous result holds for $SU(n)$ with the numbers in each box restricted to be $1, 2, \ldots, n$. Consider a state of ν particles denoted by a Young tableau with ν boxes. Such a state is an irreducible tensor, (i.e., a basis tensor for an irreducible representation) of the symmetric group S_ν. We now state without proof an important theorem.

THEOREM. *If a ν-particle state is an irreducible tensor of S_ν and is constructed from one-particle states which are basis vectors of an irreducible n-dimensional representation of $SU(n)$, then the state is an irreducible tensor of $SU(n)$.*

A consequence of this theorem is that any Young tableau with n rows or less denotes the basis tensors of an irreducible representation of $SU(n)$. If there are more than n rows in a tableau, then at least one column must have more than n boxes. But a column with more than n boxes denotes a tensor which is identically zero. This follows because if the tensor is to be antisymmetric, no more than one particle can be in each state. But there are more than n particles and only n available states. When we try to antisymmetrize a function containing two particles in the same state we get zero.

To obtain the dimensionality of an irreducible representation of $SU(n)$, we simply count the number of standard arrangements of the corresponding Young tableau. When we put a number in a box, the number stands for one of the n possible states of a single particle. Therefore, the prescription for obtaining a standard arrangement of a tableau of $SU(n)$ is to put an integer from 1 to n in each box of the tableau such that the numbers increase from top to bottom in a column and do not decrease from left to right in a row. The number of possible standard arrangements is the number of components of the irreducible tensor which the tableau denotes. This, of course, equals the dimensionality of the corresponding irreducible representation. Furthermore, the dimensionality holds not only for the group $SU(n)$, but for the groups $GL(n)$, $SL(n)$, and $U(n)$. However, for the noncompact groups $GL(n)$ and $SL(n)$, the irreducible representations thus obtained are not unitary. In general the faithful unitary representations of noncompact groups are infinite dimensional. We shall not consider $GL(n)$ and $SL(n)$ further.

Each standard arrangement of a Young tableau stands for a particular irreducible tensor, and therefore has a weight. This weight can be obtained from the weights of the single-particle states by ordinary addition. Therefore, we can construct the weight diagram of a multiplet from its Young tableau.

As an example, we consider the one-particle vectors of $SU(n)$, which are denoted by a single box □. This box represents the basis vectors for the first so-called *fundamental* representation of $SU(n)$. We can obtain the dimensionality of this fundamental representation by counting the number of standard arrangements. We have already pointed out that for $SU(2)$ we have two arrangements

 and 

with weights $m = \pm\frac{1}{2}$. Clearly, for $SU(n)$ there are n standard arrangements of the tableau □, since any number from 1 to n can be put in the box. There are actually $n-1$ different fundamental representations of $SU(n)$, at most two of which have dimension n. We shall define the others later in this section.

Next consider a two-particle state. There are two possible Young tableaux

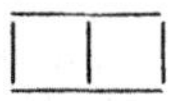 and 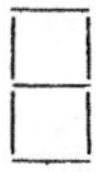

The dimensionality of these irreducible tensors depends on n. We have already considered the number of standard arrangements for $SU(2)$. For $SU(3)$ the standard arrangements are

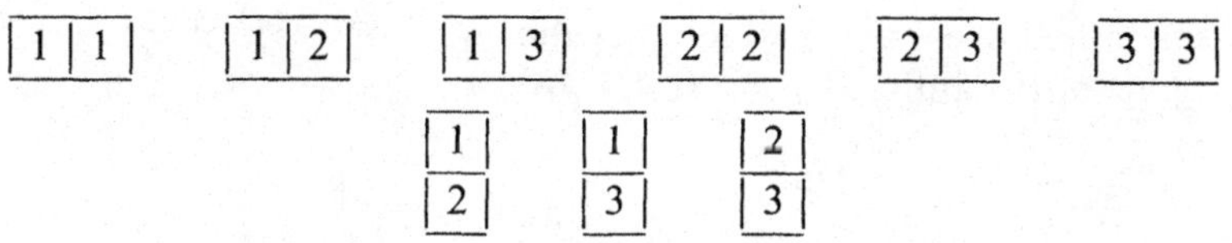

Thus the tableau □□, which we previously found was a 3-component irreducible tensor of $SU(2)$, is a 6-component irreducible tensor of $SU(3)$. Likewise the tableau

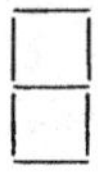

which we found to be a scalar of $SU(2)$, is a 3-component tensor of $SU(3)$.

Since the weights are additive, it is easy to obtain the weights of the standard arrangements of the tableaux

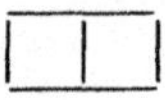 and 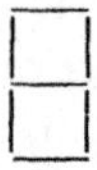

starting with the weights of the first fundamental representation. From Fig. 6.6, these weights are

$$
\begin{aligned}
\boxed{1}&: \quad \mathbf{m}(1) = \mathbf{M}^{(1)} = (1/2,\ 1/2\sqrt{3}),\\
\boxed{2}&: \quad \mathbf{m}(2) = (-1/2,\ 1/2\sqrt{3}),\\
\boxed{3}&: \quad \mathbf{m}(3) = (0,\ -1/\sqrt{3}).
\end{aligned}
\tag{7.3}
$$

Then the weights of the two-particle states are

$$
\begin{aligned}
\begin{array}{|c|c|}\hline 1 & 1 \\ \hline\end{array}&: \quad \mathbf{m}(1) = \mathbf{M} = (1,\ 1/\sqrt{3}),\\
\begin{array}{|c|c|}\hline 1 & 2 \\ \hline\end{array}&: \quad \mathbf{m}(2) = (0,\ 1/\sqrt{3}),\\
\begin{array}{|c|c|}\hline 2 & 2 \\ \hline\end{array}&: \quad \mathbf{m}(3) = (-1,\ 1/\sqrt{3}),\\
\begin{array}{|c|c|}\hline 1 & 3 \\ \hline\end{array}&: \quad \mathbf{m}(4) = (1/2,\ -1/2\sqrt{3}),\\
\begin{array}{|c|c|}\hline 2 & 3 \\ \hline\end{array}&: \quad \mathbf{m}(5) = (-1/2,\ -1/2\sqrt{3}),\\
\begin{array}{|c|c|}\hline 3 & 3 \\ \hline\end{array}&: \quad \mathbf{m}(6) = (0,\ -2/\sqrt{3}),
\end{aligned}
\tag{7.4}
$$

where we have ordered the weights according to our convention that the last component is controlling. From these values, we obtain the weight diagram of Fig. 7.1. Because the weights are additive, the weight of a state does not depend on the configuration of its Young tableau, but only on the numbers in its boxes, i.e., a weight depends only on a set of integers. Thus, the weights of the antisymmetric states

$$
\begin{array}{|c|}\hline 1 \\ \hline 2 \\ \hline\end{array} \qquad \begin{array}{|c|}\hline 1 \\ \hline 3 \\ \hline\end{array} \qquad \text{and} \qquad \begin{array}{|c|}\hline 2 \\ \hline 3 \\ \hline\end{array}
$$

are given by $\mathbf{m}(2)$, $\mathbf{m}(4)$, and $\mathbf{m}(5)$, respectively, of (7.4). These weights are shown in Fig. 7.1 connected by a solid triangle which is the weight diagram of the $\bar{\mathbf{3}}$ representation. The use of Young tableaux also enables us to obtain the multiplicity of a given weight. Since the value of a weight depends only on a set of integers, all standard arrangements of a tableau with the same integers have the same weight. Thus, we need merely count the number of these arrangements to obtain the multiplicity of the weight.

The procedure for obtaining the dimensionality of an irreducible representation by counting the standard arrangements of a Young tableau can be simplified by noting that for $SU(n)$ [and also $U(n)$] any column with n boxes has only one standard arrangement. Therefore, so far as counting

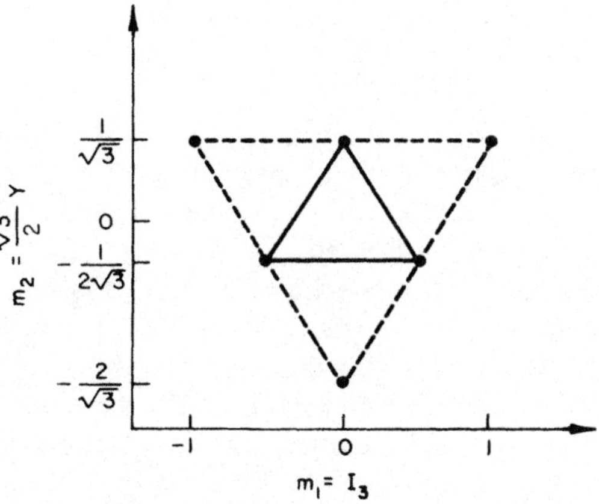

Fig. 7.1. Weight diagram of the representation (20) or **6**. The small solid triangle shows the part which is also the weight diagram of the representation (01) or $\bar{\mathbf{3}}$.

states is concerned, we can simplify a tableau by removing all such columns. For example, consider the tableau

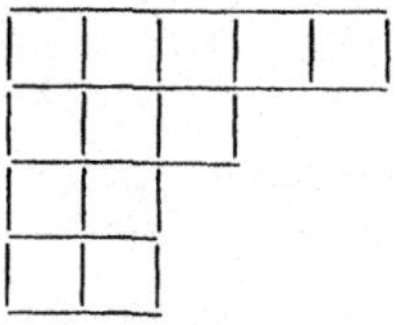

If $n = 4$, this tableau corresponds to the same irreducible representation as the tableau

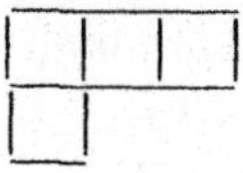

Thus, we can make the identification

$SU(4)$:

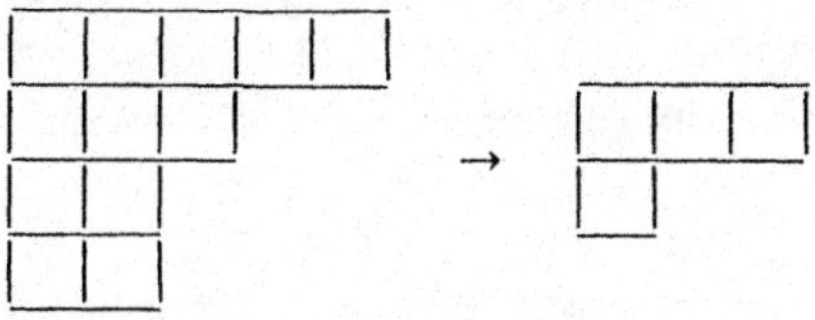

Of course, the first tableau denotes a 12-particle state while the second denotes a four-particle state; nevertheless, the dimensionalities of the representations are equal. Thus, if we are interested only in the dimensionality of a representation of $SU(n)$ associated with a particular Young tableau, but not in the number of particles, we can remove all columns with n boxes.

We need consider only tableaux specified by the $n-1$ numbers

$$\mathbf{p} = (p_1, p_2, \ldots, p_{n-1}). \tag{7.5}$$

For $SU(n)$, the p_i are the same integers that appear in Eq. (6.35), which gives the dominant weight of the multiplet specified by $\mathbf{p}$.

We next consider the effect on the weights of omitting all columns with n boxes. It can be seen from Eqs. (6.4) and (6.12) that the Pauli matrices of $SU(2)$ and the matrices of $SU(3)$ are all traceless. It is generally true that the generators of $SU(n)$ can be represented by $n^2 - 1$ traceless matrices in n dimensions. Since all the matrices are traceless, this is true in particular of the diagonal matrices $\mathbf{H}$ whose elements are the weights. Therefore the sum of all the weights of the first fundamental representation is zero.

$$\sum_{i=1}^{n} \mathbf{m}(i) = 0. \tag{7.6}$$

This means that a column of n boxes has weight $\mathbf{m} = 0$. Thus, omitting columns of n boxes has no effect on the weights. This last statement is not true for $U(n)$ since $U(n)$ has an additional generator which has a nonzero trace. (In fact it is a multiple of the unit matrix.) The weight associated with this extra generator is proportional to the number of particles, i.e., to the number of boxes.

Thus, the group properties of a representation of $SU(n)$ [but not of $U(n)$] are completely specified by the numbers p_i $(i = 1, 2, \ldots, n-1)$. All representations with the same values of p_i are equivalent. The p_i are just the numbers discussed in Section 6.3 in connection with constructing a dominant weight from the $n-1$ fundamental dominant weights of $SU(n)$. [See Eq. (6.35)].

A Young tableau of $SU(n)$ containing only a single column of n boxes is a singlet of $SU(n)$ and is denoted by

$$\mathbf{p} = (00 \cdots 0), \qquad n-1 \quad \text{zeros}. \tag{7.7}$$

In drawing such a tableau, if we remove the column, we have nothing left and might forget that we have a singlet. Thus, we shall either keep such a tableau or replace it by the symbol ①. The weight of any singlet of $SU(n)$ is

$$\mathbf{m} = (00 \cdots 0), \qquad n-1 \quad \text{zeros}. \tag{7.8}$$

If the number of boxes in the longest column of a tableau is less than $n-1$, we still use $n-1$ integers to specify it. For example, suppose we have the four-particle tableau

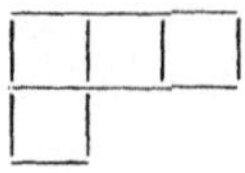

of $SU(6)$. This tableau is specified by the five integers

$$p_1 = 2, \qquad p_2 = 1, \qquad p_3 = 0, \qquad p_4 = 0, \qquad p_5 = 0,$$

or (21000). We can simplify this notation further by using a superscript for repeated integers $(21000) = (210^3)$.

Now suppose we have a tableau specified by the integers $(p_1, p_2, \ldots, p_{n-1})$. Then we define the *conjugate tableau* with respect to $SU(n)$ to be the one specified by the integers $(p_{n-1}, \ldots, p_2, p_1)$. As an example, the tableau conjugate to (210^3) is $(0^3 12)$ or

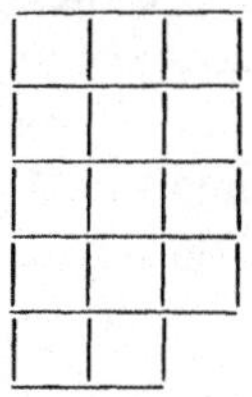

The number of boxes of a tableau is not equal, in general, to the number of boxes in its conjugate tableau. This is in contrast to conjugation with respect to S_n defined in Chapter 4. When we refer to a conjugate tableau, it should be clear from the context whether we mean with respect to $SU(n)$ or S_n.

If the same tableau is denoted by the integers $(p_1, \ldots, p_{n-1})$ and $(p_{n-1}, \ldots, p_n)$ it is said to be *self-conjugate*. If a tableau and its conjugate are not the same, they represent irreducible tensors of two *inequivalent* representations with the same dimensionality. In $SU(2)$, all tableaux are specified by a single integer p_1, and are therefore all self-conjugate. Thus, a representation of $SU(2)$ is fully specified by giving its dimensionality, but this is not true of $SU(n)$ with $n > 2$.

As we remarked previously, there are $n - 1$ fundamental representations of $SU(n)$. These correspond to all the Young tableaux with a single column with up to $n - 1$ boxes. The ith fundamental representation of $SU(n)$ is specified by the $n - 1$ integers

$$\begin{aligned} p_i &= 1, \\ p_j &= 0, \qquad j \neq i. \end{aligned} \tag{7.9}$$

Thus, for example, $SU(3)$ has the fundamental representations (10) and (01). Since they are conjugates of each other, they have the same dimensionality.

As another example, the three fundamental representations of $SU(4)$ have the tableaux

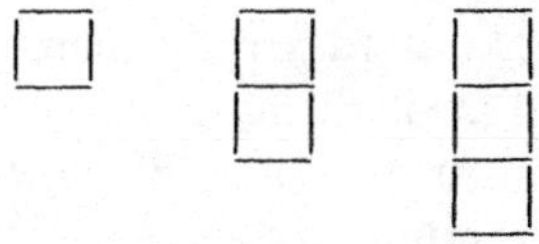

(100) (010) (001)

The representations (100) and (001) are conjugates of each other, and each has four dimensions. However the self-conjugate representation (010) has six dimensions. We have made particular use of the first fundamental representation of $SU(n)$, denoted by (10^{n-2}), as all others can be constructed from it.

7.2 Dimensionality Formulas

It can become very tedious to find the dimensionality of a Young tableau of $SU(n)$ by counting all its standard arrangements. However, we can simplify the problem by obtaining a general formula for the number of standard arrangements. This is a combinatorial problem with a straightforward solution. We shall derive the formula for $SU(2)$, and just state the result for $SU(n)$.

THEOREM. *For $SU(2)$, the dimensionality $N(p)$ of an irreducible representation corresponding to a tableau with p boxes is*

$$N(p) = p + 1. \tag{7.10}$$

Proof. Consider the tableau consisting of a single box $\square$. By direct counting, we find there are two standard arrangements $\boxed{1}$ and $\boxed{2}$. Therefore $N(1) = 2$, or $N(p) = p + 1$ if $p = 1$. We next show that if $N(p) = p + 1$, then $N(p + 1) = p + 2$. It then will follow by induction that $N(p) = p + 1$ holds for all p. Consider the standard arrangements with p boxes:

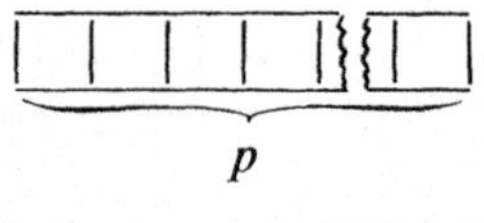

The last box may contain either a 1 or a 2. There is only one standard arrangement with the last box containing a 1: namely, the arrangement in which all boxes contain a 1. Now if we add another box, this last box can contain either a 1 or a 2:

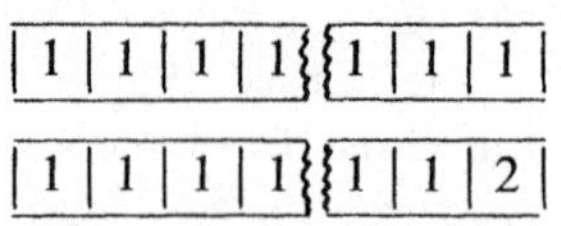

Therefore, we obtain two standard arrangements where we had one before. Now consider all other standard arrangements of p boxes in which the integer in the last box is 2. There must be p such arrangements, since by assumption, there are a total of $p + 1$ standard arrangements of p boxes, and only one of them ends with a 1. If we add one more box to these tableaux, we must put the integer 2 into this box. So the number of standard arrangements arising from the source remains at p. Finally, all together we have $p + 2$ standard arrangements of $p + 1$ boxes, which completes the proof.

The dimensionalities for $SU(3)$ and $SU(4)$ are given by

$$N_3(p_1 p_2) = \tfrac{1}{2}(p_1 + 1)(p_1 + p_2 + 2)(p_2 + 1), \tag{7.11}$$

or

$$N_3(p_1 p_2) = \tfrac{1}{2}(p_1 + p_2 + 2)(p_2 + 1)N_2(p_1). \tag{7.12}$$

$$\begin{aligned} N_4(p_1 p_2 p_3) = \frac{1}{2!\,3!}&(p_1 + 1)(p_1 + p_2 + 2)(p_1 + p_2 + p_3 + 3) \\ &\times (p_2 + 1)(p_2 + p_3 + 2)(p_3 + 1), \end{aligned} \tag{7.13}$$

or

$$N_4(p_1 p_2 p_3) = (p_1 + p_2 + p_3 + 3)(p_2 + p_3 + 2)(p_3 + 1)N_3(p_1 p_2)/3! \tag{7.14}$$

The dimensionality for $SU(n + 1)$ is given by

$$\begin{aligned} &N_{n+1}(p_1 \cdots p_n) \\ &= \frac{1}{2!\,3!\cdots n!}(p_1 + 1)(p_1 + p_2 + 2)\cdots(p_1 + p_2 + \cdots + p_n + n) \\ &\quad\times (p_2 + 1)(p_2 + p_3 + 2)\cdots(p_2 + \cdots + p_{n-1} + n - 1)\cdots(p_n + 1) \end{aligned} \tag{7.15}$$

or

$$\begin{aligned} &N_{n+1}(p_1 \cdots p_n) \\ &= (p_n + 1)(p_n + p_{n-1} + 2)\cdots(p_n + p_{n-1} + \cdots + p_1 + n)N_n(p_1 \cdots p_{n-1})/n!. \end{aligned} \tag{7.16}$$

It is also a combinatorial problem to find the multiplicity of any weight associated with a given tableau of $SU(n)$. This is just the number of standard arrangements of a tableau with a given set of integers. For $SU(3)$, the solution is that the maximum multiplicity v of any tableau is given by Eq. (6.43)

$$v = \tfrac{1}{2}(p_1 + p_2) - \tfrac{1}{2}|p_1 - p_2| + 1.$$

We shall not prove this, but shall prove the special case that all weights of a rectangular Young tableau of $SU(3)$ are simple.

First we prove that all weights of $SU(2)$ are simple. The proof follows at once from the fact that any tableau consists of a single row. If we have a given set of 1's and 2's, the only standard arrangement consists of all the 1's followed by all the 2's. The generalization to a single-row tableau of $SU(n)$ follows immediately, since all the 1's, 2's, 3's, ..., n's must go in consecutive order. For $SU(3)$, all rectangular Young tableaux have either one or two rows. The tableaux with two rows are conjugates of the one-row tableaux, and therefore have the same multiplicity. This completes the proof.

For tableaux of $SU(n)$, the combinational problem is harder. We shall not give the general result for the maximum multiplicity of a Young tableau of $SU(n)$. We merely state that even rectangular tableaux may have multiple weights for $n \geq 4$.

There is still another method for obtaining the dimensionality of a given Young tableau of $SU(n)$. For brevity, we write $\mathbf{p}$ for $p_1, p_2, \ldots, p_{n-1}$. Also we write the expression for the dimensionality $N_n(\mathbf{p})$ as a quotient:

$$N_n(\mathbf{p}) = a_n(\mathbf{p})/b(\mathbf{p}) \tag{7.17}$$

where $a_n(\mathbf{p})$ depends on the particular configuration of the Young tableau and on n, but $b(\mathbf{p})$ depends only on the configuration of the tableau.

We now give a prescription for obtaining $a_n(\mathbf{p})$ and $b(\mathbf{p})$ which, for large n, is more convenient than using the general formula (7.15). Suppose we have a tableau of $SU(n)$. We find $a_n(\mathbf{p})$ as follows: We write an n in the upper left corner of the tableau and then increase the numbers by unity in going from left to right in every row, and decrease them by unity in going down in every column. Then the product of the integers in the tableau is $a_n(\mathbf{p})$. To obtain $b(\mathbf{p})$, we proceed as follows. We put a point in the center of a given box of the tableau and draw two lines from the point, the first to the right and the second down. The number of boxes intersected by the lines is written in the given box, and the procedure is repeated for all other boxes. The product of these numbers is then $b(\mathbf{p})$.

We illustrate the method by an example. Consider the tableau

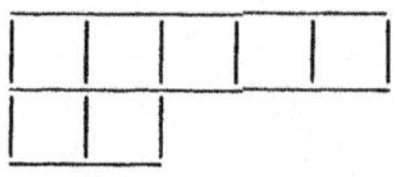

for $SU(3)$. The procedure to obtain the numerator $a_3(32)$ is to fill in the boxes of the tableau with numbers according to the prescription described for $a_n(\mathbf{p})$ with $n = 3$. We obtain the arrangement

3	4	5	6	7
2	3			

Then $a_3(320)$ is given by

$$a_3(32) = 3 \cdot 4 \cdot 5 \cdot 6 \cdot 7 \cdot 2 \cdot 3.$$

We illustrate the procedure to find $b(\mathbf{p})$ by drawing the two lines for one of the boxes. The lines intersect five boxes, so we put a 5 in the box

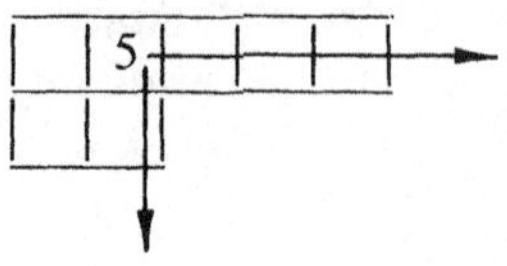

Repeating this procedure for the other boxes, we obtain

$$\begin{array}{|c|c|c|c|c|}\hline 6 & 5 & 3 & 2 & 1 \\ \hline 2 & 1 \\ \cline{1-2}\end{array}$$

Then $b(\mathbf{p})$ is given by

$$b(32) = 6 \cdot 5 \cdot 3 \cdot 2 \cdot 1 \cdot 2 \cdot 1.$$

Finally the dimensionality is given by

$$N_3(32) = \frac{a_3(32)}{b(32)} = \frac{3 \cdot 4 \cdot 5 \cdot 6 \cdot 7 \cdot 2 \cdot 3}{6 \cdot 5 \cdot 3 \cdot 2 \cdot 1 \cdot 2 \cdot 1} = 42.$$

To find the dimensionality of this tableau for $SU(4)$, we need only compute $a_4(320)$. This is

$$\begin{array}{|c|c|c|c|c|}\hline 4 & 5 & 6 & 7 & 8 \\ \hline 3 & 4 \\ \cline{1-2}\end{array}$$

Then $N_4(320)$ is given by

$$N_4(320) = \frac{4 \cdot 5 \cdot 6 \cdot 7 \cdot 8 \cdot 3 \cdot 4}{6 \cdot 5 \cdot 3 \cdot 2 \cdot 1 \cdot 2 \cdot 1} = 224,$$

since $b(\mathbf{p})$ for a given tableau is the same for all $SU(n)$.

7.3 Multiplets of the $SU(n-1)$ Subgroup of $SU(n)$

If we look at a weight diagram of a given multiplet of $SU(3)$, we see that it contains a number of multiplets of $SU(2)$. For example, the $SU(3)$ triplet contains a doublet and singlet of $SU(2)$, as can be seen from the weight diagram of Fig. 6.6. A more complicated case is that of the $SU(3)$ octet, which contains two doublets, a triplet, and a singlet of $SU(2)$, as can be seen from Fig. 6.8. These results can also be obtained directly from Young tableaux in a way that can be generalized to any multiplet of $SU(n)$.

We illustrate with the $SU(3)$ octet. The basis functions for the eight-dimensional representation of $SU(3)$ are summarized in the Young tableau

$$SU(3): \quad \begin{array}{|c|c|}\hline & \\ \hline \\ \cline{1-1}\end{array}$$

The eight standard arrangements of this tableau are

$$\begin{array}{|c|c|}\hline 1 & 1 \\ \hline 2 \\ \cline{1-1}\end{array} \quad \begin{array}{|c|c|}\hline 1 & 1 \\ \hline 3 \\ \cline{1-1}\end{array} \quad \begin{array}{|c|c|}\hline 1 & 2 \\ \hline 2 \\ \cline{1-1}\end{array} \quad \begin{array}{|c|c|}\hline 1 & 2 \\ \hline 3 \\ \cline{1-1}\end{array} \quad \begin{array}{|c|c|}\hline 1 & 3 \\ \hline 2 \\ \cline{1-1}\end{array} \quad \begin{array}{|c|c|}\hline 1 & 3 \\ \hline 3 \\ \cline{1-1}\end{array} \quad \begin{array}{|c|c|}\hline 2 & 2 \\ \hline 3 \\ \cline{1-1}\end{array} \quad \begin{array}{|c|c|}\hline 2 & 3 \\ \hline 3 \\ \cline{1-1}\end{array}$$

Now if a Young tableau refers to $SU(2)$, it can contain only the integers 1 and 2. So let us first look at those standard arrangements of the octet diagram in which 3 does not appear. There are two such arrangements

$$\begin{array}{|c|c|}\hline 1 & 1 \\ \hline 2 & \multicolumn{1}{c}{} \\ \cline{1-1}\end{array} \qquad \begin{array}{|c|c|}\hline 1 & 2 \\ \hline 2 & \multicolumn{1}{c}{} \\ \cline{1-1}\end{array}$$

The column

$$\begin{array}{|c|}\hline 1 \\ \hline 2 \\ \hline\end{array}$$

corresponds to no free indices, because in $SU(2)$ we can have only one antisymmetric combination of 1 and 2. Thus these arrangements correspond to an $SU(2)$ doublet:

$$\left.\begin{array}{c} \begin{array}{|c|c|}\hline 1 & 1 \\ \hline 2 & \multicolumn{1}{c}{} \\ \cline{1-1}\end{array} \\ \\ \begin{array}{|c|c|}\hline 1 & 2 \\ \hline 2 & \multicolumn{1}{c}{} \\ \cline{1-1}\end{array} \end{array}\right\} \to \left.\begin{array}{c} \begin{array}{|c|}\hline 1 \\ \hline\end{array} \\ \\ \begin{array}{|c|}\hline 2 \\ \hline\end{array} \end{array}\right\} \to \begin{array}{|c|}\hline \\ \hline\end{array}$$

Next let us consider the arrangements in which a 3 appears only in the right-hand box; there is only one such arrangement

$$\begin{array}{|c|c|}\hline 1 & 3 \\ \hline 2 & \multicolumn{1}{c}{} \\ \cline{1-1}\end{array}$$

The only part of this tableau on which $SU(2)$ can act is the part where 3 does not occur, namely

$$\begin{array}{|c|}\hline 1 \\ \hline 2 \\ \hline\end{array},$$

which is a singlet. Next, consider the tableaux in which 3 appears only at the bottom; there are three such arrangements

$$\begin{array}{|c|c|}\hline 1 & 1 \\ \hline 3 & \multicolumn{1}{c}{} \\ \cline{1-1}\end{array} \qquad \begin{array}{|c|c|}\hline 1 & 2 \\ \hline 3 & \multicolumn{1}{c}{} \\ \cline{1-1}\end{array} \qquad \begin{array}{|c|c|}\hline 2 & 2 \\ \hline 3 & \multicolumn{1}{c}{} \\ \cline{1-1}\end{array}$$

Again the box containing the 3 is irrelevant for $SU(2)$, and we can remove it. What is left are the arrangements

$$\begin{array}{|c|c|}\hline 1 & 1 \\ \hline\end{array} \qquad \begin{array}{|c|c|}\hline 1 & 2 \\ \hline\end{array} \qquad \begin{array}{|c|c|}\hline 2 & 2 \\ \hline\end{array}$$

which form a triplet. Finally, we consider the two tableaux containing two boxes with the number 3

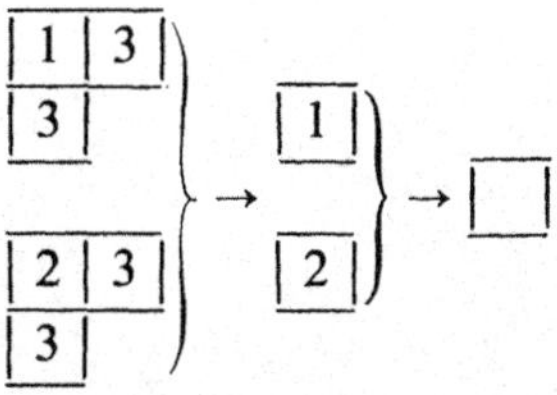

Since $SU(2)$ does not act on the index 3, this is another doublet. Thus, we find that the octet representation of $SU(3)$ contains two $SU(2)$ doublets, a triplet, and a singlet. It can be seen that we do not need to fill in the 1's and 2's to get this result, only the 3's. The formula for the multiplicity of a tableau of $SU(2)$ will then enable us to obtain the multiplicities.

By a generalization of this procedure, we can find the $SU(n-1)$ multiplets contained in any tableau of $SU(n)$. The prescription is to consider all possible ways of putting n's in the boxes consistent with standard arrangements from no n's to the maximum allowed number. We then remove all boxes with n's. The remaining tableaux give the $SU(n-1)$ multiplets.

For example suppose we have the tableau

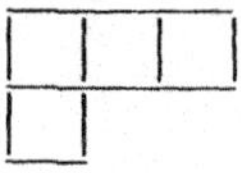

Following the prescription for putting n's in boxes, we obtain

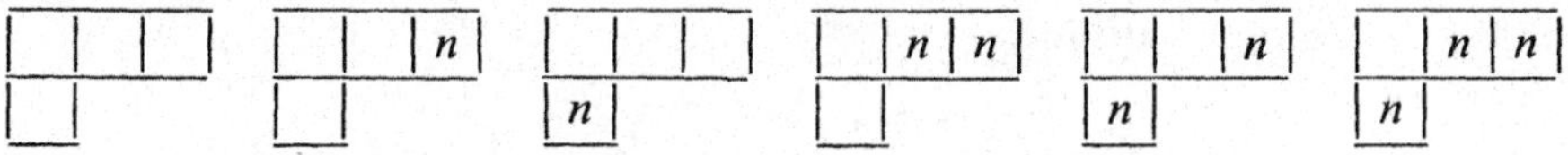

Each of these tableaux specifies a representation of $SU(n-1)$. Since $SU(n-1)$ does not care about the index n, we drop the boxes which contain an n. We thus obtain

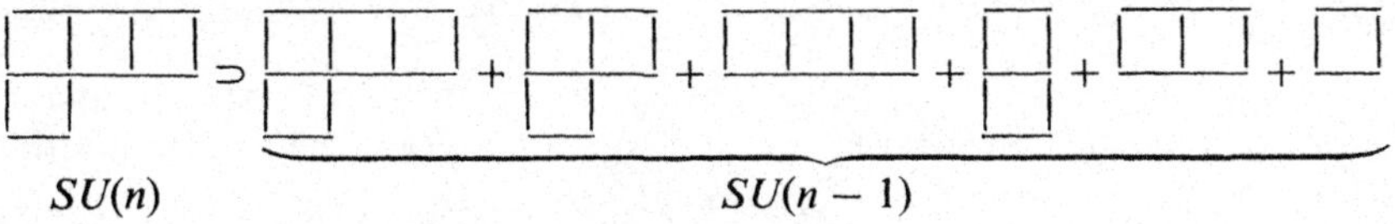

We can proceed further, obtaining the $SU(n-2)$ multiplets contained in these tableaux of $SU(n-1)$, and then continue the process until we have obtained the multiplets of all the $SU(n-n')$ subgroups of $SU(n)$, where $1 \leq n' \leq n-2$.

We can obtain the value of the weight m_2 which is characteristic of an entire $SU(2)$ multiplet contained in a multiplet of $SU(3)$. The single-particle states of $SU(3)$ have the following values of m_2

$$m_2(1) = m_2(2) = 1/(2\sqrt{3}), \qquad m_2(3) = -1/\sqrt{3}. \tag{7.18}$$

Then, since the weights are additive, a standard arrangement of a Young diagram with ν boxes and ν' 3's has a value of m_2 given by

$$m_2 = \tfrac{1}{2}(\nu - 3\nu')/\sqrt{3}, \qquad \nu \geq \nu' = 0, 1, 2, \ldots .$$

If m_2 is associated with the hypercharge as in the usual interpretation of $SU(3)$, then

$$Y = 2m_2/\sqrt{3} = \tfrac{1}{3}\nu - \nu'.$$

But the number of boxes of an $SU(3)$ Young diagram is given by $\nu = p_1 + 2p_2$. Therefore the hypercharge is given by

$$Y = 2m_2/\sqrt{3} = \tfrac{1}{3}(p_1 + 2p_2) - \nu'. \tag{7.19}$$

As an example of this procedure, let us find the hypercharge of each $SU(2)$ isospin multiplet contained in the $SU(3)$ decuplet or (30) representation. We have

$$\boxed{}\boxed{}\boxed{} \supset \boxed{}\boxed{}\boxed{} + \boxed{}\boxed{}\boxed{3} + \boxed{}\boxed{3}\boxed{3} + \boxed{3}\boxed{3}\boxed{3}$$

or

$$\boxed{}\boxed{}\boxed{} \supset \boxed{}\boxed{}\boxed{} + \boxed{}\boxed{} + \boxed{} + \textcircled{1}$$

Thus, the decomposition is

$$\mathbf{10} \supset \mathbf{4} \oplus \mathbf{3} \oplus \mathbf{2} \oplus \mathbf{1}.$$

Using Eq. (7.19), we have the following connection between $SU(2)$ multiplicity N_2 and hypercharge:

$$\begin{array}{lrrrr} N_2: & 4 & 3 & 2 & 1, \\ Y: & 1 & 0 & -1 & -2. \end{array}$$

We can obtain the weight diagrams of $SU(3)$ in an easy way by making use of the $SU(2)$ subgroups of $SU(3)$. First we note that the isospin multiplet with the highest value of m_2 (or equivalently of the hypercharge Y) has Y given by Eq. (7.19) with $\nu' = 0$. Also, the isospin of this multiplet is given by the Young tableau obtained after removing the columns with two boxes. Thus this multiplet has isospin and hypercharge given by

$$I_1 = \tfrac{1}{2}p_1, \qquad Y = \tfrac{1}{3}(p_1 + 2p_2). \tag{7.20}$$

If the weight diagram is rotated by 60° in either direction, the multiplet at the top of the diagram in its new position has $SU(2)$ multiplicity N_2 given by

$$N_2 = p_2 + 1. \tag{7.21}$$

If we fill in that portion of the weight diagram which we obtain from Eqs. (7.20) and (7.21), we obtain either one side of a triangle (for a rectangular tableau) or three sides of a hexagon (for any other tableau). Then, using 120° rotational symmetry and the knowledge that $SU(2)$ subgroups have weights that differ by integral values, we can complete the diagram except for the problem of multiple weights. This problem can be solved as follows: If the weight diagram is a triangle, i.e., $N_2 = 1$, the diagram is finished. If the diagram is not a triangle, then to every weight of the interior, add an identical weight. If these double weights form a triangle, the diagram is finished. If not, then to every weight in the interior of the double weights, add an identical weight. Proceed until there is a triangle of weights or a central weight with the highest multiplicity. These rules can all be justified by counting standard arrangements of Young tableaux.

As an example, we consider the weight diagram of the representation (41) or **35**. From Eqs. (7.20) and (7.21), we obtain that portion of the weight

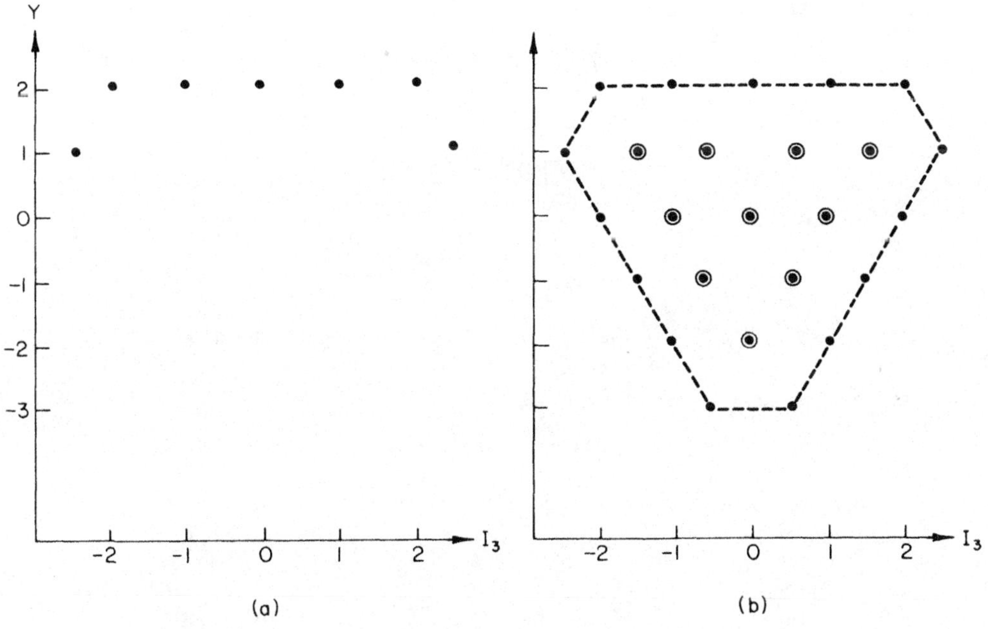

Fig. 7.2. Construction of the weight diagram of the (31) or **35** representation by a method given in Section 7.3.

diagram shown in Fig. 7.2a. Making use of 120° symmetry, plus the knowledge that $SU(2)$ weights differ by integers, and the rule for the multiplicities of weights, we complete the diagram as shown in Fig. 7.2b. The hypercharge Y, rather than m_2, is plotted as the ordinate in Fig. 7.2, but the vertical scale is shrunk so that the visual 120° symmetry remains.

7.4 Decomposition of Products of Irreducible Representations

Suppose we have two irreducible unitary representations of $SU(n)$ (or for that matter, of any compact simple group). Denote these representations by $D^{(\alpha)}$ and $D^{(\beta)}$, where $D^{(\alpha)}$ and $D^{(\beta)}$ have dimension n_α and n_β, respectively. Let the basis vectors for $D^{(\alpha)}$ and $D^{(\beta)}$ be x_i and y_i. Then the transformed basis vectors x_i' and y_i' are given by

$$x_i' = \sum_{j=1}^{n_\alpha} D_{ji}^{(\alpha)} x_j,$$

$$y_i' = \sum_{l=1}^{n_\beta} D_{li}^{(\beta)} y_l.$$

We now ask how the two-index quantity $x_i y_k$ transforms. From the transformations on the vectors we obtain

$$x_i' y_k' = \sum_{jl} D_{ji}^{(\alpha)} D_{lk}^{(\beta)} x_j y_l. \tag{7.22}$$

We can introduce the notation

$$D_{ji}^{(\alpha)} D_{lk}^{(\beta)} = D_{jl,\,ik}^{(\alpha \times \beta)}, \tag{7.23}$$

and obtain

$$x_i' y_k' = \sum_{jl} D_{jl,\,ik}^{(\alpha \times \beta)} x_j y_l. \tag{7.24}$$

The numbers $D_{jl,\,ik}^{(\alpha \times \beta)}$ are matrix elements of the representation $D^{(\alpha \times \beta)}$, which is called the *Kronecker product* of the representations $D^{(\alpha)}$ and $D^{(\beta)}$. This product representation is written

$$D^{(\alpha \times \beta)} = D^{(\alpha)} \times D^{(\beta)}. \tag{7.25}$$

We should note that the products $x_i y_k$ can be considered either as two-index tensors, or, by relabeling

$$z_j = x_i y_k, \qquad j = 1, 2, \ldots, n_\alpha n_\beta,$$

as basis vectors of a representation of $n_\alpha n_\beta$ dimensions. It is usually more convenient to consider the $x_i y_k$ as tensors. The representation $D^{(\alpha \times \beta)}$ is in general reducible. If the group is finite or simple and compact, $D^{(\alpha\times\beta)}$ decomposes into a direct sum of irreducible representations:

$$D^{(\alpha\times\beta)} = \sum_i \Gamma_i D^{(\gamma_i)}, \tag{7.26}$$

where the number Γ_i denotes how many times the irreducible representation $D^{(\gamma_i)}$ appears in the sum. This decomposition is called the *Clebsch–Gordan series* and is sometimes written

$$\alpha \times \beta = \sum_i \Gamma_i \gamma_i . \tag{7.27}$$

If none of the Γ_i is greater than unity, the product representation is said to be *simply reducible.* For a simple Lie group of rank one, a theorem says that the Kronecker product of any two irreducible representations is simply reducible. This is the principal requirement for calling a group simply reducible. Actually four requirements are necessary for a group to be simply reducible. They are:

(1) The direct sum contained in the Kronecker product of two irreducible representations contains each irreducible representation no more than once.

(2) Any representation D is either real (i.e., can be brought to real form by a similarity transformation), or is equivalent to D^*, the complex conjugate representation (i.e., the representation by matrices which are the complex conjugates of the original matrices).

(3) If a group element belongs to a certain class, then its inverse belongs to the same class.

(4) The characters (traces) of the matrices representing the group elements are real.

It is an important problem to construct from the $x_i y_k$ those *irreducible* tensors which form the basis functions of the irreducible unitary representations $D^{(\gamma_i)}$, as these tensors are the multiplets. We shall consider this problem for the group $SU(n)$, using the method of Young tableaux.

We consider as an example the tensors of a representation which is the Kronecker product of the fundamental representation of $SU(2)$ with itself. Suppose we have a system of two particles of spin $\frac{1}{2}$ (two doublets). Four linearly independent two-particle Kronecker product states can be constructed from the single-particle states u_1 and u_2. They are

$$u_1(1)u_1(2), \qquad u_1(1)u_2(2), \qquad u_2(1)u_1(2), \qquad u_2(1)u_2(2), \tag{7.28}$$

where the numbers in parentheses refer to particles number 1 and 2. If we adopt the convention of writing the state vector of the first particle to the left, we can omit the parentheses. Then the two-particle states of Eq. (7.28) are

$$u_1u_1, \qquad u_1u_2, \qquad u_2u_1, \qquad u_2u_2 . \tag{7.29}$$

The states $u_i u_j$ are the basis tensors of the Kronecker product representation. The irreducible tensors are the symmetric triplet states

$$u_1 u_1, \qquad (u_1 u_2 + u_2 u_1)/\sqrt{2}, \qquad \text{and} \qquad u_2 u_2, \tag{7.30}$$

and the antisymmetric singlet state

$$(u_1 u_2 - u_2 u_1)/\sqrt{2}. \tag{7.31}$$

Thus, the product representation of two doublets of $SU(2)$ decomposes into a triplet and a singlet. We can write this symbolically with Young tableaux as

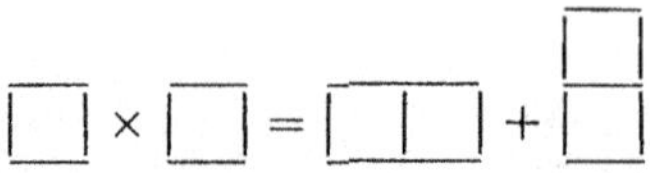

(7.32)

This is the Clebsch–Gordan series for decomposition of the Kronecker product of two doublets. There are several other notations for this series. One notation is in terms of the dimensionality of the multiplets as follows:

$$\mathbf{2} \otimes \mathbf{2} = \mathbf{3} \oplus \mathbf{1}. \tag{7.33}$$

We put circles around the multiplication and addition signs to distinguish them from ordinary arithmetic operations.

We can also denote each representation by the integer p which stands for the configuration of its Young tableau. Then the Clebsch–Gordan series is

$$(1) \times (1) = (2) + (0). \tag{7.34}$$

Next consider an example from $SU(3)$. It is easy to find the Clebsch–Gordan series for the product of two triplets of $SU(3)$ with the aid of Young tableaux. In fact, it is just given by Eq. (7.32), which holds for any $SU(n)$. The dimensionalities are different, however, but can be found from any of the methods discussed in Sections 7.1 and 7.2. In particular, applying Eq. (7.11) to the tableaux of (7.32), we find that for $SU(3)$ we have

$$\mathbf{3} \otimes \mathbf{3} = \mathbf{6} \oplus \bar{\mathbf{3}}. \tag{7.35}$$

Using the p-notation, this is

$$(10) \times (10) = (20) + (01). \tag{7.36}$$

We shall now give the general rules for reducing the Kronecker product of two representations by means of Young tableaux in order to obtain the representations of the Clebsch–Gordan series. We draw the two Young tableaux of the representations, marking each box of the second diagram with the number of the row to which it belongs. We then attach the boxes of the second tableau in all possible ways to the first tableau, subject to the following rules of the combined tableaux:

(1) Each tableau should be a proper tableau; that is, no row is longer than any row above it.

(2) No tableau should have a column with more than n boxes if the group is $SU(n)$.

(3) We can make a path by counting each row from the right, starting with the top row. At each point of the path the number of boxes encountered with the number i must be less or equal to the number of boxes with $i - 1$.

(4) The numbers must not decrease in going from left to right across a row.

(5) The numbers must increase in going from top to bottom in a column.

As an example, let us obtain the reduction of the Kronecker product of the first and second fundamental representations of $SU(n)$. We obtain

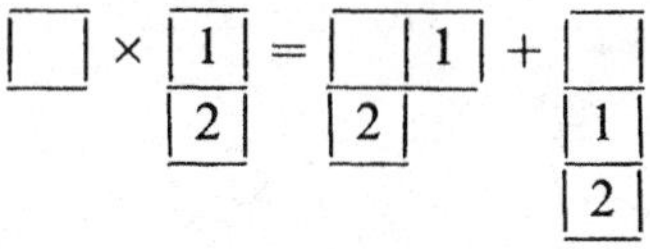

or

$$(10^{n-2}) \times (010^{n-3}) = (110^{n-3}) + (0010^{n-4}).$$

For $SU(3)$ this is the reduction of the product of the fundamental representation and its conjugate. The $SU(3)$ numerology is

$$\mathbf{3} \otimes \bar{\mathbf{3}} = \mathbf{8} \oplus \mathbf{1}. \tag{7.37}$$

Note that the reduction of $\mathbf{3} \otimes \bar{\mathbf{3}}$ [Eq. (7.37)] is different from the reduction of $\mathbf{3} \otimes \mathbf{3}$ [Eq. (7.35)], since $\mathbf{3}$ and $\bar{\mathbf{3}}$ are inequivalent representations. In general, when we take the product of the first fundamental representation of any $SU(n)$ and its conjugate we get

$$\mathbf{n} \otimes \bar{\mathbf{n}} = (\mathbf{n}^2 - 1) \oplus \mathbf{1}. \tag{7.38}$$

The representation of dimension $n^2 - 1$ has the Young tableau specified by the integers $(10^{n-3}1)$ or $p_1 = p_{n-1} = 1$, all other $p_i = 0$. It is a self-conjugate representation and is often called the *adjoint representation.*

As another example, we find the irreducible representation contained in the Kronecker product of the representations (11) and (11) of $SU(3)$. Following the rules, we obtain

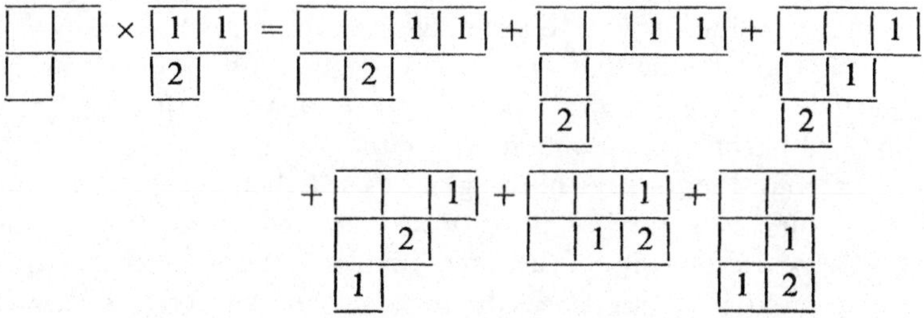

Omitting the superfluous columns with three boxes we get

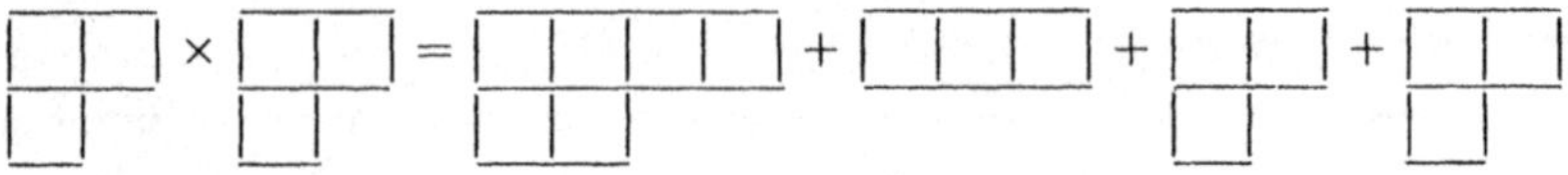

(11) × (11) = (22) + (30) + (11) + (11)

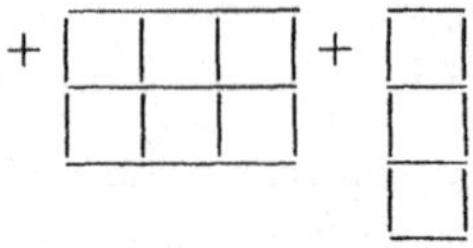

(03) + (00)

or

$$\mathbf{8} \otimes \mathbf{8} = \mathbf{27} \oplus \mathbf{10} \oplus \mathbf{8} \oplus \mathbf{8} \oplus \overline{\mathbf{10}} \oplus \mathbf{1}. \tag{7.39}$$

We see that the Clebsch–Gordan series for the Kronecker product $\mathbf{8} \otimes \mathbf{8}$ contains two equivalent representations, namely the eight-dimensional self-conjugate representations. Thus, this product is not simply reducible. This is in contrast to the case for $SU(2)$, in which the Kronecker product of any two representations is simply reducible. In fact the result for $SU(2)$ is a special case of a particular class of Kronecker products of $SU(n)$. The relevant theorem states that the Kronecker product of any representation and a representation specified by a rectangular tableau (i.e., a representation with only one p_i different from zero) is simply reducible. Since all tableaux of $SU(2)$ are rectangular (consisting of a single row), the theorem states that any Kronecker product representation of $SU(2)$ is simply reducible.

If two or more equivalent representations appear in the reduction of a Kronecker product of $SU(n)$, the use of group theory alone is not sufficient to enable us to label the states. We must know something of the physics of the problem in order to obtain a useful labeling. This problem has been encountered in $SU(2)$ in the Kronecker product of *three* representations. For example, consider the Kronecker product of three doublets of $SU(2)$

$$\mathbf{2} \otimes \mathbf{2} \otimes \mathbf{2} = (\mathbf{3} \oplus \mathbf{1}) \otimes \mathbf{2} = \mathbf{4} \oplus \mathbf{2} \oplus \mathbf{2}. \tag{7.40}$$

We see that in addition to a quartet two different doublets appear in the reduction. Let us assume that our example represents the coupling of three spin-$\frac{1}{2}$ particles. The first two particles can combine to form spin 1 or 0, and then the third particle can combine with either of these states to give a total spin of $\frac{1}{2}$. Group theory says nothing about whether the first two particles, in fact, combine to form spin 1 or 0 or a linear combination of the two. Only if we have a knowledge of the forces involved can we decide this question. At the level of $SU(n)$, with $n > 2$, the problem arises in the reduction of the Kronecker product of two representations.

7.5 Classes of Representations

A representation of $SU(2)$ can be put into one of two classes according to whether the number of boxes of its Young diagram is even or odd. The number of boxes p (all of which are in a single row) is related to the quantum number j by $p = 2j$. Thus, if p is even, j is an integer, while if p is odd, j is half of an odd integer. The representations of $SU(2)$ are called *integral* or *half integral*, according to whether j is integral or half integral.

We shall not prove the statement that a simply reducible group has only integral and half-integral unitary irreducible representations. The integral representations are those that can be made real by a similarity transformation, while the half-integral representations cannot be made real but are equivalent to their complex-conjugate representations. The group $R(3)$ has only integral vector representations, and they are just the same as the integral representations of $SU(2)$.

Suppose we have a product representation of $SU(2)$ whose states are denoted by the Young diagrams p and p'. We can use the method of Young diagrams to find the values of p'' of the irreducible representations contained in the product. Assuming $p > p'$, we have

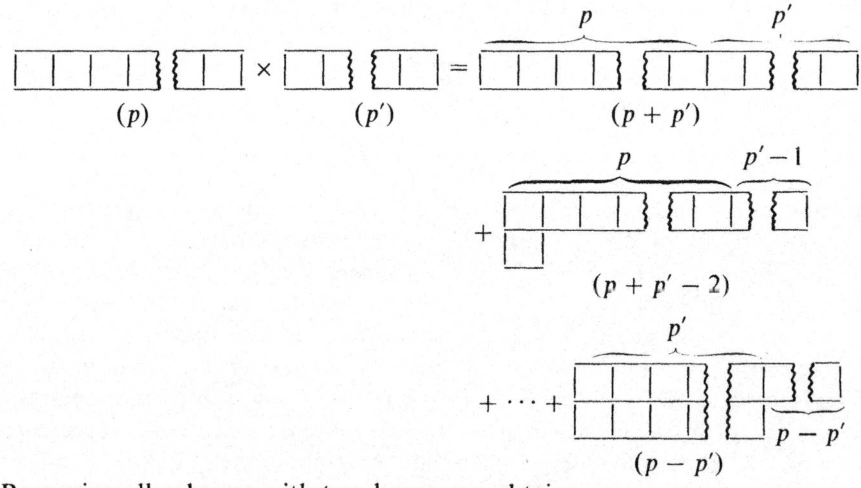

Removing all columns with two boxes, we obtain

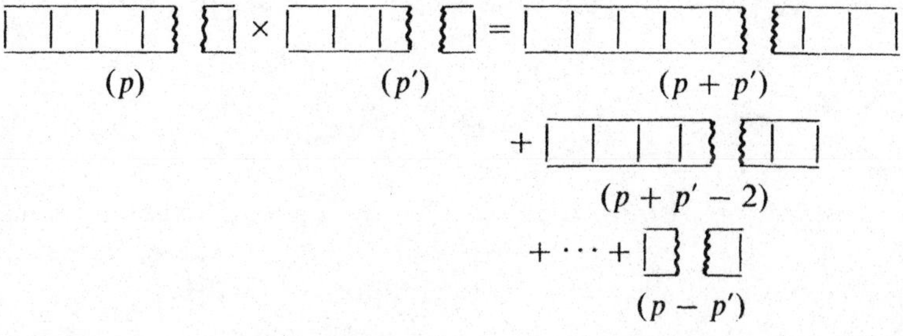

Denoting the representations by their values of j, we have

$$j'' = j + j', j + j' - 1, \ldots, j - j'. \tag{7.41}$$

According to Eq. (7.41), a product of two integral or two half-integral representations contains only integral irreducible representations in its direct sum, while a product of an integral and half-integral representation contains only half-integral irreducible representations.

Although the representations of $SU(n)$, $n > 2$, are not in general equivalent to their complex conjugates, we can nevertheless generalize the idea of the class of a representation. We subdivide the representations of $SU(n)$ into n classes according to the number of boxes contained in their Young tableaux. Let the number of boxes ν of any tableau be given by

$$\nu = ni + k, \tag{7.42}$$

where i and k are integers satisfying

$$i \geq 0, \qquad 0 \leq k \leq n - 1. \tag{7.43}$$

Each of the different possible values of k denotes a representation of a different class. Note that removing a column of n boxes from a tableau does not change the class. If we decompose the product of two representations of classes k_1 and k_2, then the irreducible representations contained in the product are of class k where

$$k = k_1 + k_2, \qquad \text{modulo } n. \tag{7.44}$$

From this formula it is clear that the Kronecker product of two representations of class $k = 0$ contains only irreducible representations of the same class.

In Chapter 2, we defined Z_2, a two-element invariant subgroup of $SU(2)$ [see Eq. (2.19)]. The group Z_2 consists of the two square roots of unity multiplied by the 2×2 unit matrix. Similarly, we can define Z_n as the group of the n nth roots of unity times the $n \times n$ unit matrix. Then Z_n is an invariant subgroup of $SU(n)$. The factor group $SU(n)/Z_n$ has vector representations only of class $k = 0$. However, if we also include multiple-valued representations of $SU(n)/Z_n$, then we include all the representations of $SU(n)$.

Baird and Biedenharn (1964) have introduced the term triality τ to distinguish the three classes of representations of $SU(3)$. They define

$$\begin{aligned} \tau &= 0 \qquad \text{if} \quad k = 0, \\ \tau &= 1 \qquad \text{if} \quad k = 1, \\ \tau &= -1 \quad\ \ \text{if} \quad k = 2. \end{aligned} \tag{7.45}$$

The representations of zero triality are the only vector representations of the group $SU(3)/Z_3$. However, if we include the triple-valued representations of this group, we include all the representations of $SU(3)$.

7.6 Multiplets of $U(n)$

The Lie algebra of $U(n)$ is the same as the Lie algebra of $U(1) \times SU(n)$. Since $U(1)$ is Abelian, all its unitary irreducible representations are one-dimensional. Therefore, all the considerations of this chapter regarding the dimensionality of representations can be directly taken over to $U(n)$. However, in the case of $U(n)$, each Young tableau is associated with an additional quantum number. For example, in the case of $U(2)$, the $SU(2)$ subgroup can be interpreted as the isospin group and the $U(1)$ subgroup as the group of the hypercharge. In the case of $U(3)$, $SU(3)$ can be interpreted as a subgroup including both isospin and hypercharge, and $U(1)$ as the subgroup of the baryon number B.

The generator associated with $U(1)$ is an additive quantum number. Because of this, Young tableaux with different numbers of boxes correspond to different values of the quantum number, and are inequivalent. Thus, in considering the Young tableaux of $U(n)$, we can lose information if we omit columns of n boxes. However, it is convenient to remove these n-box columns in any case, keeping track of the $U(1)$ quantum number separately.

Another complication with $U(n)$ is that its different fundamental representations can have values of the extra quantum number which are not related in the usual way. Again we consider $U(3)$ as an example. Let the diagram $\square$ denote a one-particle state with baryon number $B = 1$ belonging to the first fundamental representation of $U(3)$. Then the two-particle states

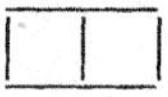 and 

both have $B = 2$. Now, however, consider the antiparticle to the state $\square$. The antiparticle is defined to be a state with the same energy as the particle but with its weight reversed. Thus, antiparticle of the state $\square$ has baryon number -1 and $SU(3)$ weight $-\mathbf{m}$, but the three states with weights $-\mathbf{m}$ belong to the conjugate $\bar{\mathbf{3}}$ representation of $SU(3)$ with Young tableau

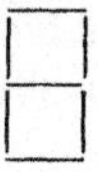

Therefore, starting from the state $\square$, we are led to two inequivalent representations of $SU(3)$, one with $B = 2$ and the other with $B = -1$, but both representations have the same Young tableau

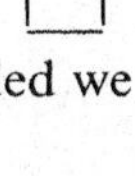

This creates no real difficulty provided we keep track of the quantum number B in a systematic way.

CHAPTER 8

CLEBSCH–GORDAN COEFFICIENTS

8.1 Some Properties of the Coefficients

Suppose we have the basis vectors of two unitary irreducible representations of a compact simple Lie group. We denote these vectors by $\psi_\mu^{(\alpha)}$ and $\psi_\nu^{(\beta)}$, where α and β stand for all numbers necessary to specify the first and second representations, and μ and ν stand for all numbers which differentiate among the different states within these representations. Then the basis tensors of the Kronecker product representation are given by $\psi_\mu^{(\alpha)}\psi_\nu^{(\beta)}$.

In general, these product basis tensors are not the basis tensors of an irreducible representation. However, the basis tensors of any irreducible representation contained in the product can be written as a linear combination of the product tensors. The coefficients in the sum are called *Clebsch–Gordan coefficients*. The irreducible tensors contained in the direct product may be written with a single index $\psi_m^{(j)}$, in which case they can be regarded as vectors. Since we shall have occasion to write them in both ways, we shall usually just call them basis functions. Likewise, the product tensors can be written as a linear combination of the irreducible functions. The coefficients in this sum are also called Clebsch–Gordan coefficients.

Let us illustrate with $SU(2)$ as an example. In this case, each of the symbols α, β, μ, and ν stands for a single number. Then we can write

$$\psi_m^{(j)} = \sum_{\mu\nu} (\alpha\beta\mu\nu \mid \alpha\beta jm)\psi_\mu^{(\alpha)}\psi_\nu^{(\beta)} \tag{8.1}$$

where the coefficients $(\alpha\beta\mu\nu \mid \alpha\beta jm)$ are the $SU(2)$ Clebsch–Gordan coefficients, often called Wigner coefficients. It is apparant from this expression that $\psi_m^{(j)}$ depends on α and β, since these indices are not summed over. However, it is customary to suppress these indices on $\psi_m^{(j)}$. Likewise, the product basis functions can be written as linear combinations of the basis functions of the irreducible representations:

$$\psi_\mu^{(\alpha)}\psi_\nu^{(\beta)} = \sum_{jm} (\alpha\beta jm \mid \alpha\beta\mu\nu)\psi_m^{(j)}. \tag{8.2}$$

The coefficients $(\alpha\beta jm \mid \alpha\beta\mu\nu)$ are also Clebsch–Gordan coefficients. It is seen that the indices $\alpha\beta$ are repeated in the symbol for a Clebsch–Gordan coefficient. Sometimes this redundancy is omitted, i.e., we write

$$(\alpha\beta jm \mid \alpha\beta\mu\nu) = (jm \mid \alpha\beta\mu\nu). \tag{8.3}$$

There is further redundancy in this expression, since the weights are additive, i.e., $m = \mu + \nu$. There are many other symbols for the Clebsch–Gordan coefficients.

A number of other relations can be proved to hold among these coefficients. For example, by taking the scalar product of Eq. (8.1) with $\psi_\mu^{(\alpha)}\psi_\nu^{(\beta)}$ we find, on using the notation of Eq. (8.3) and dropping the primes, that

$$(\psi_\mu^{(\alpha)}\psi_\nu^{(\beta)}, \psi_m^{(j)}) = (\alpha\beta\mu\nu \mid jm), \tag{8.4}$$

since the $\psi_\mu^{(\alpha)}\psi_\nu^{(\beta)}$ are orthonormal. Likewise, by taking the scalar product of Eq. (8.2) with $\psi_m^{(j)}$, we find

$$(\psi_m^{(j)}, \psi_\mu^{(\alpha)}\psi_\nu^{(\beta)}) = (jm \mid \alpha\beta\mu\nu). \tag{8.5}$$

Therefore, comparing Eq. (8.4) with Eq. (8.5), we find

$$(\alpha\beta\mu\nu \mid jm) = (jm \mid \alpha\beta\mu\nu)^*. \tag{8.6}$$

Without loss of generality, the phases of the basis functions can be chosen so that the Clebsch–Gordan coefficients are real. Then we obtain

$$(\alpha\beta\mu\nu \mid jm) = (jm \mid \alpha\beta\mu\nu). \tag{8.7}$$

We can also obtain summed relations among the coefficients. Putting the expression for $\psi_m^{(j)}$ of Eq. (8.1) into that for $\psi_\mu^{(\alpha)}\psi_\nu^{(\beta)}$ in Eq. (8.2), we obtain

$$\sum_{jm} (jm \mid \alpha\beta\mu\nu)(\alpha\beta\mu'\nu' \mid jm) = \delta_{\mu\mu'}\delta_{\nu\nu'}. \tag{8.8}$$

Similarly, substituting the expression for $\psi_\mu^{(\alpha)}\psi_\nu^{(\beta)}$ into that for $\psi_m^{(j)}$, we get

$$\sum_{\mu\nu} (\alpha\beta\mu\nu \mid jm)(j'm' \mid \alpha\beta\mu\nu) = \delta_{mm'}\delta_{jj'}. \tag{8.9}$$

The Clebsch–Gordan coefficients have many other interesting properties which we shall not discuss. See, for example, Wigner (1959), Hamermesh (1962), Rose (1957), and Edmonds (1957).

The expressions relating the basis tensors of product representations in terms of the basis functions of irreducible representations can be generalized to many other groups. We restrict ourselves to groups whose unitary representations can be decomposed into a direct sum of unitary irreducible representations. Furthermore, we consider the basis functions only of irreducible unitary representations and of representations formed from the Kronecker product of such representations. Then we can write an expression for the basis functions of a product representation in terms of the basis functions of irreducible representations in a form which looks very similar to the one for $SU(2)$. We write

$$\psi_\mu^{(\alpha)}\psi_\nu^{(\beta)} = \sum_{jm\gamma} (jm\gamma \mid \alpha\beta\mu\nu)\psi_m^{(j\gamma)}, \tag{8.10}$$

and

$$\psi_m^{(j\gamma)} = \sum_{\mu\nu} (\alpha\beta\mu\nu \mid jm\gamma)\psi_\mu^{(\alpha)}\psi_\nu^{(\beta)}. \tag{8.11}$$

There are two differences between these expressions and the analogous ones for $SU(2)$. The first difference is that, since the groups are not simply reducible, there is a sum over an additional index γ. This index distinguishes between irreducible representations with the same values of j and m.

The second difference is in the interpretation of the indices α, β, μ, ν, j, m. The upper index α on $\psi_\mu^{(\alpha)}$ again specifies the irreducible representation of which this function is a member, and the lower index μ specifies a particular basis function of that representation. But the indices α, μ can now each stand for a whole set of indices. For example, in $SU(n)$, α stands for $n-1$ integers $(p_1, p_2, \ldots, p_{n-1})$, and μ stands for a weight of $n-1$ dimensions, plus all other indices necessary to distinguish members of a representation. Furthermore, the expression $\sum_{\mu\nu}$ means a sum over all weights $\mu\nu$ appearing in the representation. A given weight may appear more than once in a given representation, and it is summed as often as it appears. Of course, one can still keep the interpretation that α is a single index by using a scheme of ordering the representations with a single number. Likewise, by ordering the basis functions within a representation, μ can retain its meaning as a single number.

In anology with the case of $SU(2)$ we can show in a straightforward way that the following relations hold among the Clebsch–Gordan coefficients:

$$\sum_{\mu\nu} (\alpha\beta\mu\nu \mid jm\gamma)(j'm'\gamma' \mid \alpha\beta\mu\nu) = \delta_{jj'}\,\delta_{mm'}\,\delta_{\gamma\gamma'}, \tag{8.12}$$

$$\sum_{jm\gamma} (jm\gamma \mid \alpha\beta\mu\nu)(\alpha\beta\mu'\nu' \mid jm\gamma) = \delta_{\mu\mu'}\,\delta_{\nu\nu'}, \tag{8.13}$$

$$(\alpha\beta\mu\nu \mid jm\gamma) = (jm\gamma \mid \alpha\beta\mu\nu), \tag{8.14}$$

where without loss of generality we have assumed that the Clebsch–Gordan coefficients are real.

For a group which is not simply reducible, the values of the Clebsch–Gordan coefficients are not completely specified by the group properties. Clebsch–Gordan coefficients depend on how the representations j are chosen whenever γ is greater than unity. We may make this choice on the basis of simplicity, but any choice is satisfactory from the standpoint of group theory.

We illustrate this point by comparing $SU(2)$ and $SU(3)$. In $SU(2)$, a reducible representation formed by the Kronecker product of two irreducible representations is specified by the numbers $\alpha\beta$ of the irreducible representations. The different basis tensors of a product representation $\alpha\beta$ are distinguished by the one-dimensional weights $\mu\nu$. Now consider the irreducible representations into which we decompose the product. These representations are characterized by the numbers $j\alpha\beta$ and the different basis functions of these representations are distinguished by the weight m. Therefore, four quantum numbers $\alpha\beta\mu\nu$ distinguish the basis tensors of various Kronecker product representations, and four quantum numbers $\alpha\beta jm$ distinguish basis tensors or irreducible representations obtained by decomposing Kronecker products.

Now let us consider $SU(3)$. We again specify a representation by α, but now α stands for two numbers, either $(p_1 p_2)$ or the quantum numbers (fg) of the Casimir operators. Alternatively, when no ambiguity arises, α can stand for the dimensionality of the representation. The different basis tensors of a representation are distinguished by three numbers: the two-dimensional weight μ and the eigenvalue of a Casimir operator of an $SU(2)$ subgroup for example, the isospin I. Therefore the basis tensors of a Kronecker product representation are fully specified by ten numbers $\alpha\beta\mu\nu I_1 I_2$ ($\alpha\beta\mu\nu$ each stands for two numbers). Now consider the irreducible representations of the direct sum. A given representation is specified by j and a weight by m. Again α and β are good quantum numbers, since they are not summed over. We also have the eigenvalue I of an $SU(2)$ subgroup to distinguish among tensors. Thus we have at our disposal only the nine quantum numbers $\alpha\beta jmI$ ($\alpha\beta jm$ each stands for two numbers) which are fully specified by group theory. In order to specify an irreducible tensor completely, we must add a tenth quantum number, denoted by γ, which must be defined in some convenient way without making reference to group properties.

We illustrate one way of defining γ with a specific example from $SU(3)$. The Kronecker product of two eight-dimensional representations has the Clebsch–Gordan series

$$\mathbf{8} \otimes \mathbf{8} = \mathbf{1} \oplus \mathbf{8}_s \oplus \mathbf{8}_a \oplus \mathbf{10} \oplus \overline{\mathbf{10}} \oplus \mathbf{27}. \tag{8.15}$$

In the direct sum, we have put a subscript "s" on one **8** and a subscript "a" on the other to indicate that the basis functions of the two eight-dimensional representations are formed from the symmetric and antisymmetric combinations, respectively of the product tensors. The Clebsch–Gordan coefficients will have these respective symmetries. It is customary, following Gell-Mann, to call the symmetric combination *D coupling* and the antisymmetric combination *F coupling*. However, the choice of symmetric and antisymmetric combinations is based on convenience, and is not dictated by group theory. We can take any linear combination of $\mathbf{8}_s$ and $\mathbf{8}_a$ to form two new representations with different Clebsch–Gordan coefficients.

The choice of symmetric and antisymmetric combinations is particularly convenient in this example. However, in other examples in which the Kronecker product representation is formed from two irreducible representations with different Young tableaux, symmetric and antisymmetric combinations are not possible, in general. Then the choice of Clebsch–Gordan coefficients is by no means obvious.

We encounter the problem of quantum numbers lying outside the group in $SU(2)$ when we consider the Kronecker product of *three* irreducible representations. For example, suppose we have three representations specified by j_1, j_2, and j_3. We can first decompose the Kronecker product of j_1 and j_2 to give an intermediate irreducible representation j'. Then we can decompose the Kronecker product of j' and j_3 to obtain an irreducible representation j. But alternatively, we can first decompose the Kronecker product of j_2 and j_3 to give an intermediate j'' and then decompose j'' and j_1 to give j. In general, we can take a linear combination of these possibilities, and the Clebsch–Gordan coefficients will depend on just what combination we choose. The choice may be dictated by our knowledge of the physics of the problem. For example, in atomic and nuclear physics, we usually couple the spins and orbital angular momenta of a many-particle system by either LS or jj coupling.

8.2 Raising and Lowering Operators

An expression for the Clebsch–Gordan coefficients of $SU(2)$ has been obtained in closed form by Wigner (see Wigner, 1959). However, we shall not use this method, but shall show how to obtain Clebsch–Gordan coefficients of simple compact groups by means of so-called lowering operators. To do this we first quote without proof several theorems about the basis functions of simple compact groups. In writing down these basis functions, we shall omit the superscripts specifying the representations to which the functions belong. Also, the subscripts will stand only for the weights, and we omit indices distinguishing between different functions with the same weight.

Consider two irreducible representations of a simple compact group. Let

ψ_μ be the basis vector belonging to the highest weight of the first representation and let ψ_ν be the basis vector belonging to the highest weight of the second representation. Then the basis tensor $\psi_\mu \psi_\nu$ of the Kronecker product representation is also a basis function of an irreducible representation of the group. Its weight is simple and is given by $\mathbf{M} = \boldsymbol{\mu} + \boldsymbol{\nu}$. Thus in the sum (8.11) there is only one Clebsch–Gordan coefficient different from zero and it is equal to one.

Once we have found one basis function of the irreducible representation, we can obtain other basis functions by using the nondiagonal standard operators E_α of the group. In fact we have the following

THEOREM. *If we operate on a basis function of an irreducible representation with a nondiagonal operator E_α, we obtain another basis function of the same irreducible representation or else get zero. Furthermore, by repeated operation on the basis function with the operators E_α, $E_\beta \cdots E_\alpha E_\alpha$, $E_\alpha E_\beta \cdots E_\alpha E_\alpha E_\alpha \cdots$ we obtain a complete set of basis functions. They may not be orthogonal or normalized, however.*

We can also find the weight of the basis function $E_\alpha \psi_m$ if we know the weight of ψ_m from the following theorem:

THEOREM. *If we have a basis function ψ_m with weight* $\mathbf{m}$, i.e.,

$$\mathbf{H}\psi_m = \mathbf{m}\psi_m \tag{8.16}$$

then the basis function $E_\alpha \psi_m$ has weight $\mathbf{m}'$ given by

$$\mathbf{m}' = \mathbf{m} + \boldsymbol{\rho}(\alpha). \tag{8.17}$$

We prove the theorem as follows.

Proof. From Eq. (5.62) we have

$$[\mathbf{H}, E_\alpha] = \boldsymbol{\rho}(\alpha) E_\alpha, \tag{8.18}$$

where $\rho(\alpha)$ is a nonvanishing root. Then

$$\mathbf{H} E_\alpha \psi_m = E_\alpha \mathbf{H} \psi_m + \boldsymbol{\rho}(\alpha) E_\alpha \psi_m. \tag{8.19}$$

Then, using Eq. (8.16) we obtain

$$\mathbf{H} E_\alpha \psi_m = E_\alpha \mathbf{m} \psi_m + \boldsymbol{\rho}(\alpha) E_\alpha \psi_m,$$

or

$$\mathbf{H} E_\alpha \psi_m = [\mathbf{m} + \boldsymbol{\rho}(\alpha)] E_\alpha \psi_m = \mathbf{m}' E_\alpha \psi_m, \tag{8.20}$$

which proves the theorem.

If the weight $\mathbf{m} + \boldsymbol{\rho}(\alpha)$ is higher than the weight $\mathbf{m}$, then E_α is said to be a *raising operator*; if $\mathbf{m} + \boldsymbol{\rho}(\alpha)$ is lower than $\mathbf{m}$, E_α is a *lowering operator*. Since $\boldsymbol{\rho}(-\alpha) = -\boldsymbol{\rho}(\alpha)$, if E_α is a raising operator, then $E_{-\alpha}$ is a lowering operator. Since our procedure to obtain the Clebsch–Gordan coefficients starts with the highest weight, we use lowering operators to obtain the other weights. However, this is a matter of convention. We can equally well start with the lowest weight and use raising operators. For $SU(3)$, all (3 or 6) weights at the corners of a weight diagram are equivalent and can serve as well as the highest weight for a starting point. In general, if we operate with a raising or lowering operator on a basis function, we do not obtain a normalized basis function. However, it is not too difficult to normalize the function. There is a theorem that helps in this regard. It says that if

$$E_\alpha \psi_m = 0, \tag{8.21}$$

then the function

$$[\boldsymbol{\rho}(\alpha) \cdot \mathbf{m}]^{1/2} E_{-\alpha} \psi_m, \tag{8.22}$$

is normalized.

We are now ready to give a prescription for decomposing the basis tensors of a representation which is the Kronecker product of two irreducible representations.

(1) We select from the basis tensors of the Kronecker product representation the one with the highest weight. This tensor is also a basis function of an irreducible representation contained in the decomposition of the Kronecker product.

(2) We apply the operators E_α, E_β, $E_\alpha E_\beta \cdots$ to the product basis tensor. By repeated application of the operators, we obtain a complete set of basis functions of the irreducible representation to which the product tensor belongs.

(3) The resulting functions will be orthogonal if they have different weights. This is clear because different eigenfunctions of the same hermitian operator with different eigenvalues are necessarily orthogonal. We can then normalize these basis functions. If two or more basis functions have the same weight, as can happen with any group of rank greater than one, we take orthogonal linear combinations of them. We have already given an example of how to make functions orthogonal to one another in Section 4.3 when we obtained the basis functions of the symmetric group S_3. This method is called the Schmidt orthogonalization procedure.

(4) In the subspace orthogonal to that generated from the highest weight we select the tensor with the highest weight. In general, such a tensor will be a linear combination of product tensors. This tensor will be a basis function of another irreducible representation. Furthermore, this function will have

the highest weight of the representation. We then repeat the process by operating on this function with the lowering operators $E_\alpha \cdots$. We repeat over and over until we have obtained the basis functions of all the irreducible representations contained in the direct sum of decomposition of the Kronecker product.

Thus, we obtain the irreducible functions as linear combinations of tensors of the Kronecker product. The coefficients obtained in this manner are the Clebsch–Gordan coefficients.

This procedure applies to any simple compact group. Therefore, we can apply it not only to the groups $SU(n)$, but also, for example, to the groups C_2 and G_2 discussed in Section 6.3. We can use this method to obtain the weight diagrams of C_2 and G_2 shown in Figs. 6.4 and 6.5, in addition to the Clebsch–Gordan coefficients of these groups. We leave this as an exercise for the interested reader.

8.3 Matrix Representation of the Algebra of $SU(n)$

In order to obtain the Clebsch–Gordan coefficients explicitly, we need to express the basis tensors $E_\alpha \psi_\mu \psi_\nu$, $E_\beta \psi_\mu \psi_\nu \cdots$ in terms of linear combinations of the product tensors. We do not need a matrix representation of the Lie algebra to do this, as we can work with the operator properties. For example, if we know that $E_\alpha^{(1)}$ is a lowering operator for $\psi_\mu^{(1)}$ and $E_\alpha^{(2)}$ is a lowering operator for $\psi_\nu^{(2)}$, then

$$E_\alpha = E_\alpha^{(1)} + E_\alpha^{(2)}, \tag{8.23}$$

is a lowering operator for $\psi_\mu^{(1)}\psi_\nu^{(2)}$. We can generalize further and state that if the product tensor $\psi_\mu^{(1)}\psi_\nu^{(2)} \cdots \psi_\lambda^{(k)}$ is also an irreducible tensor, then the lowering operator for this tensor is

$$E_\alpha = \sum_{i=1}^{k} E_\alpha^{(i)}. \tag{8.24}$$

However, it is helpful to have a matrix representation of the algebra so that our considerations are less abstract. We shall obtain the matrix generators of $SU(n)$. The lowest-dimensional nontrivial matrix representation of the algebra of $SU(n)$ is n-dimensional. Now consider a representation of the algebra of $SU(n)$ which operates on the basis functions of an N-dimensional irreducible representation of the group. The representation of the algebra may be by N-dimensional matrices if the basis functions are considered as vectors, or by a sum of n-dimensional matrices [Eq. (8.24)] if the basis functions are considered as tensors. We shall adopt the latter approach.

We have already written down in standard form the matrix generators of $SU(2)$ and $SU(3)$. These are given in Eqs. (6.7) and (6.22). Also, in Eqs. (6.29) and (6.30) we have written down in standard form the matrix elements of the generators of $SU(n)$. The normalization of these generators is useful to make use of a number of theorems about roots and weights, for example, the theorems whose major content is expressed in Eqs. (8.17) and (8.22). However, for our purposes the normalization of the generators given in Eqs. (6.29) and (6.30) is cumbersome because of the square root factors. Therefore we introduce $n \times n$ matrix generators of $SU(n)$ in standard form except for normalization, which we shall choose in such a way as to be most convenient. We denote the diagonal matrices by h_i $(i = 1, 2, \ldots, n-1)$ and the nondiagonal ones by e_{ij} $(i \neq j = 1, 2, \ldots, n)$. The diagonal $n \times n$ matrices are defined as

$$(h_i)_{\mu\nu} = \begin{cases} \delta_{\mu\nu}, & \mu < i+1 \\ -i\delta_{\mu\nu}, & \mu = i+1 \\ 0, & \mu > i+1. \end{cases} \tag{8.25}$$

The nondiagonal matrices e_{ij} are defined by

$$(e_{ij})_{\mu\nu} = \delta_{i\mu}\,\delta_{j\nu}. \tag{8.26}$$

It is easy to see that

$$e_{ij} = e_{ji}^{\dagger}. \tag{8.27}$$

If e_{ij} is a raising operator, then e_{ji} is a lowering operator. With our ordering of weights according to the last component rather than the first, e_{ij} is a raising operator if $j > i$.

As an example, for $SU(2)$ the generators are

$$h_1 = \begin{pmatrix} 1 & 0 \\ 0 & -1 \end{pmatrix}, \qquad e_{12} = \begin{pmatrix} 0 & 1 \\ 0 & 0 \end{pmatrix}, \qquad e_{21} = \begin{pmatrix} 0 & 0 \\ 1 & 0 \end{pmatrix}. \tag{8.28}$$

These are just the Pauli matrices: $h_1 = \sigma_3$, $e_{12} = \sigma_+$, $e_{21} = \sigma_-$. For $SU(3)$ the diagonal generators are

$$h_1 = \begin{pmatrix} 1 & 0 & 0 \\ 0 & -1 & 0 \\ 0 & 0 & 0 \end{pmatrix}, \qquad h_2 = \begin{pmatrix} 1 & 0 & 0 \\ 0 & 1 & 0 \\ 0 & 0 & -2 \end{pmatrix}. \tag{8.29}$$

The lowering operators are

$$e_{21} = \begin{pmatrix} 0 & 0 & 0 \\ 1 & 0 & 0 \\ 0 & 0 & 0 \end{pmatrix}, \qquad e_{31} = \begin{pmatrix} 0 & 0 & 0 \\ 0 & 0 & 0 \\ 1 & 0 & 0 \end{pmatrix}, \qquad e_{32} = \begin{pmatrix} 0 & 0 & 0 \\ 0 & 0 & 0 \\ 0 & 1 & 0 \end{pmatrix}. \tag{8.30}$$

The raising operators are obtained from Eq. (8.30) by using Eq. (8.27). The basis vectors u_i on which the operators h_i and e_{ij} act are the column vectors

$$(u_i)_j = \delta_{ij}, \qquad i, j = 1, 2, \ldots, n. \tag{8.31}$$

For $SU(2)$ these vectors are

$$u_1 = \begin{pmatrix} 1 \\ 0 \end{pmatrix}, \qquad u_2 = \begin{pmatrix} 0 \\ 1 \end{pmatrix}.$$

We do not need to write out the generators of $SU(n)$ explicitly. What is more relevant is a knowledge of how the generators act on the basis vectors u_j of the first fundamental representation. From Eqs. (8.25) and (8.31) we see that u_j is an eigenfunction of h_i

$$h_i u_j = h_{ij} u_j. \tag{8.32}$$

The eigenvalue h_{ij} is given by

$$h_{ij} = \begin{cases} 1, & j < i+1 \\ -i, & j = i+1 \\ 0, & j > i+1 \end{cases} \tag{8.33}$$

or in compact form

$$h_{ij} = \sum_{k=1}^{i} \delta_{kj} - i\delta_{i+1,j}. \tag{8.34}$$

From Eqs. (8.26) and (8.31) we have

$$e_{ij} u_k = u_i \delta_{jk}. \tag{8.35}$$

From Eq. (8.35) we see that application of the raising or lowering operators to basis vectors of the first fundamental representation yields normalized basis vectors. This is not true in general for raising and lowering operators normalized in the standard way.

It is also useful to obtain expressions for the products of the generators. From the definitions (8.25) and (8.26) we obtain that

$$h_i e_{jk} = h_{ij} e_{jk} \qquad e_{jk} h_i = h_{ik} e_{jk} \tag{8.36}$$

where we have used Eq. (8.35). Similarly we obtain

$$e_{ij} e_{kl} = \delta_{jk} e_{il}, \qquad i \neq l. \tag{8.37}$$

The product $e_{ij} e_{ki}$ cannot be written as a linear combination of the generators, since its trace is different from zero. The product is given in terms of a projection operator P_i

$$e_{ij} e_{ki} = \delta_{jk} P_i, \tag{8.38}$$

where

$$(P_i)_{\mu\nu} = \delta_{i\mu} \delta_{\mu\nu}. \tag{8.39}$$

From Eqs. (8.36)–(8.39) we obtain the commutation relations of the generators. They are

$$[h_i, e_{jk}] = (h_{ij} - h_{ik}) e_{jk}, \tag{8.40}$$

$$[e_{ij}, e_{jl}] = e_{il}, \qquad i \neq l, \tag{8.41}$$

$$[e_{ij}, e_{ji}] = P_i - P_j. \tag{8.42}$$

The operators $P_i - P_j$ are traceless and therefore can be expressed as linear combinations of the h_i. We shall not do so, however, except for $SU(2)$ and $SU(3)$. Using Eqs. (8.40)–(8.42), we see that for $SU(2)$ we have only the following nonvanishing commutation relations

$$[h_1, e_{12}] = e_{12}, \qquad [h_1, e_{21}] = -e_{21}, \qquad [e_{12}, e_{21}] = h_1, \tag{8.43}$$

since $P_1 - P_2 = h_1$. For $SU(3)$ the commutation relations of Eq. (8.42) can be expressed in terms of the h_i as follows:

$$[e_{12}, e_{21}] = h_1, \qquad [e_{13}, e_{31}] = \tfrac{1}{2}(h_1 + h_2), \qquad [e_{23}, e_{32}] = \tfrac{1}{2}(h_2 - h_1). \tag{8.44}$$

The other commutation relations of $SU(3)$ are compactly given in Eqs. (8.40) and (8.41).

For convenience, we introduce still a bit more notation. We keep the convention that in writing a tensor as the product of v vectors, each vector referring to a different particle, we write the vector of the first particle on the left, next the vector of the second particle etc. Then we can omit the integers specifying the particles. For example, a three-particle tensor is written

$$u_i(1)u_j(2)u_k(3) = u_i u_j u_k. \tag{8.45}$$

The irreducible tensor with the highest weight of a product of v single-particle vectors is clearly

$$u_1 u_1 \cdots u_1 \qquad (v \quad \text{times}). \tag{8.46}$$

Also, we denote the raising or lowering operators which operate on product tensors by e_{ij}, where

$$\mathsf{e}_{ij} = \sum_{k=1}^{v} e_{ij}(k). \tag{8.47}$$

Similarly, we denote the diagonal operators by

$$\mathsf{h}_i = \sum_{k=1}^{v} h_i(k). \tag{8.48}$$

These operators are the generators of $SU(n)$ which act on the irreducible functions formed from linear combinations of the product tensors. The generators h_i and e_{ij} do not satisfy Eqs. (8.36)–(8.38) except in the case that $v = 1$. This is because for $v > 1$, these operators contain cross terms, for example $h_i(1)e_{jk}(2)$, which act on the state vectors of different particles. However, since any operator referring to one particle commutes with any operator referring to another particle, the h_i and e_{ij} do satisfy the commutation relations of Eqs. (8.40)–(8.42). In the case of $SU(3)$, we show in Fig. 8.1 the effect of operating on an irreducible function ψ_i with e_{ij}.

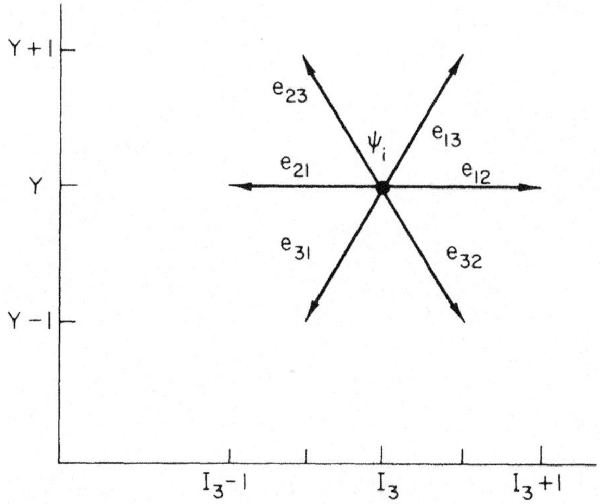

Fig. 8.1. Properties of the raising and lowering operators of $SU(3)$ when acting on a state ψ_i with weight $(I_3 Y)$.

We now have all the tools necessary to obtain the Clebsch–Gordan coefficients of $SU(n)$. We shall do this for $SU(2)$ and $SU(3)$ by the straightforward, if tedious, procedure of building up from the first fundamental representation. Such a procedure does not require knowledge of all the commutation relations of the generators. However, the procedure can be applied quite generally to $SU(n)$.

If we know all the commutation relations of the group, plus the expressions for the Casimir operators in terms of the generators, we can use more powerful methods to obtain the Clebsch–Gordan coefficients. However, since we have not given the Casimir operators of $SU(n)$, we shall not discuss this method.

8.4 Clebsch–Gordan Coefficients of $SU(2)$

We shall illustrate a procedure to obtain the Clebsch–Gordan coefficients of $SU(2)$ by a technique which does not make use of the Casimir operator J^2. We begin by decomposing the basis tensors of the Kronecker product of two fundamental representations. These four basis tensors are of the form $u_i u_j$ $(i, j = 1, 2)$. By a theorem quoted in Section 8.2 the basis tensor with the highest weight $u_1 u_1$ is already an irreducible function. This should be obvious, as there is only one way to obtain a function with the highest weight: namely, the way given in Eq. (8.46). Let us order the irreducible functions according to their weights, using a single subscript which denotes, *not* the weight, but the *order* of the weight. (This is to avoid fractions.) Then

$$\psi_1 = u_1 u_1. \tag{8.49}$$

Operating on ψ_1 with the lowering operator e_{21}, defined by Eq. (8.47), and, using Eq. (8.35), we obtain

$$\mathsf{e}_{21}\psi_1 = u_1 u_2 + u_2 u_1. \tag{8.50}$$

Normalizing, we obtain the irreducible function ψ_2 with the second highest weight

$$\psi_2 = (u_1 u_2 + u_2 u_1)/\sqrt{2}. \tag{8.51}$$

Again using e_{21}, we obtain

$$\mathsf{e}_{21}\psi_2 = \sqrt{2}\, u_2 u_2, \tag{8.52}$$

or normalizing,

$$\psi_3 = u_2 u_2. \tag{8.53}$$

If we operate on ψ_3 with e_{21} we obtain zero. Thus we have obtained the irreducible functions of a triplet representation. The coefficients multiplying the product tensors in Eqs. (8.49), (8.51), and (8.53) are $SU(2)$ Clebsch–Gordan coefficients.

From Eqs. (8.50) and (8.52), we see that for the triplet representation (which we now denote by a superscript),

$$\mathsf{e}_{21}\,\psi_i^{(3)} = \sqrt{2}\,\psi_{i+1}^{(3)}, \tag{8.54}$$

if $\psi_{i+1}^{(3)}$ is nonvanishing. Here the subscripts i and $i+1$ denote the order, not the weight.

The triplet function $\psi_2^{(3)}$ has the second-highest weight. Following the procedure outlined in Section 8.2, we next construct a function orthogonal to $\psi_2^{(3)}$ from the product tensors $u_1 u_2$ and $u_2 u_1$. Such a function is

$$\psi_1 = (u_1 u_2 - u_2 u_1)/\sqrt{2}. \tag{8.55}$$

If we operate on this function with e_{21}, we get zero. Thus ψ_1 is a singlet $\psi_1^{(1)}$, a fact we should know immediately, since there are only four product tensors $u_i u_j$.

The overall sign of $\psi_1^{(1)}$ is not obtained by this method. In fact, we must adopt a convention to determine it. The convention we use is that of Condon and Shortley (1935) which states that the coefficient

$$(\alpha\beta\alpha, j-\alpha \,|\, jj)$$

is positive. With this convention we have

$$\begin{aligned}(\alpha\beta\mu\nu \mid jm) &= (-)^{j-\alpha-\beta}(\beta\alpha\nu\mu \mid jm) = (-)^{j-\alpha-\beta}(\alpha\beta, -\mu, -\nu \mid j, -m) \\ &= (-1)^{\alpha-\mu}[(2j+1)/(2\beta+1)]^{1/2}(\alpha j\mu, -m \mid \beta, -\nu). \qquad (8.56)\end{aligned}$$

The triplet and singlet irreducible functions can be obtained without the lowering operators by Young tableaux. Since the tableau $\boxed{}\boxed{}$ stands for a symmetric combination, we immediately have (after normalization)

$$\begin{aligned}\psi_1^{(3)} &= \boxed{1}\boxed{1} = u_1 u_1, \\ \psi_2^{(3)} &= \boxed{1}\boxed{2} = (u_1 u_2 + u_2 u_1)/\sqrt{2}, \qquad (8.57) \\ \psi_3^{(3)} &= \boxed{2}\boxed{2} = u_2 u_2 .\end{aligned}$$

Likewise, since

$$\begin{array}{c}\boxed{} \\ \boxed{}\end{array}$$

is antisymmetric, we have

$$\psi_1^{(1)} = \begin{array}{c}\boxed{1} \\ \boxed{2}\end{array} = \frac{1}{\sqrt{2}}(u_1 u_2 - u_2 u_1). \qquad (8.58)$$

We can extend these procedures to obtain any Clebsch–Gordan coefficients. For example, suppose we wish to obtain the irreducible functions contained in the Kronecker product of $\psi_i^{(3)} u_j$. Since the Kronecker product has the decomposition

$$\mathbf{3} \otimes \mathbf{2} = \mathbf{4} \oplus \mathbf{2},$$

we know that the irreducible functions contained in $\psi_i^{(3)} u_j$ belong to a quartet and a doublet. The product function with the highest weight must belong to the quartet:

$$\psi_1^{(4)} = \psi_1^{(3)} u_1. \qquad (8.59)$$

We operate on this function with e_{21}, using Eqs. (8.35) and (8.54). We then obtain, omitting the superscript (3)

$$\mathsf{e}_{21}\psi_2^{(4)} = \sqrt{2}\,\psi_2 u_1 + \psi_1 u_2 . \qquad (8.60)$$

Normalizing, we obtain

$$\psi_2^{(4)} = \sqrt{\tfrac{2}{3}}\,\psi_2 u_1 + \sqrt{\tfrac{1}{3}}\,\psi_1 u_2 . \qquad (8.61)$$

We can obtain the other members of this quartet state from Eqs. (8.60) and (8.61) either by Eq. (8.56) or by continued use of the lowering operators. The irreducible functions are

$$\begin{aligned}\psi_3^{(4)} &= \sqrt{\tfrac{2}{3}}\,\psi_2 u_2 + \sqrt{\tfrac{1}{3}}\,\psi_3 u_1,\\ \psi_4^{(4)} &= \psi_3^{(3)} u_2.\end{aligned} \tag{8.62}$$

Alternatively, we can use Young tableaux. We have

$$\begin{aligned}\psi_1^{(4)} &= \boxed{1\,|\,1\,|\,1} = u_1 u_1 u_1,\\ \psi_2^{(4)} &= \boxed{1\,|\,1\,|\,2} = (u_1 u_1 u_2 + u_1 u_2 u_1 + u_2 u_1 u_1)/\sqrt{3},\\ \psi_3^{(4)} &= \boxed{1\,|\,2\,|\,2} = (u_1 u_2 u_2 + u_2 u_1 u_2 + u_2 u_2 u_1)/\sqrt{3},\\ \psi_4^{(4)} &= \boxed{2\,|\,2\,|\,2} = u_2 u_2 u_2.\end{aligned} \tag{8.63}$$

Then by substituting Eqs. (8.49), (8.51), and (8.53) into Eqs. (8.63), we immediately get the normalized functions of Eqs. (8.59), (8.61), and (8.62).

We can obtain the other multiplet by constructing a function which is orthogonal to the function $\psi_2^{(4)}$, which has the second highest weight. It is

$$\psi_1^{(2)} = -\sqrt{\tfrac{1}{3}}\,\psi_2 u_1 + \sqrt{\tfrac{2}{3}}\,\psi_1 u_2. \tag{8.64}$$

By using the lowering operator we find the other member of this doublet, which is

$$\psi_2^{(2)} = \sqrt{\tfrac{1}{3}}\,\psi_2 u_2 - \sqrt{\tfrac{2}{3}}\,\psi_3 u_1. \tag{8.65}$$

Thus we have accomplished the decomposition of the product $\psi_i^{(3)} u_j$, obtaining all the relevant Clebsch–Gordan coefficients.

As a last example of this procedure, we shall obtain the decomposition of $\psi_i^{(3)}\psi_j^{(3)}$. The decomposition of the product is

$$\mathbf{3} \otimes \mathbf{3} = \mathbf{5} \oplus \mathbf{3} \oplus \mathbf{1}. \tag{8.66}$$

In this case we can obtain many of the Clebsch–Gordan coefficients by symmetry arguments. For example, we must have, omitting the superscript 3,

$$\begin{aligned}\psi_1^{(5)} &= \psi_1 \psi_1,\\ \psi_2^{(5)} &= (\psi_1 \psi_2 + \psi_2 \psi_1)/\sqrt{2}.\end{aligned} \tag{8.67}$$

Symmetry arguments are not sufficient to obtain $\psi_3^{(5)}$ as there exist two independent symmetric functions with the relevant weight. We therefore operate on $\psi_2^{(5)}$ with e_{21}, using Eq. (8.54). We obtain

$$e_{21}\psi_2^{(5)} = \psi_1\psi_3 + 2\psi_2\psi_2 + \psi_3\psi_1. \tag{8.68}$$

Normalizing, we obtain

$$\psi_3^{(5)} = (\psi_1\psi_3 + 2\psi_2\psi_2 + \psi_3\psi_1)/\sqrt{6}. \tag{8.69}$$

The remaining functions $\psi_4^{(5)}$ and $\psi_5^{(5)}$ are easily obtained either by symmetry arguments or by the lowering operator. We obtain $\psi_1^{(3)}$ by constructing a function orthogonal to $\psi_2^{(5)}$. It is

$$\psi_1^{(3)} = (\psi_1\psi_2 - \psi_2\psi_1)/\sqrt{2}. \tag{8.70}$$

Since this function is antisymmetric, the other members of the triplet must also be antisymmetric. We thus have

$$\begin{aligned} \psi_2^{(3)} &= (\psi_1\psi_3 - \psi_3\psi_1)/\sqrt{2} \\ \psi_3^{(3)} &= (\psi_2\psi_3 - \psi_3\psi_2)/\sqrt{2} \end{aligned} \tag{8.71}$$

The singlet function $\psi_1^{(1)}$ must be orthogonal to $\psi_3^{(5)}$ and $\psi_2^{(3)}$. Therefore it is given by

$$\psi_1^{(1)} = (\psi_1\psi_3 - \psi_2\psi_2 + \psi_3\psi_1)/\sqrt{3} \tag{8.72}$$

Thus, we have obtained all the Clebsch–Gordan coefficients relevant in decomposing the product $\psi_i^{(3)}\psi_j^{(3)}$.

It should be clear from these examples that in many cases we can obtain Clebsch–Gordan coefficients by considerations of symmetry. However, in general, we use the lowering operator e_{21}. We have seen that e_{21} operating on $\psi_m^{(j)}$ gives a constant multiplied by $\psi_{m-1}^{(j)}$. From our examples, it can be seen that this constant, which we denote by $c(jm)$, depends in general on j and m. In order to obtain the Clebsch–Gordan coefficients, we need to obtain an expression for the constant $c(jm)$ in terms of j and m.

We have described a procedure for obtaining the constants $c(jm)$ by building up irreducible functions from the fundamental irreducible vectors. This procedure is straightforward and can be generalized to any $SU(n)$.

Unfortunately, the method we have described can become rather tedious if the dimensionality of the multiplets is large. There exist a number of short-cut methods which make use of the Casimir operator J^2. These are described in many places, for example, Rose (1957); Edmonds (1957), and Gottfried (1966). We shall not reproduce these methods. A number of $SU(2)$ Clebsch–Gordan coefficients are given in Table 8.1 with the phases of Condon and Shortley (1935). In Table 8.2 are given some spherical harmonics.

TABLE 8.1

SOME $SU(2)$ CLEBSCH-GORDAN COEFFICIENTS WITH CONDON AND SHORTLEY (1935) PHASES. THE FORMAT FOLLOWS THAT OF THE PARTICLE DATA GROUP (1969).

Note: A $\sqrt{\ }$ is to be understood over every coefficient, e.g., for -8/15 read $-\sqrt{8/15}$.

Notation:

		J	J	...
		M	M	...
m_1	m_2	Coefficients		
m_1	m_2			
.	.			
.	.			

1/2 × 1/2

		1
		+1
+1/2	+1/2	1

		1	0
		0	0
+1/2	-1/2	1/2	1/2
-1/2	+1/2	1/2	-1/2

		1
		-1
-1/2	-1/2	1

1 × 1/2

		3/2
		+3/2
+1	+1/2	1

		3/2	1/2
		+1/2	+1/2
+1	-1/2	1/3	2/3
0	+1/2	2/3	-1/3

		3/2	1/2
		-1/2	-1/2
0	-1/2	2/3	1/3
-1	+1/2	1/3	-2/3

		3/2
		-3/2
-1	-1/2	1

2 × 1

		3
		+3
+2	+1	1

		3	2
		+2	+2
+2	0	1/3	2/3
+1	+1	2/3	-1/3

		3	2	1
		+1	+1	+1
+2	-1	1/15	1/3	3/5
+1	0	8/15	1/6	-3/10
0	+1	6/15	-1/2	1/10

		3	2	1
		0	0	0
+1	-1	1/5	1/2	3/10
0	0	3/5	0	-2/5
-1	+1	1/5	-1/2	3/10

		3	2	1
		-1	-1	-1
0	-1	6/15	1/2	1/10
-1	0	8/15	-1/6	-3/10
-2	+1	1/15	-1/3	3/5

		3	2
		-2	-2
-1	-1	2/3	1/3
-2	0	1/3	-2/3

		3
		-3
-2	-1	1

1 × 1

		2
		+2
+1	+1	1

		2	1
		+1	+1
+1	0	1/2	1/2
0	+1	1/2	-1/2

		2	1	0
		0	0	0
+1	-1	1/6	1/2	1/3
0	0	2/3	0	-1/3
-1	+1	1/6	-1/2	1/3

		2	1
		-1	-1
0	-1	1/2	1/2
-1	0	1/2	-1/2

		2
		-2
-1	-1	1

2 × 1/2

		5/2
		+5/2
+2	1/2	1

		5/2	3/2
		3/2	+3/2
+2	-1/2	1/5	4/5
+1	+1/2	4/5	-1/5

		5/2	3/2
		+1/2	+1/2
+1	-1/2	2/5	3/5
0	+1/2	3/5	-2/5

		5/2	3/2
		-1/2	-1/2
0	-1/2	3/5	2/5
-1	+1/2	2/5	-3/5

		5/2	3/2
		-3/2	-3/2
-1	-1/2	4/5	1/5
-2	+1/2	1/5	-4/5

		5/2
		-5/2
-2	-1/2	1

3/2 × 1/2

		2
		+2
+3/2	+1/2	1

		2	1
		+1	+1
+3/2	-1/2	1/4	3/4
+1/2	+1/2	3/4	-1/4

		2	1
		0	0
+1/2	-1/2	1/2	1/2
-1/2	+1/2	1/2	-1/2

		2	1
		-1	-1
-1/2	-1/2	3/4	1/4
-3/2	+1/2	1/4	-3/4

		2
		-2
-3/2	-1/2	1

3/2 × 1

		5/2
		+5/2
+3/2	+1	1

		5/2	3/2
		+3/2	+3/2
+3/2	0	2/5	3/5
+1/2	+1	3/5	-2/5

		5/2	3/2	1/2
		+1/2	+1/2	+1/2
+3/2	-1	1/10	2/5	1/2
+1/2	0	3/5	1/15	-1/3
-1/2	+1	3/10	-8/15	1/6

		5/2	3/2	1/2
		-1/2	-1/2	-1/2
+1/2	-1	3/10	8/15	1/6
-1/2	0	3/5	-1/15	-1/3
-3/2	+1	1/10	-2/5	1/2

		5/2	3/2
		-3/2	-3/2
-1/2	-1	3/5	2/5
-3/2	0	2/5	-3/5

		5/2
		-5/2
-3/2	-1	1

$$\langle j_1 j_2 m_1 m_2 | j_1 j_2 J M \rangle = (-1)^{J - j_1 - j_2} \langle j_2 j_1 m_2 m_1 | j_2 j_1 J M \rangle$$

TABLE 8.2

SOME SPHERICAL HARMONICS WITH THE PHASES OF CONDON AND SHORTLEY (1935)

$$Y_0{}^0 = \sqrt{\frac{1}{4\pi}}$$

$$Y_1{}^0 = \sqrt{\frac{3}{4\pi}} \cos\theta; \qquad Y_1{}^1 = -\sqrt{\frac{3}{8\pi}} \sin\theta\, e^{i\phi}$$

$$Y_2{}^0 = \sqrt{\frac{5}{4\pi}} \left(\frac{3}{2}\cos^2\theta - \frac{1}{2}\right); \qquad Y_2{}^1 = -\sqrt{\frac{15}{8\pi}} \sin\theta \cos\theta\, e^{i\phi}$$

$$Y_2{}^2 = \frac{1}{4}\sqrt{\frac{15}{2\pi}} \sin^2\theta e^{2i\phi}$$

$$Y_3{}^0 = \sqrt{\frac{7}{4\pi}} \left(\frac{5}{2}\cos^3\theta - \frac{3}{2}\cos\theta\right); \qquad Y_3{}^1 = -\frac{1}{4}\sqrt{\frac{21}{4\pi}} \sin\theta\,(5\cos^2\theta - 1)e^{i\phi}$$

$$Y_3{}^2 = \frac{1}{4}\sqrt{\frac{105}{2\pi}} \sin^2\theta \cos\theta\, e^{2i\phi}; \qquad Y_3{}^3 = -\frac{1}{4}\sqrt{\frac{35}{4\pi}} \sin^3\theta\, e^{3i\phi}$$

$$(Y_j{}^m)^* = (-1)^m\, Y_j{}^{-m}$$

8.5 Clebsch–Gordan Coefficients of $SU(3)$

The procedure outlined in the last section can be used with only a few modifications which are necessary because $SU(3)$ is not a simply reducible group. The basis functions of the first fundamental representation of $SU(3)$ are u_1, u_2 and u_3 with values of (I_3, Y) given by $(\frac{1}{2}, \frac{1}{3})$, $(-\frac{1}{2}, \frac{1}{3})$, $(0, -\frac{2}{3})$, respectively. (Alternatively, we can specify the weights shown in Fig. 6.6.)

As in the case of $SU(2)$, we shall distinguish the functions belonging to a given representation by a single subscript, which denotes the *order* of the function. We shall adopt a convention which orders the states by the values of Y, I, and I_3, respectively.

Let us consider the Clebsch–Gordan coefficients which arise in reducing the Kronecker product $\mathbf{3} \otimes \mathbf{3}$. The basis tensors of the Kronecker product representation are the nine two-particle states

$$u_i u_j, \qquad i, j = 1, 2, 3. \tag{8.73}$$

We know that the product representation can be composed into a symmetric representation with Young tableau  and an antisymmetric representation with Young tableau

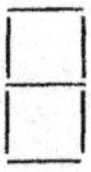

By writing down symmetric and antisymmetric combinations, we immediately get the relevant Clebsch–Gordan coefficients, except for possible sign ambiguities which must be decided by convention. In the symmetric case the normalized irreducible functions are (suppressing a superscript which denotes the dimensionality of the representation):

$$
\begin{aligned}
\begin{array}{|c|c|}\hline 1 & 1 \\ \hline\end{array} &= u_1 u_1 = \psi_1, \\
\begin{array}{|c|c|}\hline 1 & 2 \\ \hline\end{array} &= \frac{1}{\sqrt{2}}(u_1 u_2 + u_2 u_1) = \psi_2, \\
\begin{array}{|c|c|}\hline 2 & 2 \\ \hline\end{array} &= u_2 u_2 = \psi_3, \\
\begin{array}{|c|c|}\hline 1 & 3 \\ \hline\end{array} &= \frac{1}{\sqrt{2}}(u_1 u_3 + u_3 u_1) = \psi_4, \\
\begin{array}{|c|c|}\hline 2 & 3 \\ \hline\end{array} &= \frac{1}{\sqrt{2}}(u_2 u_3 + u_3 u_2) = \psi_5, \\
\begin{array}{|c|c|}\hline 3 & 3 \\ \hline\end{array} &= u_3 u_3 = \psi_6 .
\end{aligned}
\tag{8.74}
$$

The Clebsch–Gordan coefficients are just the coefficients in these sums. In the antisymmetric case, the irreducible functions are

$$
\begin{aligned}
\begin{array}{|c|}\hline 1 \\ \hline 2 \\ \hline\end{array} &= \frac{1}{\sqrt{2}}(u_1 u_2 - u_2 u_1) = \bar{u}_3, \\
-\begin{array}{|c|}\hline 1 \\ \hline 3 \\ \hline\end{array} &= \frac{1}{\sqrt{2}}(u_3 u_1 - u_1 u_3) = \bar{u}_2, \\
\begin{array}{|c|}\hline 2 \\ \hline 3 \\ \hline\end{array} &= \frac{1}{\sqrt{2}}(u_2 u_3 - u_3 u_2) = \bar{u}_1,
\end{aligned}
\tag{8.75}
$$

with a particular sign convention which we shall discuss shortly.

We have introduced the notation $\bar{u}_1$, $\bar{u}_2$, and $\bar{u}_3$ for the basis functions of the antisymmetric representation because these functions also serve as basis vectors for the second fundamental representation of $SU(3)$. These states can be interpreted as the ones which denote the antiparticles of the states u_1, u_2, and u_3. This can be seen first by noting that $\bar{u}_i$ has a weight which is opposite in sign to the weight of u_i:

$$\mathsf{h}_i \bar{u}_j = -h_{ij} \bar{u}_j . \tag{8.76}$$

Furthermore, if we operate on $\bar{u}_k$ with $\mathbf{e}_{ij}$ we find, using the definitions of Eq. (8.75), that

$$\mathbf{e}_{ij}\bar{u}_k = -\bar{u}_j\,\delta_{ik}. \qquad (8.77)$$

Thus, comparing (8.76) and (8.77) with Eqs. (8.32) and (8.35) we see that the generators of the $\bar{\mathbf{3}}$ representation (denoted by $\bar{h}_i$ and $\bar{e}_{ij}$) satisfy

$$\bar{h}_i = -h_i, \qquad \bar{e}_{ij} = -e_{ji}, \qquad (8.78)$$

in agreement with Eq. (6.16).

Although the sign convention of Eq. (8.75) is satisfactory, two other sign conventions have been used more extensively in the literature. One of these has been used by de Swart (1963) and is the convention used by the Particle Data Group (1976) in its Table of Particle Properties. In this convention the matrix elements of $\mathbf{e}_{21}$ and $\mathbf{e}_{31}$ are required to be positive. Unfortunately, although the de Swart convention is most widely used, it is not the one most suitable for generalization to $SU(n)$. Therefore, we adopt still a different convention, the one used by Baird and Biedenharn (1963). In this convention the matrix elements of $\mathbf{e}_{21}$ and $\mathbf{e}_{32}$ are positive. This convention is readily generalized to $SU(n)$ by requiring that the matrix elements of $\mathbf{e}_{i+1,i}$ be positive.

To conform to the Baird–Biedenharn convention, we introduce the vectors v_1, v_2, and v_3 which are ordered according to Y, I, and I_3 and are defined by

$$v_1 = \bar{u}_3, \qquad v_2 = -\bar{u}_2, \qquad v_3 = \bar{u}_1. \qquad (8.79)$$

If we operate on the states v_1, v_2, and v_3 with the lowering operators, we obtain

$$\mathbf{e}_{32}v_1 = v_2, \qquad \mathbf{e}_{21}v_2 = v_3, \qquad \mathbf{e}_{31}v_1 = -v_3. \qquad (8.80)$$

Thus, we see that by adopting the sign convention of Eq. (8.79), we can force $\mathbf{e}_{32}$ and $\mathbf{e}_{21}$ to have only positive matrix elements. The matrix elements of $\mathbf{e}_{31}$ are then determined by the $SU(3)$ commutation relations. In order to deal with matrix elements with only positive signs, we shall not use the operator $\mathbf{e}_{31}$, but only the lowering operators $\mathbf{e}_{32}$ and $\mathbf{e}_{21}$ (and their corresponding raising operators $\mathbf{e}_{23}$ and $\mathbf{e}_{12}$). More generally, for $SU(n)$, we use only the lowering operators $\mathbf{e}_{i+1,i}$ and their corresponding raising operators.

We are now ready to construct the Clebsch–Gordan coefficients of other irreducible representations. Let us consider as an example the coefficients arising in the decomposition

$$\mathbf{3} \otimes \bar{\mathbf{3}} = \mathbf{8} \oplus \mathbf{1}. \qquad (8.81)$$

By symmetry, the six states at the periphery of the octet weight diagram can be obtained at once. They are (suppressing a superscript on $\psi_i^{(8)}$)

$$\begin{aligned} &\psi_1 = u_1 v_1, \qquad \psi_2 = u_2 v_1, \qquad \psi_3 = u_1 v_2, \\ &\psi_5 = u_2 v_3, \qquad \psi_7 = u_3 v_2, \qquad \psi_8 = u_3 v_3. \end{aligned} \tag{8.82}$$

The signs are correct, as can be checked by operating with the $\mathbf{e}_{i+1,i}$. The states of Eq. (8.82) are all eigenstates of isospin. We can obtain another eigenstate of isospin by operating on ψ_3 with $\mathbf{e}_{21}$:

$$\mathbf{e}_{21}\psi_3 = u_1 v_3 + u_2 v_2.$$

Normalizing, we get

$$\psi_4 = (u_1 v_3 + u_2 v_2)/\sqrt{2}. \tag{8.83}$$

The functions ψ_3, ψ_4, and ψ_5 are an isospin triplet with $Y = 0$, ψ_1 and ψ_2 are a doublet with $Y = 1$, and ψ_7 and ψ_8 are a doublet with $Y = -1$.

Thus, we have obtained seven functions of the octet. The remaining function ψ_6 has the same weight as ψ_4, but differs from ψ_4 in that it has $I = 0$ rather than $I = 1$. To obtain ψ_6, we first obtain a function φ by operating on ψ_2 with the lowering operator $\mathbf{e}_{32}$. We get

$$\varphi = \mathbf{e}_{32}\psi_2 = u_3 v_1 + u_2 v_2.$$

However φ is not an eigenstate of the isospin I, but is an eigenstate of another $SU(2)$ subgroup of $SU(3)$. In general, we obtain an eigenstate of I when we operate on an I-eigenstate with $\mathbf{e}_{21}$, but not when we operate with $\mathbf{e}_{32}$. We get an I-eigenstate by operating with $\mathbf{e}_{32}$ only if we reach a simple weight as a result of the operation. We must now obtain a function orthogonal to ψ_4 by the Schmidt procedure. We let

$$\psi_6 = c(\varphi - a\psi_4),$$

where c is a normalization constant and a is a parameter chosen so that $(\psi_6, \psi_4) = 0$. We obtain

$$\psi_6 = (2u_3 v_1 + u_2 v_2 - u_1 v_3)/\sqrt{6}, \tag{8.84}$$

where the overall sign is determined by the requirement that the matrix element $(\psi_6, \mathbf{e}_{32}\psi_2)$ be positive. In Eqs. (8.82)–(8.84) we have exhibited the $SU(3)$ Clebsch–Gordan coefficients which arise in constructing an octet from $\mathbf{3} \otimes \bar{\mathbf{3}}$.

We obtain a singlet by constructing a state function which is orthogonal to Eqs. (8.83) and (8.84). It is

$$\psi = (u_1 v_3 - u_2 v_2 + u_3 v_1)/\sqrt{3}. \tag{8.85}$$

The overall sign of this state is chosen by the requirement that the coefficient with the highest value of the weight **m** of the first function be positive.

As another example, let us calculate the Clebsch–Gordan coefficients relevant to the decomposition

$$\mathbf{6} \otimes \mathbf{3} = \mathbf{10} \oplus \mathbf{8}. \tag{8.86}$$

It is easiest to obtain the Clebsch–Gordan coefficients for the 10 by using Young tableaux. In terms of the one-particle states u_i, the **10** is completely symmetric and the states $\psi_i^{(10)}$ of the **10** can be written down by inspection. They are:

$$\begin{aligned}
\boxed{1\,|\,1\,|\,1} &= u_1u_1u_1 = \psi_1^{(10)},\\
\boxed{1\,|\,1\,|\,2} &= (u_1u_1u_2 + u_1u_2u_1 + u_2u_1u_1)/\sqrt{3} = \psi_2^{(10)},\\
\boxed{1\,|\,2\,|\,2} &= (u_1u_2u_2 + u_2u_1u_2 + u_2u_2u_1)/\sqrt{3} = \psi_3^{(10)},\\
\boxed{2\,|\,2\,|\,2} &= u_2u_2u_2 = \psi_4^{(10)},\\
\boxed{1\,|\,1\,|\,3} &= (u_1u_1u_3 + u_1u_3u_1 + u_3u_1u_1)/\sqrt{3} = \psi_5^{(10)},\\
\boxed{1\,|\,2\,|\,3} &= (u_1u_2u_3 + u_1u_3u_2 + u_2u_1u_3\\
&\quad + u_2u_3u_1 + u_3u_1u_2 + u_3u_2u_1)/\sqrt{6} = \psi_6^{(10)},\\
\boxed{2\,|\,2\,|\,3} &= (u_2u_2u_3 + u_2u_3u_2 + u_3u_2u_2)/\sqrt{3} = \psi_7^{(10)},\\
\boxed{1\,|\,3\,|\,3} &= (u_1u_3u_3 + u_3u_1u_3 + u_3u_3u_1)/\sqrt{3} = \psi_8^{(10)},\\
\boxed{2\,|\,3\,|\,3} &= (u_2u_3u_3 + u_3u_2u_3 + u_3u_3u_2)/\sqrt{3} = \psi_9^{(10)},\\
\boxed{3\,|\,3\,|\,3} &= u_3u_3u_3 = \psi_{10}^{(10)}.
\end{aligned} \tag{8.87}$$

Next, we substitute the states of Eq. (8.74) for the first two particles of Eq. (8.87), obtaining

$$\begin{aligned}
\psi_1^{(10)} &= \psi_1u_1,\\
\psi_2^{(10)} &= (\psi_1u_2 + \sqrt{2}\psi_2u_1)/\sqrt{3},\\
\psi_3^{(10)} &= (\sqrt{2}\psi_2u_2 + \psi_3u_1)/\sqrt{3},\\
\psi_4^{(10)} &= \psi_3u_2,\\
\psi_5^{(10)} &= (\psi_1u_3 + \sqrt{2}\psi_4u_1)/\sqrt{3},\\
\psi_6^{(10)} &= (\psi_2u_3 + \psi_4u_2 + \psi_5u_1)/\sqrt{3},\\
\psi_7^{(10)} &= (\psi_3u_3 + \sqrt{2}\psi_5u_2)/\sqrt{3},
\end{aligned} \tag{8.88}$$

$$\psi_8^{(10)} = (\sqrt{2}\,\psi_4 u_3 + \psi_6 u_1)/\sqrt{3},$$
$$\psi_9^{(10)} = (\sqrt{2}\,\psi_5 u_3 + \psi_6 u_2)/\sqrt{3},$$
$$\psi_{10}^{(10)} = \psi_6 u_3 .$$

We obtain the highest member of the **8** by constructing a function $\psi_1^{(8)}$ orthogonal to $\psi_2^{(10)}$. It is given by

$$\psi_1^{(8)} = (\sqrt{2}\,\psi_1 u_2 - \psi_2 u_1)/\sqrt{3}.$$

By symmetry we can obtain the other members of the octet which are at the corners of the octet weight diagram. However, such a procedure does not give us the overall signs of the octet functions relative to $\psi_1^{(8)}$. Therefore, we use the lowering operators $\mathbf{e}_{21}$ and $\mathbf{e}_{32}$ and the raising operator $\mathbf{e}_{12}$ to obtain all the remaining octet functions from $\psi_1^{(8)}$. The functions are

$$\begin{aligned}
\psi_1^{(8)} &= (\sqrt{2}\,\psi_1 u_2 - \psi_2 u_1)/\sqrt{3},\\
\psi_2^{(8)} &= (\psi_2 u_2 - \sqrt{2}\,\psi_3 u_1)/\sqrt{3},\\
\psi_3^{(8)} &= (\sqrt{2}\,\psi_1 u_3 - \psi_4 u_1)/\sqrt{3},\\
\psi_4^{(8)} &= (2\psi_2 u_3 - \psi_5 u_1 - \psi_4 u_2)/\sqrt{6},\\
\psi_5^{(8)} &= (\sqrt{2}\,\psi_3 u_3 - \psi_5 u_2)/\sqrt{6},\\
\psi_6^{(8)} &= (\psi_4 u_2 - \psi_5 u_1)/\sqrt{2},\\
\psi_7^{(8)} &= (\psi_4 u_3 - \sqrt{2}\,\psi_6 u_1)/\sqrt{3},\\
\psi_8^{(8)} &= (\psi_5 u_3 - \sqrt{2}\,\psi_6 u_2)/\sqrt{3}.
\end{aligned} \tag{8.89}$$

In obtaining $\psi_6^{(8)}$, we have used the Schmidt orthogonalization procedure.

It is instructive to compare the Clebsch–Gordan coefficients of $SU(2)$ and $SU(3)$. If we do so, we see that some of them are the same in both cases. For example, the coefficients of Eq. (8.63) are the same as those in the first four equations of (8.87). Such coefficients do not need to be recalculated for $SU(3)$ after they have been obtained for $SU(2)$.

In order to take advantage of tables of $SU(2)$ Clebsch–Gordan coefficients, it is useful to introduce the concept of *unitary scalar factors* of $SU(2)$, also called $SU(2)$ *singlet factors* or *isoscalar factors* (Edmonds, 1962). We begin with the Kronecker product tensors $\psi_\mu^{(\alpha)}\psi_\nu^{(\beta)}$, where μ and ν each stands for three numbers: the hypercharge, the isospin, and the third component of the isospin, as these are the quantities which distinguish members of a multiplet. More generally, for $SU(n)$, μ and ν stand for the weight of $n-1$ dimensions and for quantum numbers which distinguish the subgroups of $SU(n)$. The latter may be, for example, the eigenstates of the

Casimir operators of the subgroups, or the dimensionalities of the representations.

Returning to $SU(3)$, we introduce a somewhat less compact notation which has the advantage of making the quantum numbers explicit. We let

$$\mu = yii_3, \qquad \nu = y'i'i_3', \qquad m = YII_3, \qquad j = N, \tag{8.90}$$

where i, i', and I stand for the isospin, y, y', and Y for the hypercharge, i_3, i_3', and I_3 for the third component of the isospin, and N for the dimensionality of the representation (where that is sufficient to specify it), and γ to distinguish between equivalent representations. For example, in reduction of the product $\mathbf{8} \otimes \mathbf{8}$, there arise $\mathbf{8}_s$ and $\mathbf{8}_a$, and γ distinguishes between them. We next form linear combinations of the product tensors $\psi^{(\alpha)}_{yii_3}\psi^{(\beta)}_{y'i'i_3'}$, which correspond to eigenstates of the total isospin by summing over i_3 and i_3'. Denoting these isospin eigenstates by $\chi^{(I)}_{yiy'i'}$, we have

$$\chi^{(I)}_{yiy'i'} = \sum_{i_3 i_3'} (ii'i_3 i_3' | II_3)\psi^{(\alpha)}_{yii_3}\psi^{(\beta)}_{y'i'i_3'}, \tag{8.91}$$

where the coefficients $(ii'i_3 i_3' | II_3)$ are $SU(2)$ Clebsch–Gordan coefficients. We then obtain the eigenstates $\psi^{(N\gamma)}_{YII_3}$ by taking linear combinations of the $\chi^{(I)}_{yiy'i'}$ as follows:

$$\psi^{(N\gamma)}_{YII_3} = \sum_{yiy'i'} (\alpha\beta yiy'i' | NYI\gamma)\chi^{(I)}_{yiy'i'}. \tag{8.92}$$

The coefficients $(\alpha\beta yiy'i' | NIY\gamma)$ are called *isoscalar factors* because they depend on the magnitude of the isospin but not on the value of its third component. These coefficients are more generally called unitary scalar factors, because the $SU(2)$ group does not have to be isospin. If we substitute Eq. (8.91) in Eq. (8.92) we find that an $SU(3)$ Clebsch–Gordan coefficient is given by

$$(\alpha\beta\mu\nu | Nm\gamma) = (\alpha\beta yiy'i' | NYI\gamma)(ii'i_3 i_3' | II_3), \tag{8.93}$$

where $\mu\nu m$ are given by Eq. (8.90). Some tables of isoscalar factors are given by Rabl *et al.* (1975) and by Haacke *et al.* (1976). Earlier tables of isoscalar factors were given by de Swart (1963) with a different convention. It is the de Swart convention which is used in the tables of the Particle Data Group (1976). McNamee and Chilton (1964) give tables of $SU(3)$ Clebsch–Gordan coefficients based on the de Swart convention.

Let us obtain the isoscalar factors which arise in constructing an $\mathbf{8}$ from $\mathbf{3} \otimes \bar{\mathbf{3}}$. We already have obtained the $SU(3)$ Clebsch–Gordan coefficients; they are given in Eqs. (8.82)–(8.84). The isoscalar factors of Eq. (8.82) are all unity because both the $SU(2)$ and $SU(3)$ Clebsch–Gordan coefficients are unity. The function ψ_4 of Eq. (8.83) is constructed from two isospin doublets. Therefore, the coefficient $1/\sqrt{2}$ is an $SU(2)$ coefficient, and the isoscalar factor is unity. This can also be seen in another way. Because the functions ψ_3, ψ_4, and ψ_5 form an isospin triplet, the isoscalar factor must be the same for all of them. But we have already seen from Eq. (8.82) that the relevant isoscalar factor for ψ_3 is unity, and therefore, in constructing ψ_4, we use the same value unity for the isoscalar factor. The function ψ_6 of Eq. (8.84) is an isosinglet. It is composed partly from the function χ_1 defined by

$$\chi_1 = u_3 v_1,$$

which is composed of two isosinglets, and partly from the function φ_1 defined by

$$\varphi_1 = (u_1 v_3 - u_2 v_2)/\sqrt{3},$$

which is made from two isodoublets. Thus, we can write Eq. (8.84) as

$$\psi_6 = (\sqrt{2}\chi_1 - \varphi_1)/\sqrt{3}.$$

The coefficients $\sqrt{2/3}$ and $-\sqrt{1/3}$ are the relevant isoscalar factors. In the notation of Eq. (8.92) we write

$$\begin{aligned}(3\,\bar{3},\, -\tfrac{2}{3}\,0\,\tfrac{2}{3}\,0|8\,0\,0) &= \sqrt{2/3},\\ (3\,\bar{3}\,\tfrac{1}{3}\,\tfrac{1}{2},\, -\tfrac{1}{3}\,\tfrac{1}{2}|8\,0\,0) &= -\sqrt{1/3}\end{aligned} \tag{8.94}$$

where we have omitted the index γ.

Next we consider the $SU(2)$ singlet function ψ of Eq. (8.85). This can be written

$$\psi = (\chi_1 + \sqrt{2}\varphi_1)/\sqrt{3}.$$

Therefore, the relevant isoscalar factors are $\sqrt{1/3}$ and $\sqrt{2/3}$.

As another example, consider the $\psi_i^{(10)}$ of Eq. (8.88). The functions $\psi_i^{(10)}$ ($i = 1, \ldots, 4$), form an isospin quartet. Since the isoscalar factor of $\psi_1^{(10)}$ is obviously 1, all the isoscalar factors of the quartet are unity, and the coefficients are Clebsch–Gordan coefficients of $SU(2)$. The function $\psi_5^{(10)}$ is composed partly of the isospin triplet state $\chi_3 = \psi_1 u_3$, which is made of an isospin triplet and singlet, and partly of $\varphi_3 = \psi_4 u_1$, which is made of two isospin doublets. We have

$$\psi_4^{(10)} = (\chi_3 + \sqrt{2}\varphi_3)/\sqrt{3}.$$

Therefore isoscalar factors of the triplet $\psi_i^{(10)}$ ($i = 4, 5, 6$) are $1/\sqrt{3}$ and $\sqrt{2/3}$. The other isoscalar factors of this $SU(3)$ multiplet are obtained in an analogous manner.

8.6 Clebsch–Gordan Coefficients of $SU(4)$ and Beyond

The procedure we have outlined in the last section generalizes quite readily to $SU(n)$. We take the matrix elements of $\mathbf{e}_{i+1,i}$ to be positive. We shall illustrate with some examples from $SU(4)$.

A particularly simple case is the reduction of the Kronecker product $\mathbf{4} \otimes \mathbf{4} = \mathbf{10} \oplus \mathbf{6}$. We can make use of our earlier work for $SU(3)$. The first 6 functions of the symmetric **10** are given in Eq. (8.74). The remaining ones are easily obtained by symmetry (but can be checked by using the raising and lowering operators). These functions are

$$\begin{aligned}
\begin{array}{|c|c|}\hline 1 & 4 \\ \hline\end{array} &= (u_1u_4 + u_4u_1)/\sqrt{2} = \psi_7, \\
\begin{array}{|c|c|}\hline 2 & 4 \\ \hline\end{array} &= (u_2u_4 + u_4u_2)/\sqrt{2} = \psi_8, \\
\begin{array}{|c|c|}\hline 3 & 4 \\ \hline\end{array} &= (u_3u_4 + u_4u_3)/\sqrt{2} = \psi_9, \\
\begin{array}{|c|c|}\hline 4 & 4 \\ \hline\end{array} &= u_4u_4 = \psi_{10}.
\end{aligned} \tag{8.95}$$

The first 3 functions of the antisymmetric **6** are given in Eq. (8.75). If we operate with the $\mathbf{e}_{i+1,i}$, we find that the others are

$$\begin{aligned}
\begin{array}{|c|}\hline 1 \\ \hline 4 \\ \hline\end{array} &= (u_1u_4 - u_4u_1)/\sqrt{2}, \\
\begin{array}{|c|}\hline 2 \\ \hline 4 \\ \hline\end{array} &= (u_2u_4 - u_4u_2)/\sqrt{2}, \\
\begin{array}{|c|}\hline 3 \\ \hline 4 \\ \hline\end{array} &= (u_3u_4 - u_4u_1)/\sqrt{3}.
\end{aligned} \tag{8.96}$$

We shall not construct other examples.

As n gets larger, tables of $SU(n)$ Clebsch–Gordan coefficients become increasingly unwieldy. We believe it is most convenient not to make use of tables for $n > 3$ but to compute the coefficients as we need them. The next best thing is to make use of unitary scalar factors. Eq. (8.93) can be schematically written

$$C_3 = f_2 C_2, \tag{8.97}$$

where C_3 and C_2 are Clebsch–Gordan coefficients of $SU(3)$ and $SU(2)$, respectively, and f_2 is an $SU(2)$ scalar factor. This equation is readily generalized to $SU(n+1)$ by defining $SU(n)$ singlet factors f_n. Then we can write

$$C_{n+1} = f_n C_n = f_n f_{n-1} \cdots f_2 C_2 . \tag{8.98}$$

Tables of $SU(3)$ and $SU(2)$ scalar factors are given by Rabl *et al.* (1975) and Haacke *et al.* (1976). We emphasize again that the sign convention used in these tables is not the most common one, but is rather the one which generalizes most readily to $SU(n)$.

By now it can be seen that our methods allow us to calculate the Clebsch–Gordan coefficients and unitary scalar factors of any $SU(n)$. The only part of the procedure we have not completely specified is the phase convention. Any phase convention we choose is all right, provided we use it consistently. Many of the sign ambiguities are resolved by the convention that the matrix elements of $\mathbf{e}_{i+1,i}$ be positive. To go beyond this and state a phase convention complete enough to be applicable to all coefficients is quite complicated. We prefer to choose the signs in any convenient way. It should be clear that no physical results can depend upon the choice of an arbitrary phase convention.

8.7 Other Matrix Representations of the Algebra of $SU(n)$

In Section 8.3 we gave a scheme for obtaining the $n^2 - 1$ generators of $SU(n)$ as $n \times n$ matrices. In fact we were able to write down the matrix elements of all the $n^2 - 1$ generators of $SU(n)$. However, this is only one representation, and it is not always the most useful one. The representation of Section 8.3 is useful for obtaining all the $SU(n-1)$ multiplets contained in a multiplet of $SU(n)$. More generally, it is a scheme which is good for the following hierarchy of symmetries:

$$SU(n) \supset SU(n-1) \supset SU(n-2) \supset \cdots \supset SU(2). \tag{8.99}$$

However, it is also useful for the purposes of elementary particles to consider $SU(n)$ broken down with respect to other subgroups. A particular example is the case (letting $n \rightarrow 2n$)

$$SU(2n) \supset SU(n) \times SU(2), \tag{8.100}$$

where the direct-product group $SU(n) \times SU(2)$ is emphasized. In the application to particle physics, $SU(n)$ is taken to be an internal symmetry group, and $SU(2)$ is the group of ordinary spin.

Let us begin with $SU(4)$ as an example. We shall construct the generators of $SU(4)$ from the generators of the $SU(2)$ isospin group, the generators of the $SU(2)$ spin group, and the unit matrices in these spaces. One representation for the isospin generators is by the 2×2 matrices τ_1, τ_2, τ_3. We add to these the 2×2 unit matrix τ_0. The same representation for spin is by the

σ matrices, to which we add σ_0. Now consider the product generators $\varphi_{\mu\nu}$ given by

$$\varphi_{\mu\nu} = \sigma_\mu \tau_\nu, \qquad \mu, \nu = 1, \ldots, 4. \tag{8.101}$$

The 16 matrices $\varphi_{\mu\nu}$ are the generators of $U(4)$. If we eliminate the unit matrix $\sigma_0 \tau_0$, the remaining 15 are the generators of $SU(4)$. We can, of course, obtain another representation by using τ_+, τ_-, σ_+, and σ_- instead of τ_1, τ_2, σ_1, and σ_2.

In a similar fashion we can construct the generators of $SU(6)$, but instead of using the τ_μ we use the λ_μ of Gell-Mann. We define $\beta_{\mu\nu}$ to be

$$\beta_{\mu\nu} = \sigma_\mu \lambda_\nu, \qquad \mu = 0, 1, \ldots, 4, \quad \nu = 0, 1, \ldots, 9. \tag{8.102}$$

Then the 36 $\beta_{\mu\nu}$ are the generators of $U(6)$. Only the unit generator β_{00} has a trace different from zero. This element generates $U(1)$. The remaining 35 $\beta_{\mu\nu}$ are the generators of $SU(6)$. These generators are written in a product notation. However, we can obtain a 6×6 representation of these matrices by taking the outer product of the 3×3 with the 2×2 matrices.

The matrix elements of the *outer product* of two matrices A and B are defined in the following way:

$$(A \times B)_{ij,kl} = A_{ik} B_{jl}. \tag{8.103}$$

We can order these matrix elements in a square array as follows:

$$A \times B = \begin{pmatrix} A_{11}B_{11} & A_{11}B_{12} & \cdots & A_{11}B_{1m} & A_{12}B_{11} & \cdots & A_{1n}B_{1m} \\ A_{11}B_{21} & A_{11}B_{22} & \cdots & A_{11}B_{2m} & & & \vdots \\ \vdots & & & & & & \\ A_{11}B_{m1} & & & & & & \\ A_{21}B_{11} & & & & & & \\ \vdots & & & & & & \vdots \\ A_{n1}B_{m1} & \cdots & & & & \cdots & A_{nn}B_{mm} \end{pmatrix}.$$

In other words, each matrix element of the matrix A is multiplied on the right by the entire matrix B:

$$A \times B = \begin{pmatrix} A_{11}B & A_{12}B & \cdots & A_{1n}B \\ \vdots & & & \vdots \\ A_{n1}B & \cdots & \cdots & A_{nn}B \end{pmatrix},$$

where

$$A_{ij}B = A_{ij}\begin{pmatrix} B_{11} & B_{12} & \cdots & B_{1m} \\ \vdots & & & \vdots \\ B_{m1} & & \cdots & B_{mm} \end{pmatrix}.$$

An alternative definition which leads to a different representation is that each matrix element of B is multiplied on the left by the entire matrix A. Using the first definition, for example, the matrix $\beta_{18} = \sigma_1 \lambda_8$ written as an outer product is given by

$$\beta_{18} = \sqrt{\frac{1}{3}} \left(\begin{array}{ccc|ccc} & 0 & & 1 & 0 & 0 \\ & & & 0 & 1 & 0 \\ & & & 0 & 0 & -2 \\ \hline 1 & 0 & 0 & & & \\ 0 & 1 & 0 & & 0 & \\ 0 & 0 & -2 & & & \end{array} \right).$$

There are many other useful representations for the generators of $SU(6)$. For example, instead of using the nondiagonal λ_μ and σ_ν, we can use the raising and lowering operators of $SU(3)$ and $SU(2)$. Then the 30 nondiagonal $SU(6)$ generators are just the e_{ij} of $SU(6)$ with matrix elements

$$(e_{ij})_{\alpha\beta} = \delta_{i\alpha}\, \delta_{j\beta}, \tag{8.104}$$

which satisfy

$$e_{ij}^\dagger = e_{ji}. \tag{8.105}$$

The five diagonal generators in this representation are the same as in the representation by the $\beta_{\mu\nu}$ matrices. This representation of the generators is most useful for obtaining the Clebsch–Gordan coefficients of $SU(6)$. The six states of $SU(6)$ on which the generators act have the following interpretation: The states $u_1\ u_2\ u_3$ of the first fundamental representation are the states with the same weights as the three states of $SU(3)$ and also with $J_3 = \frac{1}{2}$, while the states $u_4\ u_5\ u_6$ correspond to the $SU(3)$ states with $J_3 = -\frac{1}{2}$.

The preceding remarks about $SU(6)$ can be generalized to $SU(8) \supset SU(4) \times SU(2)$ by making use of the $SU(4)$ generators of Eq. (6.27) and λ_0 of Eq. (6.28), instead of the corresponding $U(3)$ generators.

We illustrate how to obtain Clebsch–Gordan coefficients with an example from $SU(4)$. There are 10 state functions belonging to the symmetric representation $\boxed{}\boxed{}$. These functions are

$$\begin{aligned} \boxed{i}\boxed{i} &= u_i u_i, \qquad i = 1, \ldots, 4 \\ \boxed{i}\boxed{j} &= (u_i u_j + u_j u_i)/\sqrt{2}, \qquad i \neq j. \end{aligned} \tag{8.106}$$

We now wish to emphasize the $SU(2) \times SU(2)$ content of these functions. We therefore write

$$u_1 = x_1 y_1, \qquad u_2 = x_2 y_1, \qquad u_3 = x_1 y_2, \qquad u_4 = x_2 y_2, \tag{8.107}$$

where x_1 and x_2 are new symbols for the $SU(2)$ internal symmetry group, and y_1 and y_2 denote spin up ($J_3 = \frac{1}{2}$) and down ($J_3 = -\frac{1}{2}$), respectively. (Here we are writing the isospin function first; the choice is arbitrary.) Then, for example, the function $\begin{array}{|c|c|}\hline 1 & 4 \\ \hline\end{array}$ becomes

$$\begin{array}{|c|c|}\hline 1 & 4 \\ \hline\end{array} = (x_1 y_1 x_2 y_2 + x_2 y_2 x_1 y_1)/\sqrt{2}.$$

This expression can be written

$$\begin{array}{|c|c|}\hline 1 & 4 \\ \hline\end{array} = \tfrac{1}{2}[(x_1 x_2 + x_2 x_1)(y_1 y_2 + y_2 y_1) + (x_1 x_2 - x_2 x_1)(y_1 y_2 - y_2 y_1)]/\sqrt{2}$$

where we have added and subtracted some terms. The interpretation of $\begin{array}{|c|c|}\hline 1 & 4 \\ \hline\end{array}$ is that it is a linear combination of two particles, the first having spin and isospin 1 and the second having spin and isospin 0. Likewise, the state $\begin{array}{|c|c|}\hline 2 & 3 \\ \hline\end{array}$ is another such linear combination, which is readily found to be

$$\begin{array}{|c|c|}\hline 2 & 3 \\ \hline\end{array} = \tfrac{1}{2}[(x_1 x_2 + x_2 x_1)(y_1 y_2 + y_2 y_1) - (x_1 x_2 - x_2 x_1)(y_1 y_2 - y_2 y_1)]/\sqrt{2}.$$

The states ψ_{IJ} that correspond to particles of definite spin and isospin are then given by

$$\begin{aligned}
\psi_{11} &= \left(\begin{array}{|c|c|}\hline 1 & 4 \\ \hline\end{array} + \begin{array}{|c|c|}\hline 2 & 3 \\ \hline\end{array}\right)/\sqrt{2}, \\
\psi_{00} &= \left(\begin{array}{|c|c|}\hline 1 & 4 \\ \hline\end{array} - \begin{array}{|c|c|}\hline 2 & 3 \\ \hline\end{array}\right)/\sqrt{2}.
\end{aligned} \tag{8.108}$$

Tables of unitary scalar factors appropriate to the reduction of $SU(6)$ to $SU(3) \times SU(2)$ have been given by Cook and Murtaza (1965). Papers making use of $SU(8)$ Clebsch–Gordan coefficients appropriate to the subgroup $SU(4) \times SU(2)$ have appeared in the literature, but we have not seen any extensive tables as of this writing. In any case, we believe it is just as easy to work these out when needed as to consult complicated tables.

8.8 Wigner–Eckart Theorem

In Section 6.9 we discussed irreducible tensor operators $T_m^{(j)}$ for a simple compact Lie group. Let us consider the matrix element of $T_m^{(j)}$ between two states $\psi_\mu^{(\alpha)}$ and $\psi_\nu^{(\beta)}$ which are basis vectors of two irreducible representations of the group. The Wigner–Eckart theorem states that such a matrix element $(\psi_\mu^{(\alpha)}, T_m^{(j)}\psi_\nu^{(\beta)})$ can be written

$$(\psi_\mu^{(\alpha)}, T_m^{(j)}\, \psi_\nu^{(\beta)}) = \sum_\gamma (\alpha\beta\mu\nu \,|\, jm\gamma)(\alpha \,\|\, T_m^{(j)} \,\|\, \beta)_\gamma, \tag{8.109}$$

where $(\alpha\beta\mu\nu \,|\, jm\gamma)$ is a Clebsch–Gordan coefficient of the group and the $(\alpha \,\|\, T_m^{(j)} \,\|\, \beta)_\gamma$ are independent of the quantum numbers μ and ν which distinguish different basis vectors of the representations α and β. The quantities $(\alpha \| T_m^{(j)} \| \beta)_\gamma$ are known as *reduced matrix elements*. In Eq. (8.109), the sum over γ includes as many terms as the irreducible representation j is contained in the decomposition of the kronecker product $\alpha \otimes \beta$. Since $SU(2)$ is simply reducible, the right-hand side of Eq. (8.109) consists of only a single term in this case.

The proof of the theorem for the rotation group is given in Wigner (1959, p. 244). The method given by Wigner can be generalized in a straightforward manner to a simple compact Lie group, although if the group is not simply reducible there is a certain arbitrariness in how the γ are chosen and how the states are labeled. The values of the Clebsch–Gordan coefficients depend of course on this arbitrary choice. The proof of the Wigner–Eckart theorem for $SU(3)$ is sketched by de Swart (1963).

Let us consider the matrix elements of a tensor between two states belonging to the same irreducible representation of a group. If the tensor T_μ is of rank one, or an irreducible *vector*, then the matrix element is proportional to the matrix element of the irreducible vector X_μ whose components are the generators of the group. For $SU(2)$, we have

$$(\psi_{m'}^{(j)}, T_\mu \psi_m^{(j)}) = C(j)(\psi_{m'}^{(j)}, J_\mu \psi_m^{(j)}), \tag{8.110}$$

where $C(j)$ is a constant of proportionality and we have put $X_\mu = J_\mu$, the components of the angular momentum. For $SU(3)$, on the other hand, there are two independent irreducible vectors, the generators F_μ, and the symmetric combination

$$D_\mu = \tfrac{2}{3} \sum_{\alpha\beta} d_{\mu\alpha\beta} F_\alpha F_\beta .$$

Then the matrix element of T_μ is given by

$$(\psi_{m'}^{(j)}, T_\mu \psi_m^{(j)}) = C_F(j)(\psi_{m'}^{(j)}, F_\mu \psi_m^{(j)}) + C_D(j)(\psi_{m'}^{(j)}, D_\mu \psi_m^{(j)}). \tag{8.111}$$

The index γ has been replaced by the symbols F and D in this case. The fact that there are two matrix elements on the right hand side of Eq. (8.111) is related to the fact that in reducing the Kronecker product of any representation $D^{(\alpha)}$ with an **8**, the representation $D^{(\alpha)}$ appears twice in the direct sum. An exception is that if $D^{(\alpha)}$ has a rectangular Young diagram, then $D^{(\alpha)}$ appears only once in the direct sum. The ratio of C_F to C_D is called the F/D ratio. By the Wigner–Eckart theorem, equations similar to (8.110) and (8.111) can be written for the reduced matrix elements.

CHAPTER 9

THE EIGHTFOLD WAY

9.1 *SU*(3) and Hadrons

Many of the hadrons (strongly interacting particles) which have been observed so far can be conveniently classified as members of $SU(3)$ multiplets. The unclassified hadrons are also believed to belong to such multiplets, but not all members have been seen experimentally. Since additional hadrons continue to be discovered, it seems reasonable to assume that many of the known states that are now unclassified will later be found to fit in the $SU(3)$ scheme.

As we have seen in previous chapters, the group $SU(3)$ is characterized by two additive quantum numbers, which, for hadrons, can be identified with the third component of isospin I_3 and the hypercharge Y. In 1976, some new hadrons were discovered (Goldhaber *et al.*, 1976; Knapp *et al.*, 1976), which require a third additive quantum number, called charm C. In consequence, the symmetry group of hadrons has to be enlarged, and our best guess for this enlarged group is $SU(4)$. Because at present we have no theoretical principle which favors any particular internal symmetry group for the hadrons, we shall not be surprised if one day other hadrons are discovered which require more additive quantum numbers. It is primarily for this reason that we have stressed methods for any $SU(n)$, although, of course, there is no guarantee that $SU(n)$ with large n will be a relevant group.

If we look at members of an isospin or $SU(2)$ multiplet, we find that all members have very nearly the same mass, the masses of members with different values of I_3 differing by less than 10 MeV. If we look at members of $SU(3)$ multiplets, we find that the masses of states with different values of Y differ from one another by as much as a few hundred MeV. The fact that not all members of a multiplet have the same mass shows that the symmetry is only approximate. The group $SU(4)$ is even more badly broken than $SU(3)$, with masses of members of a multiplet with different values of C differing by around 1 GeV.

Because of this large $SU(4)$ symmetry breaking, we can confine ourselves to $SU(3)$ when looking at hadrons of relatively low mass. It is these low-mass $SU(3)$ multiplets which we treat in this chapter.

A characteristic of the $SU(3)$ multiplets with $C = 0$ is that they all have zero triality. The lowest-dimensional representation of $SU(3)$ with zero triality (except for the trivial one-dimensional representation) has eight dimensions. In fact, all zero-triality representations of $SU(3)$ can be obtained from the reduction of Kronecker products of eight-dimensional representations. Thus, the **8**, rather than the **3**, is basic to this approach, as first postulated by Gell-Mann (1961) and Ne'eman (1961). This idea led Gell-Mann to name the symmetry scheme "the eightfold way." The book by Gell-Mann and Ne'eman (1964) contains reprints of the original papers on the eightfold way, as well as reprints of a number of interesting papers giving applications.

As we have previously pointed out, the Clebsch–Gordan series for the Kronecker product of two **8**'s is

$$\mathbf{8} \otimes \mathbf{8} = \mathbf{1} \oplus \mathbf{8} \oplus \mathbf{8} \oplus \mathbf{10} \oplus \overline{\mathbf{10}} \oplus \mathbf{27}. \tag{9.1}$$

We can then take the Kronecker product of an **8** with a **10** to obtain

$$\mathbf{8} \otimes \mathbf{10} = \mathbf{8} \oplus \mathbf{10} \oplus \mathbf{27} \oplus \mathbf{35}. \tag{9.2}$$

In similar fashion, we can build up all the zero-triality representations.

If we restrict ourselves to hadronic states with baryon number $0 \leq B \leq 1$, and $C = 0$, then nearly all states observed so far belong to multiplets of 1, 8, or 10 states, that is, to singlets, octets, or decuplets.[1] In the case of the

[1] There is perhaps an inconsistent nomenclature for the multiplets. Multiplets of 3, 4, ..., 8, members should be designated "triplets, quadruplets, ..., octuplets" if the similarity of the members is emphasized rather than "trios, quartets, ..., octets." However, in the old days of atomic spectroscopy, after naming multiplets of three states "triplets," physicists named multiplets of 4, 5, "quartets, quintets." Gell-Mann continued this spectroscopic designation by naming the 8 an "octet" and the 10 a "decimet." However, others (especially Glashow and Rosenfeld, 1963) have tried to switch to the other notation. The term "octet" has stuck, even though there are a few "octuplets" in the literature. The terms "decimet" and "decuplet" are both used, but we shall use "decuplet."

mesons, because $SU(3)$ is a broken symmetry, the singlets and octets are mixed into nonets. Also, heavy mesons were discovered (Aubert *et al.*, 1974; Augustin *et al.*, 1974) which enlarge the number of mesons in some multiplets from 9 to 10. There is no place in the $SU(3)$ scheme for such mesons, and we shall not consider them until we discuss $SU(4)$ in the next chapter.

Even if we omit, for the present, states which apparently require an enlarged group, there is still some inconclusive evidence for the existence of hadrons not belonging to singlets, octets, or decuplets—the so-called *exotic* states. If exotic states exist, they are evidently coupled relatively weakly to the more commonly observed hadrons. The reason that only singlets, octets, and decuplets with $|B| \leq 1$ appear as strongly-coupled states is a question that cannot be answered on the basis of the octet model alone. Not all representations of a symmetry group are required to be realized by states in nature.

Suppose we consider an $SU(3)$ multiplet which is not a singlet. In general, the state vectors of these particles have symmetry properties under transformations of coordinates other than their $SU(3)$ indices. For example, the individual states of an $SU(3)$ multiplet may be subjected to a gauge transformation corresponding to the baryon number, to an inhomogeneous Lorentz transformation, or a parity transformation. If we neglect all symmetry breaking, we may consider as the overall symmetry group of the multiplet the direct-product group of the individual symmetry groups. Then $SU(3)$ is an invariant subgroup of this larger direct-product group.

Now transformations within the $SU(3)$ subgroup do not change any of the quantum numbers of the larger group which are not part of the $SU(3)$ subgroup. These quantum numbers then remain invariant operators for transformations within the $SU(3)$ subgroup. We conclude that all members of an $SU(3)$ multiplet should have the same values of all quantum numbers outside the subgroup.

This picture runs into difficulty, however. The square of the mass of a particle is an invariant of the Poincaré group. Therefore, when we make an $SU(3)$ transformation, we should not change the mass of a particle. But we know that the states of an $SU(3)$ multiplet are not degenerate in mass. Thus, a symmetry group which is a direct product of $SU(3)$ and the Poincaré group is not an exact symmetry of the Hamiltonian unless the symmetry is spontaneously broken. We shall exclude this latter possibility for the remainder of this section.

This difficulty persists at the $SU(2)$ level, since isospin is also an approximate symmetry. It is instructive to compare the case of isospin with ordinary spin. The ordinary spin group is a subgroup of the Poincaré group, whereas the isospin group is not. If we consider the two spin states of an electron in an external magnetic field, these states lose their degeneracy. But the magnetic field breaks the Poincaré group (rotational invariance is lost), so that we do not expect mass degeneracy. On the other hand, the electromagnetic interaction which is always present breaks the mass degeneracy between proton and neutron without breaking the Poincaré group.

We have seen that the Hamiltonian describing the elementary particles cannot be invariant under the direct-product group $SU(3) \times P$, where P is the Poincaré group. Under such circumstances we may ask whether there exists a group G, containing both $SU(3)$ and P but not as a direct product, such that the masses of the members of an $SU(3)$ multiplet may differ from one another. However, If G is a finite Lie group, this possibility is ruled out, as has been discussed by many authors under various assumptions. McGlinn (1964) has given one of the early proofs that no such G exists. O'Raifeartaigh (1968) has given a proof with quite weak assumptions that an internal symmetry group and the Poincaré cannot be embedded in a finite Lie group in such a way as to break the mass degeneracy of a multiplet.

We shall assume that the Hamiltonian which describes the hadrons consists of two parts, one of which H_0 is invariant under $SU(3)$ and another H' which is not. Then we can regard eigenstates of H_0 as multiplets of the direct product of $SU(3)$ and the Poincaré group. The symmetry-breaking term H' must be of such a form as to break the mass degeneracy, but to leave other quantum numbers outside $SU(3)$, such as baryon number, spin, and parity, invariant for the entire multiplet.

9.2 Baryon Multiplets

Any state with baryon number $B = 1$ we call a baryon. Baryons have half-integral spin, and all except the proton are unstable. If a baryon has hypercharge $Y \neq 1$, it is called a hyperon. Instead of the hypercharge, the quantum number strangeness S is often used. For a hadron with baryon number B, hypercharge Y, and charm $C = 0$, the strangeness is defined as

$$S = Y - B. \tag{9.3}$$

Any hadron with $S \neq 0$ is sometimes called a strange particle.

The electric charge Q of a hadron in units of the proton charge is given by (again for $C = 0$)

$$Q = I_3 + \tfrac{1}{2}Y \qquad \text{or} \qquad Q = I_3 + \tfrac{1}{2}(B + S).$$

For the first fundamental triplet, the charge Q, as well as I_3 and Y, are given as multiples of the diagonal matrices λ_3 and λ_8 of Eq. (6.12). We have

$$I_3 = \tfrac{1}{2}\lambda_3, \qquad Y = \sqrt{\tfrac{1}{3}}\,\lambda_8, \qquad Q = \tfrac{1}{2}\lambda_3 + \tfrac{1}{2}\sqrt{\tfrac{1}{3}}\,\lambda_8.$$

However, B and S cannot be written solely in terms of λ_3 and λ_8, but require the use of λ_0 of Eq. (6.13). The expressions are

$$B = \tfrac{1}{3}\lambda_0, \qquad S = \sqrt{\tfrac{1}{3}}\,\lambda_8 - \tfrac{1}{3}\lambda_0 .$$

Thus, because of the existence of baryon number, the relevant group of hadrons is $U(3)$ rather than $SU(3)$. We continue for the most part to use $SU(3)$, however, because the multiplicities are the same as for $U(3)$.

We denote a baryon by a symbol which depends on the value of its isospin I and hypercharge Y according to the scheme of Table 9.1. The electric charge of a baryon is denoted by a superscript except in the case of the nucleon N, where we use p (for proton) instead of N^+, and n (for neutron) instead of N^0. We sometimes give the spin and parity J^P in parentheses and sometimes also the mass in MeV. (We let $c = 1$.) Thus the symbol $N(939, \tfrac{1}{2}^+)$ means the nucleon has $I = \tfrac{1}{2}$, $Y = 1$, $J^P = \tfrac{1}{2}^+$, and a mass (averaged over the different charge states) of 939 MeV. The symbol unmodified denotes the state of lowest mass with the values of I and Y specified by Table 9.1. If we do not give the mass of an excited state, we put an asterisk on the symbol. For example, the symbol Σ means $\Sigma(1193)$; Σ^* means any higher-mass baryon with $I = 1$, $Y = 0$. Baryons with $Y = 0$ are usually given the symbol Y^* in the literature, but we shall not use this notation.

A bar on the symbol for a particle denotes the antiparticle. If a particle has quantum numbers J, B, I, I_3, Y, then the antiparticle has quantum numbers $J, -B, I, -I_3, -Y$. A baryon and its antiparticle have opposite parity, while a meson and its antiparticle have the same parity. A further discussion of the notation used in the classification of hadrons is given by

TABLE 9.1

NOMENCLATURE FOR BARYONS[a]

Symbol	Isospin I	Hypercharge Y
N	$\tfrac{1}{2}$	1
Λ	0	0
Σ	1	0
Ξ	$\tfrac{1}{2}$	-1
Δ	$\tfrac{3}{2}$	1
Ω	0	-2

[a] An asterisk on the symbol for a particle denotes an excited state. If the mass of the state is given in parentheses following the symbol, the asterisk is omitted.

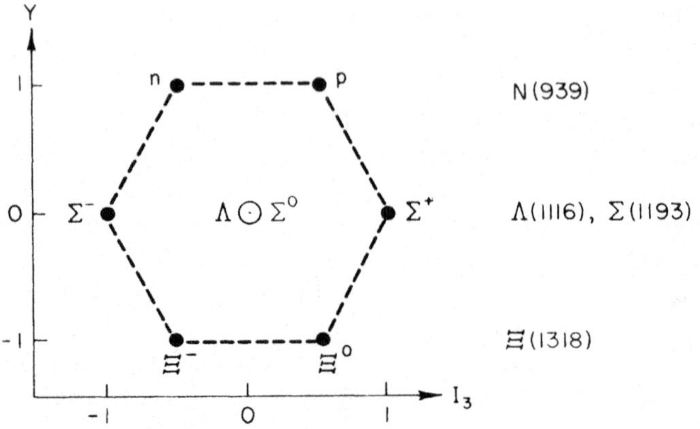

Fig. 9.1. Weight diagram of the baryon octet with spin and parity $J^P = \frac{1}{2}^+$. The average mass in MeV of each isospin multiplet is also given.

Chew *et al.* (1964). A discussion of the quantum numbers of the hadrons and how they are determined is given by Lichtenberg (1965).

The baryon of lowest mass is the proton p. This particle is a member of the nucleon isospin doublet N, which also includes the neutron n. The N belongs in a baryon octet together with another isospin doublet (the Ξ hyperon), an isospin singlet (the Λ hyperon), and an isospin triplet (the Σ hyperon). These states are shown on a weight diagram in Fig. 9.1, with their masses in MeV in parentheses, neglecting the electromagnetic mass splitting within isospin multiplets. The hypercharge is plotted on the vertical scale in this section, but to preserve the 120° symmetry, the vertical scale is contracted by a factor $\frac{1}{2}\sqrt{3}$. The masses of all members of the octet are given in Table

TABLE 9.2

SOME PROPERTIES OF THE BARYON OCTET

Symbol	Mass (MeV)	Mean life (sec)	Principal decay modes
p	938.28	Stable	—
n	939.57	918 ± 14	$pe^-\bar{\nu}$
Λ	1115.6 ± 0.1	$(2.58 \pm 0.02) \times 10^{-10}$	$p\pi^-$ 64% $n\pi^0$ 36%
Σ^+	1189.4 ± 0.1	$(0.80 \pm 0.01) \times 10^{-10}$	$p\pi^0$ 52% $n\pi^+$ 48%
Σ^0	1192.5 ± 0.1	$< 10^{-14}$	$\Lambda\gamma$
Σ^-	1197.3 ± 0.1	$(1.48 \pm 0.03) \times 10^{-10}$	$n\pi^-$
Ξ^0	1314.9 ± 0.7	$(3.0 \pm 0.1) \times 10^{-10}$	$\Lambda\pi^0$
Ξ^-	1321.3 ± 0.2	$(1.65 \pm 0.02) \times 10^{-10}$	$\Lambda\pi^-$

9.2. Also given are their lifetimes (mean lives), and principal decay models. These values are adapted from the compilation of the Particle Data Group (1976).

According to our assumptions of the previous section, an $SU(3)$ multiplet should have the same values of all quantum numbers which refer to other symmetry groups, except for the mass. Let us see how this agrees with experiment for the baryon octet. All members of this octet have $B = 1$, as they should have. Likewise each baryon has spin measured to be $J = \frac{1}{2}$. The parity of the Ξ has not been definitively measured, but the others all have positive parity. (The parity of the p, n, and Λ can be defined to be positive, but then the measurement of the parity of the Σ is a confirmation of the assignment $P = +$. See, for example, Lichtenberg, 1965.)

There is another octet consisting of the antiparticles of the baryon octet. In general, for every $SU(3)$ multiplet of baryons, there will exist an $SU(3)$ multiplet of antibaryons belonging to the conjugate weight diagram. In the case of an octet, the $SU(3)$ weight diagram is self-conjugate. However, the octet of antibaryons is distinguished from the octet of baryons by the opposite baryon number and opposite intrinsic parity.

Another baryon multiplet which has been well established experimentally is the baryon decuplet, consisting of an isospin quartet Δ, triplet Σ^* (or Y^*), doublet Ξ^*, and singlet Ω. The members of the decuplet are shown on the weight diagram of Fig. 9.2. In Table 9.3 are given the masses of the members of the decuplet. Also given are the full widths of the states at half maximum

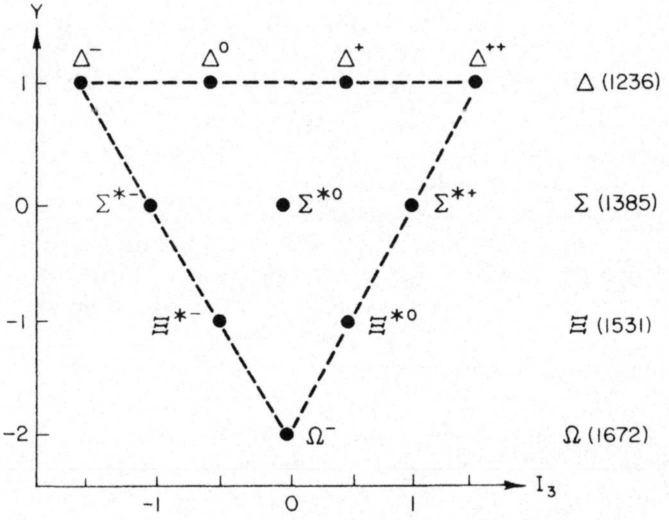

Fig. 9.2. Weight diagram of the baryon decuplet with $J^P = \frac{3}{2}^+$. The Σ^* is often called Y^*.

TABLE 9.3

SOME PROPERTIES OF THE BARYON DECUPLET

Symbol	Mass (MeV)	Full width (MeV) or mean life (sec)	Principal decay modes
Δ^{++}	1231 ± 1		
Δ^{+}	1231 ± 1	120 ± 10 MeV	$N\pi$
Δ^{0}	1232 ± 2		
Δ^{-}	1239 ± 8		
Σ^{*+}	1382 ± 1	35 ± 2 MeV	
Σ^{*0}	1381 ± 2	39 ± 6 MeV	$\Lambda\pi$ $(88 \pm 2)\%$ $\Sigma\pi$ $(12 \pm 2)\%$
Σ^{*-}	1386 ± 2	42 ± 4 MeV	
Ξ^{*0}	1532 ± 1	9 ± 1 MeV	$\Xi\pi$
Ξ^{*-}	1535 ± 1	10 ± 2 MeV	
Ω^{-}	1672 ± 1	$(1.3 \pm 0.3) \times 10^{-10}$ sec	$\Xi^{0}\pi^{-}$ $\Xi^{-}\pi^{0}$ ΛK^{-}

(except in the case of the Ω where the lifetime is given) and the principal decay modes.

If all these particles are correctly classified as members of the same $SU(3)$ multiplet, they should all have the same baryon number, spin, and parity (but not the same mass). All of them have $B = 1$. The spins and parities of the Δ, Σ^*, and Ξ^* have been measured and found to be $J^P = \frac{3}{2}^+$. The spin and parity of the Ω has not been measured, and it can be regarded as a prediction of the classification scheme that the Ω also has $J^P = \frac{3}{2}^+$. An interesting property of the baryon decuplet is that the Δ, Σ^*, and Ξ^* decay strongly, while the Ω decays only weakly. The Ω lives about 10^{13} times as long as a Δ. Thus, stability against decay by strong interactions is not a good criterion to use in classification of the hadrons.

The octet and decuplet are the two best established baryon $SU(3)$ multiplets. There also appear to be $SU(3)$ singlets among the baryons. The weight diagram corresponding to an $SU(3)$ singlet consists merely of a point with $Y = I_3 = 0$. Many of the baryon multiplets are experimentally incomplete. This is because it is easiest experimentally to establish evidence for N^* and Δ^* baryons, somewhat more difficult to produce Λ^* and Σ^* states, still harder to produce Ξ^* states, and extremely difficult to produce Ω^* baryons. The reason for this is, of course, the symmetry breaking. Because of the symmetry-breaking interaction, protons are the only stable target particles with $B = 1$ and pions are the most readily available projectiles with $B = 0$. Thus, states with hypercharge $Y = 1$ are most accessible to experiment. Since K^- mesons are also copiously produced in accelerators, states with $Y = 0$ are also fairly accessible to experiment. A further discussion of experimental problems in obtaining evidence for multiplets is beyond the scope of this book. In Table 9.4 we list some possible baryon multiplets, some of which are incomplete, adapted from the Particle Data Group (1976). The assignments to $SU(3)$ multiplets are only tentative. Presumably, when an N^* state is discovered it belongs to an octet and when a Δ^* is discovered it belongs to a decuplet. A Λ^* can belong to a singlet or octet, while a Σ^* and Ξ^* can belong to an octet or decuplet. Many baryon states have been discovered which are not listed in Table 9.4. For further information, see, for example, the compilation of the Particle Data Group (1976).

TABLE 9.4

TENTATIVE ASSIGNMENTS OF BARYON STATES TO $SU(3)$ SINGLETS, OCTETS, AND DECUPLETS[a]

$SU(3)$ Multiplicity	Spin and parity	Symbols and masses
1	$\frac{1}{2}^-$	$\Lambda(1405)$
8	$\frac{1}{2}^-$	$N(1535)$, $\Lambda(1670)$, $\Sigma(1750)$, ?
1	$\frac{3}{2}^-$	$\Lambda(1520)$
8	$\frac{3}{2}^-$	$N(1520)$, $\Lambda(1690)$, $\Sigma(1670)$, ?
8	$\frac{5}{2}^+$	$N(1690)$, $\Lambda(1815)$, $\Sigma(1915)$, $\Xi(2030)$
10	$\frac{7}{2}^+$	$\Delta(1950)$, $\Sigma(2030)$, ?, ?

[a] Some of these multiplets are experimentally incomplete. Also, not all the quantum numbers have been verified. A question mark means that a state has not been seen.

We had remarked in Section 9.1 that all prominently observed hadron multiplets with $|B| \leq 1$ appear to belong to multiplets of 1, 8, or 10 dimensions. However, multiplets of dimensionality greater than 10 may exist for baryon number $B > 1$. An argument why this should be so is as follows: The **1**, **8**, and **10** representations are restricted to have $Y \leq 1$, as can be seen from the weight diagrams of these multiplets. In general, the larger the highest value of a weight in a multiplet (i.e., the highest value of Y), the larger is the dimensionality of the multiplet. Since a nucleon has $Y = 1$, a nucleus consisting of ν nucleons has $Y = \nu$. Therefore, if nuclei such as uranium with $Y = 238$ are members of $SU(3)$ multiplets, the multiplicity will be very large indeed. Unfortunately, at the present time it is not experimentally feasible to test whether heavy nuclei do in fact belong to $SU(3)$ multiplets.

Oakes (1963) has looked at the predictions of $SU(3)$ for baryon number $B = 2$. If $SU(3)$ is a good symmetry, it is plausible that a low-mass bound state of two baryons should belong in a multiplet contained in the Kronecker product of $\mathbf{8} \otimes \mathbf{8}$. (An alternative possibility is that such a state belongs to a multiplet contained in the Kronecker product of **8** × **10**.) Consider the deuteron as an example. This two-baryon state has $I = 0$, $Y = 2$. Such a state is contained in **8** ⊗ **8**, but not in **8** ⊗ **10**. In Eq. (9.1) we have given the Clebsch–Gordan series for the decomposition of **8** ⊗ **8**. Using the methods of Chapter 6 or 7, it is easy to obtain the (IY) content of the multiplets contained in **8** ⊗ **8**. We find that a state with $(I, Y) = (0, 2)$ is contained only of the $\overline{\mathbf{10}}$ of the Clebsch–Gordan series. Thus, if a deuteron belongs to an $SU(3)$ multiplet, it must belong to a $\overline{\mathbf{10}}$. Because of symmetry-breaking, the other members of the $\overline{\mathbf{10}}$, if they exist, are expected to be unstable. There is some inconclusive experimental evidence in favor of a $p - \Lambda$ resonance, but that is all.

Symmetry-breaking effects are expected to be especially important in states with more than one baryon. This is because the binding energy ~8 MeV per nucleon is small compared to typical symmetry-breaking energies of more than 100 MeV. The question of whether $SU(3)$ is useful for systems with $B \geq 2$ has not yet been answered.

9.3 Meson Multiplets

A hadron with $B = 0$ is called a meson. All mesons have integral spin and are unstable. The meson multiplets of $SU(3)$ differ from the baryon multiplets in several important ways. In the first place, thus far all prominently observed mesons belong only to one of two multiplets, either a singlet or an octet.

A second difference between baryons and mesons is that whereas, for any multiplet of baryons there is a distinct (conjugate) multiplet of antibaryons, for those mesons observed thus far, particles and antiparticles are contained in the same $SU(3)$ multiplet. We see how this can come about as follows. Both the octet and singlet are self-conjugate multiplets, and therefore for every state in a multiplet with a given value of I_3 and Y, there is also a state with the values of these quantum numbers given by $-I_3$, $-Y$. (If $I_3 = Y = 0$, there is only one state.)

Furthermore, for mesons, unlike baryons, the particle and antiparticle have the same values of all quantum numbers outside the group, namely spin, parity, and baryon number. (The baryon number, in general, is opposite for particle and antiparticle, but mesons have $B = 0$.) Therefore, for every state of a multiplet, there exists a state in the same multiplet (either the same state or another one) with just the quantum numbers of the antiparticle. That this state is, in fact, the antiparticle is an experimental observation.

A third difference between mesons and baryons concerns the manner of $SU(3)$ breaking, although this difference may be quantitative rather than qualitative. Suppose there exist two $SU(3)$ multiplets with the same values of spin, parity, and baryon number. Further, suppose that one of the members of the first $SU(3)$ multiplet has the same values of I, I_3, and Y as a member of the second $SU(3)$ multiplet. Then the $SU(3)$ symmetry-breaking interaction can "mix" the two multiplets. For example, consider an $SU(3)$ singlet, which has $I = Y = 0$. Also one member of an $SU(3)$ octet also has $I = Y = 0$. Then it is possible that for the symmetry-breaking interaction to cause the two *physical* states to be linear combinations of the $SU(3)$ singlet and $SU(3)$ octet. We call this $SU(3)$ mixing.

It appears that $SU(3)$ mixing occurs to a greater degree in the meson states than in the baryon octet and decuplet of lowest mass, as we shall see in Section 9.6. Furthermore, every well-established $SU(3)$ meson octet appears to be accompanied by an $SU(3)$ singlet of the same spin and parity. Since this singlet appears to mix with the octet, it is sometimes more convenient to classify all members of singlet and octet together as a nonet. Such a nonet also belongs to a *reducible* representation of $SU(3)$. However, a nonet does belong to an irreducible representation of a larger group, the group $U(3) \times U(3)$.

The nomenclature for the mesons is much more disorganized than that for baryons, as a large number of the states are given individual symbols. A scheme to label the mesons by the values of I and Y has been proposed, but has not become popular. This scheme is given in Table 9.5. In much of the literature, however, the symbols π, η, and η' are reserved for mesons with $J^P = 0^-$. We shall use the symbols of Table 9.5, but shall also give on occasion the more common symbols for mesons. Again the unmodified

TABLE 9.5

NOMENCLATURE FOR THE MESONS[a]

Symbol	I	Y	$SU(3)$ multiplicity
π	1	0	8
K	$\frac{1}{2}$	1	8
η	0	0	Mixed 8 and 1, mostly 8
η'	0	0	Mixed 8 and 1, mostly 1

[a] An asterisk on the symbol for a particle denotes an excited state. If the mass is given in parentheses following the symbol, the asterisk is omitted. A subscript "n" is sometimes used on a symbol to denote normal (or natural) parity $[P = (-1)^J]$, and a subscript "a" to denote abnormal (or unnatural) parity $[P = -(-1)^J]$.

symbol means the state of lowest mass with the values of I and Y, while an asterisk indicates an excited state. The spin, parity, and mass are sometimes given in parentheses; if the mass is given, the asterisk is omitted. The neutral states with $Y = 0$ are eigenstates of the charge conjugation operator. Thus, these states have a definite C parity,[2] which we also give. For example, the π^0 has $J^{PC} = 0^{-+}$. Thus, we denote a pion by $\pi(138, 0^{-+})$.

Sometimes, instead of the C parity, G parity is used to characterize a meson. For mesons with $Y = 0$, G is given by $G = C(-1)^I$. Although C is a good quantum number for the neutral member of a $Y = 0$ meson multiplet, G is a good quantum number for all members of such a multiplet. For example, a neutral pion has $J^{PC} = 0^{-+}$, but all pions have $J^{PG} = 0^{--}$. For this reason G is sometimes a more convenient quantum number than C, especially since G appears to be conserved in strong interactions. However, we shall use only C, which appears to be conserved in both strong and electromagnetic interactions.

The meson multiplet of lowest average mass and the first to be known experimentally is the pseudoscalar[3] meson octet, with spin and parity $J^P = 0^-$. The weight diagram of this octet is shown in Fig. 9.3. The π and η^0 mesons are in the center of the weight diagram and are their own antiparticles. Since for these states, particle and antiparticle are indistinguishable and therefore

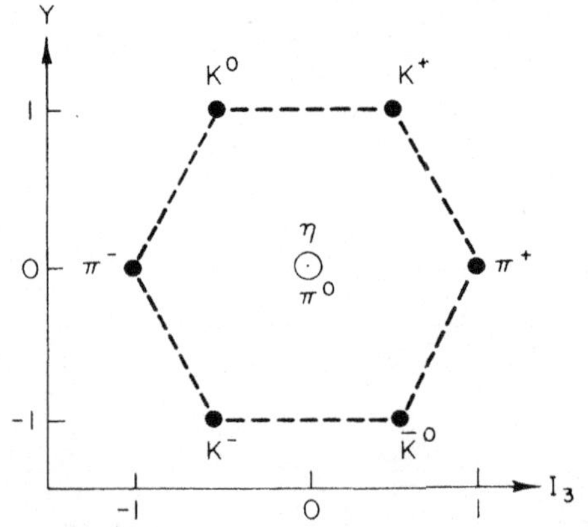

Fig. 9.3. Weight diagram of pseudoscalar meson octet. The average mass in MeV of each isospin multiplet is also given.

[2] Do not confuse the C parity (charge-conjugation parity) with charm C.

[3] A scalar particle has spin and parity $J^P = 0^+$, a pseudoscalar has $J^P = 0^-$, a vector $J^P = 1^-$, and an axial vector $J^P = 1^+$. By a tensor particle we usually mean a particle with $J^P = 2^+$ and by a pseudotensor $J^P = 2^-$. A particle with spin and parity related by $P = (-1)^J$ is sometimes said to have normal or natural parity; if $P = -(-1)^J$, then the parity is said to be abnormal or unnatural.

contained in the same multiplet, all the antiparticles must be in the same $SU(3)$ multiplet as the particles. The weight of an antiparticle of a given state is obtained by reflection the weight of the particle through the origin.

Accompanying the pseudoscalar octet is a pseudoscalar singlet, the η'. The η' is its own antiparticle. In Table 9.6 the masses, lifetimes, and principal decay modes of the pseudoscalar octet plus singlet, or nonet, are given.

TABLE 9.6

SOME PROPERTIES OF THE PSEUDOSCALAR MESON OCTET PLUS SINGLET[a] WITH $J^{PC} = 0^{-+}$

$SU(3)$ Assignment	Symbol	Mass (MeV)	Mean life (sec) or width	Principal decay modes[b]
8	π^+	139.57 ± 0.01	$(2.603 \pm 0.003) \times 10^{-8}$	$\mu\nu$
	π^0	134.96 ± 0.01	$(0.83 \pm 0.06) \times 10^{-16}$	$\gamma\gamma$
	$K^\pm$	493.71 ± 0.04	$(1.237 \pm 0.003) \times 10^{-8}$	$\mu\nu$ $(63.6 \pm 0.2)\%$
				$\pi^\pm\pi^0$ $(21.0 \pm 0.2)\%$
				$\pi^\pm\pi^+\pi^-$ $(5.6 \pm 0.1)\%$
				$\pi^\pm\pi^0\pi^0$ $(1.7 \pm 0.1)\%$
				$\mu\pi^0\nu$ $(3.2 \pm 0.1)\%$
				$e\pi^0\nu$ $(4.8 \pm 0.1)\%$
	$K^0, \bar{K}^0$	497.7 ± 0.2	$50\%\ K_L,\ 50\%\ K_S$	
	(K_S)		$(0.893 \pm 0.003) \times 10^{-10}$	$\pi^+\pi^-$ $(68.7 \pm 0.3)\%$
				$\pi^0\pi^0$ $(31.3 \pm 0.3)\%$
	(K_L)		$(5.18 \pm 0.04) \times 10^{-8}$	$3\pi^0$ $(21.4 \pm 0.7)\%$
				$\pi^+\pi^-\pi^0$ $(12.2 \pm 0.2)\%$
				$\pi\mu\nu$ $(27.1 \pm 0.5)\%$
				$\pi e\nu$ $(39.0 \pm 0.5)\%$
	η	548.8 ± 0.6	$(0.8 \pm 0.2) \times 10^{-3}$ MeV	$\gamma\gamma$ $(38 \pm 1)\%$
				$\pi^0\gamma\gamma$ $(3 \pm 1)\%$
				$3\pi^0$ $(30 \pm 1)\%$
				$\pi^+\pi^-\pi^0$ $(24 \pm 1)\%$
				$\pi^+\pi^-\gamma$ $(4.9 \pm 0.2)\%$
1	η'	957.6 ± 0.3	<1 MeV	$\eta\pi\pi$ $(68 \pm 2)\%$
				$\rho^0\gamma$ $(30 \pm 2)\%$
				$\gamma\gamma$ $(2 \pm 0.3)\%$

[a] The C-parity quantum number refers only to neutral mesons with $Y = 0$.
[b] In this table, we do not distinguish between electron neutrinos ν_e, muon neutrinos ν_μ, and their antiparticles $\bar{\nu}_e$ and $\bar{\nu}_\mu$.

The vector mesons are more conveniently classified as a nonet than as an octet plus singlet, because the evidence indicates that there is significant mixing between the octet and singlet. We shall discuss this $SU(3)$ mixing in more detail in Section 9.6. Some properties of the vector nonet are given in Table 9.7.

TABLE 9.7

SOME PROPERTIES OF THE VECTOR MESON NONET[a] WITH $J^{PC} = 1^{--}$

Symbol	Common symbol	Mass (MeV)	Width (MeV)	Principal decay modes
$\pi^{*\pm}$	$\rho^{\pm}$	766 ± 5	148 ± 5	$\pi\pi$
π^{*0}	ρ^{0}	770 ± 5	150 ± 5	$\pi\pi$
$K^{*\pm}$	$K^{*\pm}$	892 ± 1	50 ± 2	$K\pi$
K^{*0}	K^{*0}	896 ± 1	50 ± 1	$K\pi$
η^{*}	φ	1019.7 ± 0.3	4.1 ± 0.2	K^+K^- $(46 \pm 3)\%$ $K_L K_S$ $(35 \pm 2)\%$ $\pi^+\pi^-\pi^0$ $(16 \pm 2)\%$
$\eta^{*\prime}$	ω	782.7 ± 0.3	10.0 ± 0.4	$\pi^+\pi^-\pi^0$ $(90 \pm 1)\%$ $\pi^0\gamma$ $(9 \pm 1)\%$

[a] The C-parity quantum number refers only to neutral mesons with $Y = 0$.

A tensor meson nonet is also well established experimentally. Some properties of the members are given in Table 9.8. In Table 9.9 are given the masses of other meson nonets for which there is some experimental evidence.

TABLE 9.8

SOME PROPERTIES OF THE TENSOR MESON NONET[a] WITH $J^{PC} = 2^{++}$

Symbol	Common symbol	Mass (MeV)	Width (MeV)	Principal decay modes
π^{*}	A_2	1310 ± 5	102 ± 5	$\rho\pi$ $(71 \pm 2)\%$ $\eta\pi$ $(15 \pm 1)\%$ $\omega\pi\pi$ $(9 \pm 2)\%$ $K\bar{K}$ $(5 \pm 1)\%$
K^{*}	K^{*}	1421 ± 3	108 ± 10	$K\pi$ $(56 \pm 3)\%$ $K^{*}\pi$ $(30 \pm 2)\%$ $K\rho$ $(7 \pm 2)\%$ $K\omega$ $(4 \pm 2)\%$ $K\eta$ $(2 \pm 2)\%$
η^{*}	f'	1516 ± 3	40 ± 10	$K\bar{K}$
$\eta^{*\prime}$	f	1271 ± 5	180 ± 20	$\pi\pi$ $(81 \pm 1)\%$ $K\bar{K}$ $(3 \pm 1)\%$ 4π seen

[a] The C-parity quantum number refers only to neutral mesons with $Y = 0$.

TABLE 9.9

TENTATIVE ASSIGNMENTS OF MESONS TO NONETS[a]

Spin and parity	C-Parity of neutral, $Y = 0$ mesons	Symbols and masses
1^+	+	$\pi(1100)$, $K(1250)$, $\eta(1285)$, $\eta'(?)$
1^+	−	$\pi(1228)$, $K(1300)$, $\eta(?)$, $\eta'(?)$
0^+	+	$\pi(976)$, $K(1250)$, $\eta(1200)$, $\eta'(?)$

[a] Not all the quantum numbers have been verified. A question mark means a state has not yet been observed.

9.4 *U*-Spin

Thus far, we have discussed extensively the isospin subgroup of $SU(3)$, but we have not considered the possible physical significance of the other two $SU(2)$ subgroups. In Fig. 9.4 is shown the baryon octet, with the weights of three $SU(2)$ subgroups labeled as I_3, U_3, and V_3. Use of all three $SU(2)$ subgroups has been proposed by Meshkov *et al.* (1963), who have named them *I*-spin,[4] *U*-spin, and *V*-spin. These authors have discussed the three subgroups especially *U*-spin, in some detail. (See also Lipkin, 1965.)

If a weight is simple, the corresponding state vector is an eigenstate of *I*-spin, *U*-spin, and *V*-spin. However, if a weight is not simple, the eigenstates of *I*-spin having this weight are not, in general, eigenstates of *U*-spin or *V*-spin. The states Λ and Σ^0 are eigenstates of *I*-spin, but not of *U*-spin or *V*-spin. All other members of the baryon octet are eigenstates of *I*, *U*, and *V*. The generator e_{21} is a lowering operator for *I*-spin, e_{32} for *U* spin, and e_{31} for *V* spin.

It can be seen from the masses given in Fig. 9.4 that *U*-spin and *V*-spin are more badly broken than *I*-spin. For example, the different members of the *I*-spin multiplets differ in mass by at most 8 MeV, while different members of a *U*- or *V*-spin multiplet differ in mass by as much as several hundred MeV. Of course, since even *I*-spin is not exact, the physical states Λ and Σ° may not correspond to pure eigenstates of *I*-spin, but there may be a small amount of mixing between $I = 0$ and $I = 1$. Nevertheless, it is a convenient approximation for most purposes to regard the Σ^0 and Λ as eigenstates of *I*-spin.

[4] In this section only, we shall refer to isospin as "*I*-spin."

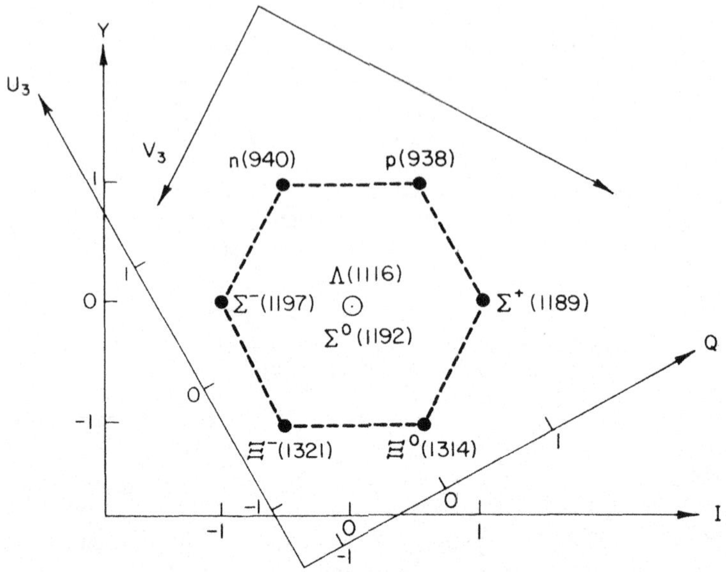

Fig. 9.4. Baryon octet, showing the I-spin, U-spin and V-spin $SU(2)$ subgroups. The states Λ and Σ^0 are eigenstates of I-spin, but not of U-spin or V-spin.

From the weight diagram of Fig. 9.4 we see that all members of a single U-spin multiplet have the same electric charge. Furthermore, the axis perpendicular to the U_3 axis is the charge Q. Thus, a U-spin eigenstate has a weight (U_3, Q), just as an I-spin eigenstate has a weight (I_3, Y). There is no such simple physical interpretation of the V-spin subgroup, and so we shall not discuss V-spin at any length.

Assuming that the Λ and Σ^0 are I-spin eigenstates, we now wish to obtain the linear combinations of Λ and Σ^0 which are eigenstates of U-spin. To do this, we shall introduce operators which act in the $SU(2)$ subspaces of $SU(3)$. We have defined the $SU(2)$ generators

$$h_1 = \begin{pmatrix} 1 & 0 \\ 0 & -1 \end{pmatrix}, \qquad e_{21} = \begin{pmatrix} 0 & 0 \\ 1 & 0 \end{pmatrix}, \qquad \text{and} \qquad e_{12} = \begin{pmatrix} 0 & 1 \\ 0 & 0 \end{pmatrix}. \tag{9.4}$$

We shall introduce the notation that these three operators will be called τ_3, τ_-, and τ_+, respectively when referred to I-spin, and σ_3, σ_-, and σ_+, respectively when referred to U-spin. We shall also define operators which act on states of more than one particle

$$\begin{aligned} I_3 &= \tfrac{1}{2}\sum_i \tau_3(i), \qquad & I_- &= \sum_i \tau_-(i), \qquad & I_+ &= \sum_i \tau_+(i), \\ U_3 &= \tfrac{1}{2}\sum_i \sigma_3(i), \qquad & U_- &= \sum_i \sigma_-(i), \qquad & U_+ &= \sum_i \sigma_+(i). \end{aligned} \tag{9.5}$$

Let us denote the two states of a fundamental doublet of I-spin by the notation α and β. We can construct the I-spin triplet Σ^+, Σ^0, Σ^- and the I-spin singlet Λ from two of these fundamental I-spin doublets. Then

$$\Sigma^+ = \alpha\alpha, \qquad \Sigma^0 = (\alpha\beta + \beta\alpha)/\sqrt{2}, \qquad \Sigma^- = \beta\beta, \qquad \Lambda = (\alpha\beta - \beta\alpha)/\sqrt{2}. \tag{9.6}$$

Let the triplet eigenstates of U-spin be n, χ, Ξ^0, and the singlet be φ, where χ and φ are orthogonal linear combinations of the I-spin eigenstates Λ and Σ^0

$$\chi = a\Sigma^0 + b\Lambda, \qquad \varphi = b\Sigma^0 - a\Lambda. \tag{9.7}$$

We want to find the coefficients a and b.

From the definitions of I_-, Σ^+, and Σ^0, we can directly verify that

$$I_-\Sigma^+ = \Sigma^0\sqrt{2}. \tag{9.8}$$

But U_- acting on the U-spin triplet eigenstate n must act analogously to I_- acting on the I-spin eigenstate Σ^+. Therefore, we have

$$U_- n = \chi\sqrt{2}. \tag{9.9}$$

Now we operate with the commutator $[U_-, I_+]$ on the state vector n of the neutron. Noting that

$$I_+ n = p, \tag{9.10}$$

and using Eq. (9.9) we obtain

$$[U_-, I_+]n = U_- p - I_+ \chi\sqrt{2}. \tag{9.11}$$

Then from Eq. (9.7) we obtain

$$[U_-, I_+]n = U_- p - I_+(a\Sigma^0 + b\Lambda)\sqrt{2}. \tag{9.12}$$

Now since

$$U_- p = \Sigma^+, \qquad I_+ \Sigma^0 = \Sigma^+\sqrt{2}, \qquad \text{and} \qquad I_+\Lambda = 0, \tag{9.13}$$

Eq. (9.12) becomes

$$[U_-, I_+]n = (1 - 2a)\Sigma^+. \tag{9.14}$$

We now point out that the commutator $[U_-, I_+]$ must vanish. This can be proved as follows: If the commutator did not vanish, it would have to be equal to a linear combination of the $SU(3)$ generators, since U_- and I_+ are members of the Lie algebra of $SU(3)$. But I_+ leads to the changes $\Delta Y = 0$, $\Delta I_3 = 1$, and U_- leads to $\Delta Y = -1$, $\Delta I_3 = \frac{1}{2}$. Therefore the product $U_- I_+$ or $I_+ U_-$ leads to a change $\Delta I_3 = \frac{3}{2}$. But no generator of $SU(3)$ changes I_3 by $\frac{3}{2}$, so the commutator must be zero. Then from Eq. (9.14) we obtain

$a = \frac{1}{2}$. Because χ is a normalized function, we must have $b = \pm\frac{1}{2}\sqrt{3}$. Then, choosing the convention that the sign of b is positive, we obtain

$$\chi = \tfrac{1}{2}\,\Sigma^0 + \tfrac{1}{2}\sqrt{3}\Lambda, \qquad \varphi = \tfrac{1}{2}\sqrt{3}\,\Sigma^0 - \tfrac{1}{2}\Lambda. \tag{9.15}$$

Our sign convention is based on the requirement that the matrix elements of $\mathbf{e}_{21}$ and $\mathbf{e}_{32}$ are positive. Because of this fact, the relative signs in Eq. (9.15) differ from those in a number of papers on the subject. We stress again that no physical differences can appear from different choices of sign convention, but, to avoid error, a single convention must be used consistently.

In a similar manner we can obtain the linear combinations of Λ and Σ° which are eigenstates of V-spin. However, we shall not do so since we shall not make use of V-spin eigenstates.

9.5 Tests of *U*-Spin Invariance

Although the electromagnetic interaction does not conserve isospin, it ought to conserve U-spin, since all members of a given U-spin multiplet have the same electric charge. This is, at best, true only to a first approximation. It holds to the extent that U-spin is conserved in strong interactions. If U-spin is not conserved in strong interactions, different members of a U-spin multiplet might not have the same spatial wave functions. Then they would have different electric and magnetic form factors, and would interact differently with the electromagnetic field. Also, when members of a given U-spin multiplet have different masses, this will also break the conservation of U-spin. We cannot get rid of the effects of the violation of U-spin by strong interactions to see whether U-spin would then be strictly conserved in electromagnetic interactions.

Nevertheless, let us regard conservation of U-spin as holding for electromagnetic interactions. Then we can make predictions about the electromagnetic properties of the baryons. For example, since p and Σ^+ belong to the same U-spin multiplet, they should have the same magnetic moment μ. Thus, we have the prediction

$$\mu(\Sigma^+) = \mu(p). \tag{9.16}$$

Similarly, since Σ^- and Ξ^- belong to a U-spin doublet, we predict that

$$\mu(\Sigma^-) = \mu(\Xi^-). \tag{9.17}$$

Likewise, n, χ, and Ξ^0 belong to a U-spin triplet. The χ is a linear combination of Λ and Σ, and is therefore not a physical state. However, we do predict that

$$\mu(n) = \mu(\Xi^\circ). \tag{9.18}$$

Making use of U-spin invariance, Rosen (1963) obtained these relations and others among the baryon magnetic moments.

Thus far, the magnetic moments of the Σ^0 and Ξ^0 have not been measured. It may be feasible to measure the Ξ^0 moment, but a measurement of the Σ^0 moment seems remote. The measured moments of p, n, Λ, Σ^+, Σ^-, and Ξ^- are (Particle Data Group, 1976)

$$\begin{aligned} \mu(p) &= 2.792846 \pm 0.000001, & \mu(n) &= -1.91315 \pm 0.00007, \\ \mu(\Lambda) &= -0.67 \pm 0.06, & \mu(\Sigma^+) &= 2.6 \pm 0.4, \\ \mu(\Sigma^-) &= -1.5 \pm 0.4, & \mu(\Xi^-) &= -1.85 \pm 0.8 \end{aligned} \tag{9.19}$$

in units of the proton magneton $e/(2M_p)$. From Eq. (9.19) we see that Eqs. (9.16) and (9.17) hold within experimental error.

We can use the same methods to obtain relations among the magnetic moments of members of the decuplet. Because all weights of the **10** are simple, the results are very easy to obtain. However, since all members of the decuplet (except for the Ω^-) decay strongly, their magnetic moments cannot be directly measured. The Ω magnetic moment has not yet been measured.

We can also use U-spin invariance to obtain relations among the electromagnetic mass splittings of baryons (Rosen, 1963). Let us assume that the mass of a baryon arises in part from a strong interaction which conserves isospin and in part from an electromagnetic interaction which conserves U-spin. Then since p, Σ^+ belong to a U-spin multiplet, the electromagnetic contributions to the masses of the proton and Σ^+ are given by

$$\delta M_p = \delta M_{\Sigma^+}. \tag{9.20}$$

Similarly we have

$$\delta M_n = \delta M_{\Xi^0}, \qquad \delta M_{\Sigma^-} = \delta M_{\Xi^-}. \tag{9.21}$$

From Eqs. (9.20) and (9.21) we obtain the relation

$$\delta M_n - \delta M_p + \delta M_{\Xi^-} - \delta M_{\Xi^0} = \delta M_{\Sigma^-} - \delta M_{\Sigma^+}. \tag{9.22}$$

But if the electromagnetic interaction is the only interaction which breaks the degeneracy of an isospin multiplet, we must have

$$\begin{aligned} \delta M_n - \delta M_p &= M_n - M_p, \\ \delta M_{\Xi^-} - \delta M_{\Xi^0} &= M_{\Xi^-} - M_{\Xi^0}, \\ \delta M_{\Sigma^-} - \delta M_{\Sigma^+} &= M_{\Sigma^-} - M_{\Sigma^+}. \end{aligned} \tag{9.23}$$

Substituting (9.23) into (9.22) we obtain

$$M_n - M_p + M_{\Xi^-} - M_{\Xi^0} = M_{\Sigma^-} - M_{\Sigma^+}. \tag{9.24}$$

This is known as the Coleman–Glashow (1961) relation. It was first obtained with somewhat different assumptions. The experimental mass splittings are (in MeV)

$$\begin{aligned} M_n - M_p &= 1.293 \pm 0.004, \\ M_{\Xi^-} - M_{\Xi^0} &= 6.4 \pm 0.7, \\ M_{\Sigma^-} - M_{\Sigma^+} &= 7.98 \pm 0.08. \end{aligned} \tag{9.25}$$

From these numbers, we see that Eq. (9.24) is well satisfied.

An equation analogous to (9.24) holds for the pseudoscalar meson octet. It is

$$M_{K^0} - M_{K^+} + M_{K^-} - M_{\bar{K}^0} = M_{\pi^-} - M_{\pi^+}.$$

However, since particle and antiparticle have the same mass, this equation says simply that $0 = 0$, and is not a useful check of U-spin conservation.

From U-spin invariance we also obtain relations along the electromagnetic mass differences of the members of the decuplet. These relations are

$$\begin{aligned} M_{\Delta^-} - M_{\Delta^0} &= M_{\Sigma^{*-}} - M_{\Sigma^{*0}} = M_{\Xi^{*-}} - M_{\Xi^{*0}}, \\ M_{\Delta^0} - M_{\Delta^+} &= M_{\Sigma^{*0}} - M_{\Sigma^{*+}}. \end{aligned} \tag{9.26}$$

The present experimental evidence is consistent with these relations, but is not accurate enough to provide a severe test of them.

Let us now see how to make use of the isospin and U-spin subgroups of $SU(3)$ to obtain relations among reaction cross sections. We consider the reactions

$$M + N \to M + D, \tag{9.27}$$

where M is a meson from the pseudoscalar octet, N is a nucleon (since no other members of the baryon octet are suitable as targets), and D is baryon from the decuplet.

First let us make use of the familiar isospin conservation. Consider the reactions

$$\pi^+ + p \to \Delta^{++} + \pi^0, \tag{9.28}$$

$$\pi^+ + p \to \Delta^+ + \pi^+. \tag{9.29}$$

The initial state has $I = \frac{3}{2}$. Therefore, if isospin is conserved, the final state must also be $I = \frac{3}{2}$. We need to use only the Clebsch–Gordan coefficients of $SU(2)$ to construct a state $\psi_{3/2}^{(3/2)}$ ($I = 3/2$, $I_3 = 3/2$) from states of $I = 1$ and $I = 3/2$. Letting the symbol for a particle denote its isospin state function, we obtain

$$\psi_{3/2}^{(3/2)} = \sqrt{\tfrac{3}{5}}\,\Delta^{++}\pi^0 - \sqrt{\tfrac{2}{5}}\,\Delta^+\pi^+. \tag{9.30}$$

The ratio of the amplitudes for reactions (9.28) and (9.29) is just given by the ratio of the Clebsch–Gordan coefficients in Eq. (9.30), and the ratio of the cross sections is given by the ratio of the squares of the Clebsch–Gordan coefficients. Denoting the cross sections by $\sigma(MN \to DM)$, we thus obtain the prediction

$$\sigma(\pi^+ p \to \Delta^{++}\pi^0)/\sigma(\pi^+ p \to \Delta^+\pi^+) = \tfrac{3}{2}. \tag{9.31}$$

Next, we consider the following set of reactions

$$\begin{aligned} \pi^- + p &\to \Sigma^{*-} + K^+, \\ \pi^- + p &\to \Delta^- + \pi^+, \\ K^- + p &\to \Xi^{*-} + K^+, \\ K^- + p &\to \Sigma^{*-} + \pi^+. \end{aligned} \tag{9.32}$$

We cannot obtain any information about the ratios of these cross sections from isospin conservation, since π, K, Δ, Ξ^*, and Σ^* all belong to different isospin multiplets. However, if *U*-spin is conserved, we can make predictions. The states

$$(\pi^-, K^-), \qquad (p, \Sigma^+), \qquad (K^+, \pi^+),$$

are all *U*-spin doublets, while

$$(\Delta^-, \Sigma^{*-}, \Xi^{*-}, \Omega^-),$$

is a *U*-spin quartet. The initial state $\pi^- p$ is pure $U = 1$, while the initial state $K^- p$ is a linear combination of $U = 1$ and $U = 0$:

$$\begin{aligned} \pi^- p &= \psi_1^{(1)}, \\ K^- p &= (\psi_0^{(0)} - \psi_0^{(1)})/\sqrt{2}. \end{aligned} \tag{9.33}$$

The superscript on ψ gives the value of U and the subscript the value of U_3. The final state is composed of a quartet and doublet, and so can have $U = 2$ or 1. However, if *U*-spin is conserved, the final state must be $U = 1$, since there is no $U = 2$ in the initial state. The relevant *U*-spin eigenstates are

$$\begin{aligned} \psi_1^{(1)} &= \sqrt{\tfrac{3}{4}}\,\Delta^-\pi^+ - \sqrt{\tfrac{1}{4}}\,\Sigma^{*-}K^+, \\ \psi_0^{(1)} &= \sqrt{\tfrac{1}{2}}\,\Sigma^{*-}\pi^+ - \sqrt{\tfrac{1}{2}}\,\Xi^{*-}K^+. \end{aligned} \tag{9.34}$$

From Eqs. (9.33) and (9.34), we see that the ratios of the cross sections are predicted to be

$$\sigma(\pi^- p \to \Sigma^{*-}K^+)/\sigma(\pi^- p \to \Delta^-\pi^+)/\sigma(K^- p \to \Xi^{*-}K^+)/\sigma(K^- p \to \Sigma^{*-}\pi^+)$$
$$= 1/3/1/1. \tag{9.35}$$

If *U*-spin is conserved, these ratios should hold at all energies and at all angles. However, because the masses of the different members of a given

U-spin multiplet may be very different, there will surely be a substantial correction near threshold. One might even do the experiment at an energy such that one process is allowed, whereas another is strictly forbidden by conservation of energy. Thus, because of the large amount of U-spin symmetry breaking, we should not expect Eq. (9.35) to hold very well. We can make a kinematic correction for the different masses involved in the reactions (9.32) by modifying Eq. (9.35) for the available phase space. However, there should also be significant dynamical corrections. We should expect (9.35) to hold best when the energy and momentum transfer are large compared to the mass splitting.

We can also assume V-spin conservation to obtain relations among cross sections. However, we do not obtain any useful relations which are independent of those obtained from isospin and U-spin conservation.

Conservation of U-spin also enables us to discuss photo-induced reactions. We consider the following two reactions:

$$\begin{aligned} \gamma + p &\to \Delta^0 + \pi^+, \\ \gamma + p &\to \Sigma^{*0} + K^+. \end{aligned} \tag{9.36}$$

If the electromagnetic interaction conserves U-spin, then a photon should have a definite U-spin multiplicity. Since there is only one photon, the photon must be a U-spin singlet.[5] We also know that the proton is one member of a U-spin doublet. Therefore, conservation of U-spin implies that the initial state of Eq. (9.36) has $U = \frac{1}{2}$. Therefore, assuming U-spin is conserved in strong interactions, the final states $\Delta^0\pi^+$ and $\Sigma^{*0} K^+$ should also have $U = \frac{1}{2}$. We see that both reactions are governed by a single amplitude since Δ^0 and Σ^{*0} both belong to a U-spin multiplet as do π^+ and K^+. The relevant $SU(2)$ Clebsch–Gordan coefficients appear in the following equation for the final state function:

$$\psi_{1/2}^{(1/2)} = \sqrt{\tfrac{2}{3}}\,\Delta^0\pi^+ - \sqrt{\tfrac{1}{3}}\,\Sigma^{*0}K^+. \tag{9.37}$$

Thus, the ratio of the cross sections is predicted to be

$$\sigma(\gamma p \to \Delta^0\pi^+)/\sigma(\gamma p \to \Sigma^* K^{0+}) = 2. \tag{9.38}$$

Again because of symmetry-breaking, this relation should hold best at large energies and large momentum transfers.

Now let us obtain a relation among electromagnetic decays from U-spin invariance. Let us consider as an example the decays

$$\pi^0 \to 2\gamma, \qquad \eta \to 2\gamma. \tag{9.39}$$

[5] We are referring only to internal degrees of freedom. Of course a photon has two spin orientations.

Since photons are U-spin singlets, we must construct a linear combination of π^0 and η which is a singlet. This has already been done for Σ and Λ, and the result is given in Eq. (9.15). Analogously, the U-spin singlet combination of π^0 and η is

$$\varphi = \tfrac{1}{2}(\sqrt{3}\,\pi^0 - \eta)$$

We thus obtain for the ratio of decay rates

$$\Gamma(\pi^0 \to 2\gamma)/\Gamma(\eta \to 2\gamma) = 3. \tag{9.40}$$

Equation (9.40) must be wrong as it stands, since there is much more available phase space for the η than for the π to decay into two photons. A more meaningful test of the theory would be to multiply both decay rates by the relative phase space and then compare these adjusted decay rates. We leave this as an exercise for the interested reader.

9.6 Gell-Mann–Okubo Mass Formula

Since $SU(3)$ is only an approximate symmetry, the problem arises as to how to identify the physical states belonging together in a single $SU(3)$ multiplet. This problem also exists for $SU(2)$, since isospin is also only approximately conserved.

However, it appears that isospin symmetry is broken only by the electromagnetic and weak interactions, and that therefore $SU(2)$ symmetry-breaking is relatively minor. Typically, the different members of an isospin multiplet with baryon number $|B| \leq 1$ have masses which differ by less than 10 MeV, a number which is small compared to the mass difference between different isospin multiplets of the same spin and parity. Therefore, we usually have no difficulty in identifying the different members of an isospin multiplet.

Even for isospin, however, the problem of identification of states is not always trivial. In heavy nuclei, for example, the Coulomb interaction is sufficiently large that states which are members of the same isospin multiplet may differ in energy by an amount which is much larger than the average spacing of energy levels in a nucleus. The recognition of the states belonging to a given isospin multiplet (called analog states) can be a major problem under these circumstances.

Returning to $SU(3)$, we find that the identification of states can be quite difficult, since the mass splitting within a multiplet may be several hundred MeV. Typically, this is comparable to the mass differences between multiplets. However, it turns out that the symmetry *breaking* of the masses exhibits a pattern in some cases, and we can use this pattern to help identify the multiplets.

Let us assume that the symmetry-breaking interaction, which is responsible for the mass splitting, although a strong interaction, is significantly weaker than the $SU(3)$-invariant interaction so that we may use perturbation theory in computing its effects.

The simplest assumption that leads to a splitting of the masses is that the mass operator M contains a term which is invariant under $SU(3)$ plus a term which is linear in the $SU(3)$ generators. If we neglect electromagnetic mass splitting, all members of an isospin multiplet have the same mass. Of the $SU(3)$ generators, only F_8 or the hypercharge Y does not change the third component of the isospin. Therefore, we write for the mass operator,

$$M = a + bY, \tag{9.41}$$

where a and b are constants within a given $SU(3)$ multiplet.

According to Eq. (9.41), there is a constant mass splitting among members of the decuplet, that is

$$M_{\Omega} - M_{\Xi^*} = M_{\Xi^*} - M_{\Sigma^*} = M_{\Sigma^*} - M_{\Delta}. \tag{9.42}$$

There is some difficulty in comparing this prediction with experiment, because of the electromagnetic mass differences among the members of the decuplet. It is plausible to assume that we can minimize the effect of the electromagnetic interaction if we consider the members of the decuplet which belong to a given U-spin multiplet. We therefore compare the masses of the negatively charged members of the decuplet, which belong to a U-spin quartet. We then have, using the experimental values of the masses from Table 9.3,

$$\begin{aligned} M_{\Omega^-} - M_{\Xi^{*-}} &= 137 \pm 2 \quad \text{MeV}, \\ M_{\Xi^{*-}} - M_{\Sigma^{*-}} &= 149 \pm 2 \quad \text{MeV}, \\ M_{\Sigma^{*-}} - M_{\Delta^-} &= 147 \pm 9 \quad \text{MeV}. \end{aligned} \tag{9.43}$$

The agreement between theory and experiment is not too bad.

However, Eq. (9.41) fails for the baryon octet, as it does not break the mass degeneracy between the Λ and Σ hyperons. Experimentally

$$M_{\Sigma^0} - M_{\Lambda} = 76.9 \pm 0.1 \quad \text{MeV}.$$

The next simplest assumption is to add quadratic functions of the generators to the mass operator. We can add a term proportional to the Casimir operator of the $SU(2)$ isospin group and a term proportional to Y^2 without breaking isospin conservation. We then obtain

$$M = a + bY + cI(I+1) + dY^2, \tag{9.44}$$

where c and d are constant for an $SU(3)$ multiplet. For arbitrary c and d, Eq. (9.44) will not give a constant mass splitting for members of the decuplet and so will ruin the good agreement with experiment. However, we can obtain a relation between c and d by requiring that the constant mass splitting still hold. We require for members of the decuplet

$$cI(I+1) + dY^2 = x + yY, \tag{9.45}$$

where x and y are constants. Substituting the values of I and Y of the Δ, Σ^*, Ξ^*, and Ω in Eq. (9.45) and eliminating x and y, we obtain two equations relating c and d. They both give

$$d = -\tfrac{1}{4}c.$$

Then Eq. (9.44) becomes

$$M = a + bY + c[I(I+1) - \tfrac{1}{4}Y^2]. \tag{9.46}$$

This is the Gell-Man–Okubo mass formula, which was derived by Okubo (1962) by requiring that the mass operator transform like the eighth component of an octet vector. This formula leads to Eq. (9.42) for the decuplet masses. For the octet masses, Eq. (9.46) leads to the relation

$$\tfrac{1}{2}(M_N + M_\Xi) = \tfrac{3}{4}M_\Lambda + \tfrac{1}{4}M_\Sigma. \tag{9.47}$$

Let us compare Eq. (9.47) with experiment. In this case we cannot compare all members of the same U-spin multiplet, as the octet does not contain a U-spin quartet. We shall arbitrarily use the masses of the neutral members of the octet, hoping in this way to minimize the effect of the electromagnetic mass splitting. Using the experimental values of the masses from Table 9.2, we obtain

$$\begin{aligned} \tfrac{1}{2}M_n + \tfrac{1}{2}M_{\Xi^0} &= 1127.2 \pm 0.7 \quad \text{MeV} \\ \tfrac{3}{4}M_\Lambda + \tfrac{1}{4}M_{\Sigma^0} &= 1134.8 \pm 0.2 \quad \text{MeV}. \end{aligned} \tag{9.48}$$

The difference between the two values is 7.6 ± 0.7 MeV, and thus, there is a definite disagreement between theory and experiment. However, the discrepancy is small when compared to a baryon mass or even to a typical mass splitting, the latter being of order 100 MeV.

A more elegant way of deriving the Gell-Mann–Okubo mass formula is to assume, following Okubo (1962), that the mass operator M consists of two terms, one of which M_0 is a scalar under $SU(3)$ transformations, while the second M_8 transforms like the eighth component of an $SU(3)$ octet. The mass M_h of a hadron is then taken to be the expectation value of M. Let us evaluate this expectation value:

$$M_h = (\psi_m^{(j)}, M\psi_m^{(j)}) = (\psi_m^{(j)}, M_0\psi_m^{(j)}) + (\psi_m^{(j)}, M_8\psi_m^{(j)}).$$

Since M_8 is a vector, we can make use of Eq. (8.111) to write

$$(\psi_m^{(j)}, M_8\psi_m^{(j)}) + C_F(j)(\psi_m^{(j)}, F_8\psi_m^{(j)}) + C_D(j)(\psi_m^{(h)}, D_8\psi_m^{(j)}).$$

In order to simplify this expression further, we use the definition of D_8 given in Eq. (6.19):

$$D_8 = \tfrac{2}{3} \sum_{\alpha\beta} d_{8\alpha\beta} F_\alpha F_\beta.$$

Using the values of $d_{8\alpha\beta}$ from Table 6.2 and the definitions

$$I^2 = \sum_{i=1}^{3} F_i^{\,2}, \qquad Y = (2/\sqrt{3}) F_8,$$

we obtain

$$D_8 = (I^2 - \tfrac{1}{3}F^2 - \tfrac{1}{4}Y^2)/\sqrt{3}.$$

Thus, we have written D_8 in terms of diagonal operators. Since M_0 and F_8 are already diagonal, we can write the expression for M_h in the form

$$M_h = a_j + b_j Y + c_j[I(I + 1) - \tfrac{1}{4}Y^2]$$

where a_j, b_j, and c_j are constants for a given multiplet, here denoted by j. The values of these constants depend on the unknown matrix elements $(\psi_m^{(j)}, M_0\, \psi_m^{(j)})$ and the unknown constants $C_F(j)$ and $C_D(j)$, and so must be regarded as a set of parameters for each multiplet.

This derivation of the Gell-Mann–Okubo mass formula depends on the assumption that the symmetry-breaking term in the mass operator transforms like the eighth component of an octet. Since the formula is not exact, there must be another symmetry-breaking term in the mass operator (unless perturbation theory breaks down). But for some reason, the part that transforms like an octet is enhanced. The fact of *octet enhancement* appears to play a role in many other aspects of elementary particle physics, for example in the electromagnetic and weak interactions of hadrons.

If the assumptions underlying the Gell-Mann–Okubo mass formula are correct, then a formula analogous to Eq. (9.47) should also hold for meson octets. In applying this formula to mesons, sometimes the squares of the masses are used rather than the masses themselves. If this is done, then the M of Eq. (9.46) should be regarded as a mass-squared operator when applied to mesons. One reason why some people use linear masses for baryons and squares of masses for mesons is that a free meson may propagate according to the Klein–Gordon equation (which is quadratic in the mass), while a free baryon may propagate according to a Dirac equation (which is linear in the mass). Other reasons have also been given for using squares of masses, and some of these reasons may be applied to baryons as well. Since none of the reasons given seems compelling, one can appeal to experiment to decide. However, as we shall see, the Gell-Mann–Okubo mass formula does not hold for mesons, independently of whether masses or squares of masses are used.

Let us neglect electromagnetic mass differences among the mesons in the following discussion. If we consider a meson octet, we note that it contains two isospin doublets, one with hypercharge $Y = 1$ and the other with $Y = -1$. Both these doublets have the same mass since they are antiparticles of each other.

If we let the mass of the isospin doublet, triplet, and singlet mesons be M_2, M_3, and M_1, the Gell-Mann–Okubo octet mass formula becomes

$$\begin{aligned} &\text{linear:} \quad && M_2 = \tfrac{1}{4}M_3 + \tfrac{3}{4}M_1, \\ &\text{quadratic:} \quad && M_2{}^2 = \tfrac{1}{4}M_3{}^2 + \tfrac{3}{4}M_1{}^2 \end{aligned} \tag{9.49}$$

We wish to compare these formulas with the experimental masses of the pseudoscalar, vector, and tensor meson octets. There is a complication in doing so, however. As we have already discussed in Section 9.2, we find experimentally nine mesons of a given spin and parity, rather than eight. Two of the mesons of a given spin and parity have quantum numbers $I = Y = 0$. If $SU(3)$ is a good classification scheme, one of these should belong to an octet and the other to a singlet, but the question is which meson is which.

Since the Gell-Mann–Okubo mass formula works rather well for baryons, it is perhaps reasonable to use it to decide which of the two $I = Y = 0$ mesons belongs to the octet. The procedure we shall follow is to use the experimental values of the masses of the isospin doublet and triplet in Eq. (9.49) to predict the value of the mass of the isospin singlet. We shall then compare this prediction to the observed values of the masses of the two isospin singlets. We obtain the results shown in Table 9.10.

TABLE 9.10

COMPARISON OF GELL-MANN–OKUBO MASS FORMULA WITH EXPERIMENTAL MASSES OF MESONS

		M_1 (MeV) predicted	
J^P	Observed mesons and masses (MeV)	Linear mass formula	Quadratic mass formula
0^-	$\eta(549)$, $\eta'(958)$	619	570
1^-	$\omega(783)$, $\varphi(1020)$	938	934
2^+	$f(1271)$, $f'(1516)$	1458	1456

We see that the masses of the isospin singlets calculated from either the linear or quadratic Gell-Mann–Okubo mass formula do not agree with any of the experimental masses. However, we see from Table 9.10 that in every case the predicted mass lies *between* the values of the masses of two mesons with the right quantum numbers. Because of this fact, we interpret the meson multiplets as nonets rather than as octets plus singlets. We assume that the $SU(3)$ symmetry-breaking interaction mixes the isosinglet member of the

TABLE 9.11

LINEAR AND QUADRATIC $SU(3)$ MIXING ANGLES

J^P	Nonet members	θ (linear)	θ (quadratic)
0^-	$\pi K\eta, \eta'$	$-24°$	$-11°$
1^-	$\rho K^*\varphi, \omega$	$36°$	$39°$
2^+	$A_2 K^*f', f$	$29°$	$31°$

octet with the $SU(3)$ singlet. A mixing angle θ can be defined in the following way. Let M_1 be the mass of the isosinglet member of the octet as predicted by the Gell-Mann–Okubo mass formula, and m and M be the masses of the two physically observed isosinglet mesons. Then we define θ by the formula

$$\begin{aligned} \text{linear:} \qquad & M_1 = m\cos^2\theta + M\sin^2\theta, \\ \text{quadratic:} \qquad & M_1^{\,2} = m^2\cos^2\theta + M^2\sin^2\theta, \end{aligned} \tag{9.50}$$

where m and M are identified with the two physical masses. If we assume that m is mostly octet, then $|\theta| \leq 45°$. The sign of θ is taken to be positive if the heavier meson is mostly octet, and otherwise negative. The values of the linear and quadratic mixing angles for the pseudoscalar, vector, and tensor nonets are given in Table 9.11. Similar values are given by the Particle Data Group (1976). The $SU(3)$ mixing angle for mesons has been discussed by Glashow and Socolow (1966).

It is clear that by introducing the angle θ we are able to obtain agreement with the experimental masses. However, at first sight it is not clear what is the use of such a fit, since we do not make any prediction about a relationship among the meson masses. Only if we can use the angle θ to make predictions about other processes is the idea of $SU(3)$ mixing useful. It turns out that we can indeed make a prediction about the decay modes of the φ and ω vector mesons, using $SU(3)$ mixing. We shall briefly discuss this question in Section 12.9 within the framework of the quark model.

Note that if a nonet is subsequently discovered with masses such that M_1 does not lie between m and M, Eq. (9.50) will still hold, but the angle θ will be imaginary. In such a case, the usefulness of the concept of $SU(3)$ mixing will be seriously diminished.

9.7 Meson–Baryon Coupling

Let us consider the coupling of a pseudoscalar meson to a baryon–antibaryon pair, assuming that the baryon, antibaryon, and meson all belong to octets. Then, to obtain an interaction which is invariant under $SU(3)$ transformations, we must couple $\mathbf{8} \otimes \mathbf{8}$ to give $\mathbf{8}$. As we have previously noted, there are two ways to do this, using F and D coupling.

We can use the Clebsch–Gordan coefficients of $SU(3)$ to obtain the relative coupling strengths in a straightforward way. However, we shall adopt a different procedure, in which we write the baryon and meson octets as 3×3 matrices instead of as eight-dimensional vectors. These matrices are

$$B = \begin{pmatrix} \Sigma^0/\sqrt{2} + \Lambda/\sqrt{6} & \Sigma^+ & p \\ \Sigma^- & -\Sigma^0/\sqrt{2} + \Lambda/\sqrt{6} & n \\ \Xi^- & \Xi^0 & -2\Lambda/\sqrt{6} \end{pmatrix}, \tag{9.51}$$

$$\bar{B} = \begin{pmatrix} \bar{\Sigma}^0/\sqrt{2} + \bar{\Lambda}/\sqrt{6} & \bar{\Sigma}^- & \bar{\Xi}^- \\ \bar{\Sigma}^+ & -\bar{\Sigma}^0/\sqrt{2} + \bar{\Lambda}/\sqrt{6} & \bar{\Xi}^0 \\ \bar{p} & \bar{n} & -2\bar{\Lambda}/\sqrt{6} \end{pmatrix}, \tag{9.52}$$

$$M = \begin{pmatrix} \pi^0/\sqrt{2} + \eta/\sqrt{6} & \pi^+ & K^+ \\ \pi^- & -\pi^0/\sqrt{2} + \eta/\sqrt{6} & K^0 \\ K^- & \bar{K}^0 & -2\eta/\sqrt{6} \end{pmatrix}. \tag{9.53}$$

The above matrices were originally written down by Ne'eman (1961). Our normalization differs from that of Ne'eman by a factor $\sqrt{2}$. In terms of these matrices, the F-type coupling C_F is given by a scalar formed with the antisymmetric combination of $\bar{B}$ and B, namely

$$C_F = -\sqrt{2}\,\mathrm{Tr}([\bar{B}, B]M), \tag{9.54}$$

where the minus sign and the $\sqrt{2}$ are inserted to conform to the convention of Gell-Mann (1961). On the other hand, the D-type coupling C_D is given by a scalar formed with the symmetric combination

$$C_D = \sqrt{2}\,\mathrm{Tr}(\{\bar{B}, B\}M). \tag{9.55}$$

Performing the matrix multiplication and carrying out the traces, we obtain

$$\begin{aligned} C_F = {} & (\bar{p}p - \bar{n}n + 2\bar{\Sigma}^+\Sigma^+ - 2\bar{\Sigma}^-\Sigma^- + \bar{\Xi}^0\Xi^0 - \bar{\Xi}^-\Xi^-)\pi^0 \\ & + (\sqrt{2}\,\bar{p}n - \sqrt{2}\,\bar{\Xi}^0\Xi^- + 2\bar{\Sigma}^0\Sigma^- - 2\bar{\Sigma}^+\Sigma^0)\pi^+ \\ & + (\sqrt{2}\,\bar{n}p - \sqrt{2}\,\bar{\Xi}^-\Xi^0 + 2\bar{\Sigma}^-\Sigma^0 - 2\bar{\Sigma}^0\Sigma^+)\pi^- \\ & + \sqrt{3}(\bar{p}p + \bar{n}n + \bar{\Xi}^0\Xi^0 - \bar{\Xi}^-\Xi^-)\eta \\ & + (-\sqrt{3}\,\bar{p}\Lambda + \sqrt{3}\,\bar{\Lambda}\Xi^- - \bar{p}\Sigma^0 - \sqrt{2}\,\bar{n}\Sigma^- + \bar{\Sigma}^0\Xi^- + \sqrt{2}\,\bar{\Sigma}^+\Xi^0)K^+ \\ & + (-\sqrt{3}\,\bar{\Lambda}p + \sqrt{3}\,\bar{\Xi}^-\Lambda - \bar{\Sigma}^0 p - \sqrt{2}\,\bar{\Sigma}^- n + \bar{\Xi}^-\Sigma^0 + \sqrt{2}\,\bar{\Xi}^0\Sigma^+)K^- \\ & + (-\sqrt{3}\,\bar{n}\Lambda + \sqrt{3}\,\bar{\Lambda}\Xi^0 + \bar{n}\Sigma^0 - \sqrt{2}\,\bar{p}\Sigma^+ - \bar{\Sigma}^0\Xi^0 + \sqrt{2}\,\bar{\Sigma}^-\Xi^-)K^0 \\ & + (-\sqrt{3}\,\bar{\Lambda}n + \sqrt{3}\,\bar{\Xi}^0\Lambda + \bar{\Sigma}^0 n - \sqrt{2}\,\bar{\Sigma}^+ p - \bar{\Xi}^0\Sigma^0 + \sqrt{2}\,\bar{\Xi}^-\Sigma^-)\bar{K}^0 \end{aligned} \tag{9.56}$$

$$
\begin{aligned}
C_D = {} & \{\bar{p}p - \bar{n}n + 2(\bar{\Sigma}^0\Lambda + \bar{\Lambda}\Sigma^0)/\sqrt{3} - \bar{\Xi}^0\Xi^0 + \bar{\Xi}^-\Xi^-\}\pi^0 \\
& + \{\sqrt{2}\,\bar{p}n + 2(\bar{\Sigma}^+\Lambda + \bar{\Lambda}\Sigma^-)/\sqrt{3} + \sqrt{2}\,\bar{\Xi}^0\Xi^-\}\pi^+ \\
& + \{\sqrt{2}\,\bar{n}p + 2(\bar{\Lambda}\Sigma^+ + \bar{\Sigma}^-\Lambda)/\sqrt{3} + \sqrt{2}\,\bar{\Xi}^0\Xi^0\}\pi^- \\
& + \{-\bar{p}p - \bar{n}n + 2(\bar{\Sigma}^+\Sigma^+ + \bar{\Sigma}^0\Sigma^0 + \bar{\Sigma}^-\Sigma^- - \bar{\Lambda}\Lambda) - \bar{\Xi}^0\Xi^0 - \bar{\Xi}^-\Xi^-\}\eta/\sqrt{3} \\
& + \{-\bar{p}\Lambda/\sqrt{3} + \bar{p}\Sigma^0 + \sqrt{2}\,\bar{n}\Sigma^- - \bar{\Lambda}\Xi^-/\sqrt{3} + \bar{\Sigma}^0\Xi^- + \sqrt{2}\,\bar{\Sigma}^+\Xi^0\}K^+ \\
& + \{-\bar{\Lambda}p/\sqrt{3} + \bar{\Sigma}^0 p + \sqrt{2}\,\bar{\Sigma}^- n - \bar{\Xi}^-\Lambda/\sqrt{3} + \bar{\Xi}^-\Sigma^0 + \sqrt{2}\,\bar{\Xi}^0\Sigma^+\}K^- \\
& + \{-\bar{n}\Lambda/\sqrt{3} - \bar{n}\Sigma^0 + \sqrt{2}\,\bar{p}\Sigma^+ - \bar{\Lambda}\Xi^0/\sqrt{3} - \bar{\Sigma}^0\Xi^0 + \sqrt{2}\,\bar{\Sigma}^-\Xi^-\}K^0 \\
& + \{-\bar{\Lambda}n/\sqrt{3} - \bar{\Sigma}^0 n + \sqrt{2}\,\bar{\Sigma}^+ p - \bar{\Xi}^0\Lambda/\sqrt{3} - \bar{\Xi}^0\Sigma^0 + \sqrt{2}\,\bar{\Xi}^-\Sigma^-\}\bar{K}^0.
\end{aligned}
\tag{9.57}
$$

In writing down the antibaryon–baryon–meson ($\bar{B}BM$) couplings C_D and C_F, we have included only the dependence on $SU(3)$ variables. In general, there will also be a dependence on spatial and spin variables as well.

In terms of C_D and C_F, a general $SU(3)$-invariant coupling C is given by

$$C = g[\alpha C_D + (1 - \alpha)C_F], \tag{9.58}$$

where g and α are constants. This expression is subject to the restriction that $\bar{p}p\pi^0$ coupling strength is g.

The quantity $\alpha/(1 - \alpha)$ is called the D/F ratio and cannot be determined from $SU(3)$-invariance alone. This ratio is important, not only in $\bar{B}BM$ coupling, but in any situation in which three octets are combined to form a scalar (or alternatively, two octets are coupled to form a third).

Because of the existence of two types of $SU(3)$-invariant couplings, the strength of the coupling of the hyperons to mesons, relative to that of nucleons to mesons, is not fixed. For example, in C_F there is no $\bar{\Lambda}\Sigma\pi$ coupling, while in C_D there is no $\bar{\Sigma}\Sigma\pi$ coupling.

We can use the matrix technique to obtain trilinear meson couplings as well. Let us use the symbol V for a vector meson octet and M for a pseudoscalar meson octet. There is no strong MMM coupling because such a coupling does not conserve parity. The strong MMV coupling must be pure F-type and the VVM coupling pure D-type because of the requirement of charge-conjugation invariance in strong interactions.

9.8 Hadron Decays

All the unstable members of the baryon octet decay into other baryons of the octet and ultimately into the proton plus leptons or photons. If $SU(3)$ were exact, all members of the octet would be mass degenerate and could not decay into other members of the same multiplet. Thus, since the baryon octet is the baryon multiplet of smallest average mass, all the octet decays are forbidden by $SU(3)$ invariance. However, this is not considered a difficulty for $SU(3)$, since no member of the baryon octet decays via strong interactions. In fact, all baryon octet decays proceed via the weak interaction except for the Σ^0 decay which goes via the electromagnetic interaction.

Although the members of the baryon octet cannot decay without $SU(3)$ violation, the members of the decuplet are allowed, in principle, to decay strongly in an $SU(3)$-conserving interaction. For example, $SU(3)$–invariance does not forbid the decay

$$D \to B + M,$$

where D is a member of the baryon decuplet, B a member of the baryon octet, and M a member of the meson pseudoscalar octet. This decay is allowed because **10** is contained in the Kronecker product of $\mathbf{8} \otimes \mathbf{8}$. Assuming $SU(3)$ invariance holds, then all decays $D \to B + M$ are governed by one amplitude. This is because $\mathbf{8} \otimes \mathbf{8}$ contains **10** only once. Thus, to obtain predictions concerning the ratio of decay rates of different members of the decuplet which are based on $SU(3)$ invariance, we need merely make use of the relevant Clebsch–Gordan coefficients of $SU(3)$.

In fact, the Δ, Σ^*, and Ξ^* of the decuplet are observed to decay strongly into a baryon and meson. Nevertheless, the decays violate $SU(3)$ invariance, at least in part because of mass differences among different members of a given $SU(3)$ multiplet. For example, the decay

$$\Sigma^* \to \Lambda + \pi$$

is allowed, but the decay

$$\Sigma^* \to N + \bar{K}$$

is forbidden by energy conservation.

As a first approximation, we may assume that the decay matrix elements are $SU(3)$ invariant, but that the kinematic or phase space factors break the symmetry. Such an assumption is in qualitative agreement with experiment in many cases.

On the other hand, the decay ratios $B^* \to B + M$, where B^* is an excited baryon octet, cannot be obtained from Clebsch–Gordan coefficients alone. This is because **8** appears twice in $\mathbf{8} \otimes \mathbf{8}$, once with F-type coupling and once with D-type coupling. One must know the D/F ratio to be able to calculate the relative decay rates. However, one can use the measured rates for two decays to determine the D/F ratio experimentally, and then predict other decay rates. This has been done, for example, by Glashow and Rosenfeld (1963).

An analogous situation occurs for the mesons. Because the pseudoscalar octet is the $SU(3)$ multiplet of lowest mass, the members of this octet are fordibben to decay; and in fact, all the decays of the octet go via electromagnetic or weak interactions. The vector meson octet, on the other hand, is allowed to decay into two pseudoscalar mesons without violating $SU(3)$ invariance, and the coupling is pure F-type. Such decays are observed to occur via the strong interactions, but again the mass splittings break the symmetry.

In this chapter so far, we have given some indication of the relevance of $SU(3)$ to hadron physics. However, we do not have the space to go into a detailed comparison of the predictions of broken $SU(3)$ with experiment. A detailed review of the spectroscopy of hadrons, including hadron decays, and a comparison with $SU(3)$ predictions, has been given by Samios *et al.* (1974). The conclusion of these authors is that $SU(3)$ is a well-established approximate symmetry of hadron physics.

9.9 Weak Hadron Decays

The weak interactions are certainly not invariant under $SU(3)$ transformations; in fact they are not even invariant under the isospin $SU(2)$ subgroup. However, as has been pointed out by Gell-Mann (1962, 1964) and by Cabibbo (1963), there appear to be certain regularities in the way that $SU(3)$ is violated in the decays of hadrons.

We shall not discuss the weak interactions in much detail, but shall point out a few salient features. It was proposed by Lee and Yang (1956) that parity might be violated in weak interactions. Subsequently Wu *et al.* (1957) verified that this was indeed the case. The $V-A$ (V = vector current, A = axial vector current) form of the weak interaction was proposed by Sudarshan and Marshak (1958), Feynman and Gell-Mann (1958), and Sakurai (1958). Feynman and Gell-Mann (1958) further proposed that the vector current is conserved in weak interactions. The axial vector current cannot be conserved or the pion decay would be forbidden (Taylor, 1958; Goldberger and Treiman, 1958). However, the extent to which the axial vector current is not conserved can be shown in some sense to be small. This concept is known as the partially conserved axial vector current. (See, for example, Marshak *et al.*, 1969, for details.)

Christenson *et al.* (1964) discovered that CP was not conserved in K_L-meson decay. Little is known about the CP-violating part of the weak interaction. For phenomenological descriptions, see for example, Wolfenstein (1964), Wu and Yang (1964), and Lee and Wu (1966). The CP-conserving part of the weak interaction H_w appears to be fairly well described phenomenologically by a current–current interaction of the form

$$H_w = G \int J_\alpha^\dagger J_\alpha \, d^3x/\sqrt{2}, \tag{9.59}$$

where J_α is a "weak current," G is a coupling constant, and the $\sqrt{2}$ is inserted to conform to the conventional normalization. The current J_α is assumed to be composed of two parts, a leptonic part $J_\alpha^{(l)}$ and a hadronic part $J_\alpha^{(h)}$. We are here interested in the $SU(3)$-transformation properties of the hadronic current $J_\alpha^{(h)}$.

Before discussing the form of $J_\alpha^{(h)}$, we consider an analogous problem for the electromagnetic current j_α. The time component j_t of the four-vector j_α is the charge density, and its integral over three-space is the total charge Q:

$$Q = \int j_t \, d^3x. \tag{9.60}$$

As we have stated earlier, the electric charge of a hadron, in units of the proton charge, is related to the hypercharge Y and the third component of the isospin I_3 by the Gell-Mann–Nishijima formula

$$Q = I_3 + \tfrac{1}{2}Y.$$

But I_3 and Y are given in terms of the $SU(3)$ generators F_3 and F_8 by

$$I_3 = F_3, \qquad Y = 2F_8/\sqrt{3}.$$

Therefore, the charge transforms as a component of an octet.

$$Q = F_3 + F_8/\sqrt{3}. \tag{9.61}$$

We now return to the weak hadronic current $J_\alpha^{(h)}$. This must contain an isospin-raising operator, since in the decay $n \to p + e^- + \bar{\nu}$, a neutron is converted into a proton. Likewise, $J_h^{(\alpha)}$ must contain a hypercharge (or strangeness) changing operator, since in the decay $\Lambda \to p + e^- + \bar{\nu}$, a Λ is converted into a proton. The isospin-raising operator can be written in terms of the $SU(3)$ generators as $F_1 + iF_2$, and the hypercharge-changing operator as $F_4 + iF_5$. The coupling strengths of these two parts are not equal, and the relative coupling is giving in terms of the Cabibbo angle θ. Putting all these ideas together we write for $J_\alpha^{(h)}$:

$$J_\alpha^{(h)} = \cos\theta(F_1 + iF_2) + \sin\theta(F_4 + iF_5). \tag{9.62}$$

Thus $J_\alpha^{(h)}$ can be considered as a component of an $SU(3)$ vector $J_{\mu\alpha}^{(h)}$, where the subscript μ refers to the $SU(3)$ index.

Now, just as we have written the electric charge as a spatial integral of

j_t, we write the weak hadronic charge as a spatial integral of $J_{\mu t}^{(h)}$

$$\int J_{\mu t}^{(h)}\, d^3x = V_\mu(t) + A_\mu(t). \tag{9.63}$$

We have divided the expression for the weak charge into two terms, one of which V_μ is the time component of a vector, and the other A_μ, which is the time component of an axial vector. Both these terms can exist because parity is not conserved in weak interactions.

Gell-Mann has assumed the equal-time commutators of the V_μ and A_μ are

$$\begin{aligned} [V_\mu, V_\nu] &= i \sum_\kappa f_{\mu\nu\kappa} V_\kappa, \\ [V_\mu, A_\nu] &= i \sum_\kappa f_{\mu\nu\kappa} A_\kappa, \\ [A_\mu, A_\nu] &= i \sum_\kappa f_{\mu\nu\kappa} V_\kappa. \end{aligned} \tag{9.64}$$

These equal-time commutation relations are the principal hypothesis of current algebra. They can be correct even if $SU(3)$ is badly broken. We can introduce two linear combinations of the V_μ and A_μ

$$2F_\mu^{\pm} = V_\mu \pm iA_\mu. \tag{9.65}$$

Then it can be directly verified that F_μ^+ and F_μ^- transform like two commuting F-spins, so that the current algebra is the algebra of the group $SU(3) \times SU(3)$. Gell-Mann (1964) has also pointed out a connection between current algebra and the group $SU(6)$.

If we look at Table 9.6, we see that the K^+ meson can decay weakly into several different final states. The final state $\mu^+\bar{\nu}_\mu$ is purely leptonic, the state $\pi^0\mu^+\bar{\nu}_\mu$ contains a hadron as well as leptons, and the state $\pi^+\pi^0$ contains only hadrons. These final states illustrate the different kinds of weak decays that hadrons undergo.

In general, none of the $SU(3)$ quantum numbers I, I_3, and Y is conserved in the weak decays of hadrons. However, there are some striking regularities in the decays, several of which we shall point out. Let us look first at decays with both hadrons and leptons in the final state. Then, the prominent strangeness-changing decays satisfy a selection rule; namely,

$$\Delta Q = \Delta S,$$

where ΔQ is the change in charge and ΔS is the change in strangeness, both of which refer to the hadrons only. This is illustrated by the decay $K^+ \to \pi^+\pi^-e^+\bar{\nu}_e$, which is rare (and therefore not listed in Table 9.6) but has been seen, and by the failure of physicists to see the decay

$K^+ \to \pi^+\pi^+e^-\nu_e$. If the latter decay is ever seen, it will only mean that the selection rule $\Delta Q = \Delta S$ is approximate rather than exact. But the selection rule has already been verified to approximately the 1% level.

We turn next to the nonleptonic decays. The known decays satisfy the selection rule

$$|\Delta S| = 0, 1.$$

In addition, there is the approximate selection rule for $|\Delta S| = 1$ decays,

$$|\Delta I| = \tfrac{1}{2}.$$

This latter rule is illustrated by a comparison of the decays

$$K^+ \to \pi^+\pi^0 \qquad \text{and} \qquad K_S \to \pi^+\pi^-.$$

The K meson has $I = \frac{1}{2}$, and therefore, according to the $|\Delta I| = \frac{1}{2}$ selection rule, the final two-pion state should have $I = 0$ or 1. But two pions in a state with $J = 0$ must have even I to satisfy Bose statistics (see, for example, Lichtenberg 1965). Now a charged state containing pions must have $I_3 \neq 0$, and therefore $I \neq 0$. It follows that the $\pi^+\pi^0$ state from the decay of the K^+ must have $I \neq 0$, 1, and therefore must violate the $|\Delta I| = \frac{1}{2}$ selection rule. But the mean life of the K^+ meson is about 100 times that of the K_S. This inhibition of the decay of the K^+ illustrates not only that there is a $|\Delta I| = \frac{1}{2}$ selection rule but shows that it is only approximate. For further discussion of this rule, see Gell-Mann and Ne'eman (1964).

A connection between $SU(3)$ and weak interactions is illustrated by a relation between the amplitudes for the decays

$$\Sigma^+ \to p\pi^0, \qquad \Lambda \to p\pi^-, \qquad \Xi^- \to \Lambda\pi^-.$$

Lee (1964) and Sugawara (1964) proposed that these amplitudes satisfy the relation

$$\sqrt{3}\, A(\Sigma^+) - A(\Lambda) = 2A(\Xi^-).$$

The experimental decay rates are consistent with this amplitude relation to a good approximation. Rosen (1965) has given a good discussion of the connection with $SU(3)$. We shall not discuss this relation further here, but simply refer the interested reader to the original papers.

Comprehensive treatments of the weak interaction have been given by Konopinski (1966) for the classical aspects and by Marshak *et al.* (1969) for aspects involving $SU(3)$. For further information about the CP-conserving interaction, see, for example, Adler and Dashen (1968) and Bernstein (1968).

Little is known about the part of the weak interaction which does not conserve CP. For a phenomenological description, see, for example, Wu and Yang (1964) and Lee and Wu (1966). We shall postpone discussion of the more recent developments leading to charm and $SU(4)$ until Chapter 12.

CHAPTER 10

CHARM

10.1 Need for a New Quantum Number

With the discovery of a heavy vector meson by Aubert *et al.* (1974) and Augustin *et al.* (1974), it became apparent to many physicists that the $SU(3)$ internal symmetry scheme needed to be enlarged. This meson, called J by Aubert *et al.* and ψ by Augustin *et al.*, is now usually called the J/ψ, but we shall call it ψ.[1] It has a mass of 3.1 GeV and a width 0.07 MeV.

Although the ψ does not carry any new quantum numbers, its unusually narrow width (less than 1 % of the width of a typical heavy meson) convinced many physicists that it could not fit into the usual $SU(3)$ scheme.

At the present time a whole family of heavy ψ mesons has been discovered. These all have C-parity -1. Also seen are heavy mesons of the same family with C-parity $+1$. We denote these mesons by χ.

Recently a new meson of mass 1.87 GeV was discovered by Goldhaber *et al.* (1976) and a new baryon of mass 2.26 GeV by Knapp *et al.* (1976). Both of these hadrons show explicit evidence for a new additive quantum number which is conserved in strong and electromagnetic interactions, but not in weak interactions. All the new particles are quite massive, and this fact shows that any enlarged symmetry is more badly broken than $SU(3)$.

The idea of a new quantum number, and an enlargement of the symmetry

[1] Our apologies to Aubert *et al.*

group from $SU(3)$ to $SU(4)$, was proposed some time ago by a number of authors (Katayama *et al.*, 1962; Tarjanne and Teplitz, 1963; Maki, 1964; Hara, 1964; Amati *et al.*, 1964; Bjorken and Glashow, 1964). Following Bjorken and Glashow (1964), we call the new quantum number *charm*. We shall briefly discuss in Chapter 12 why charm, unlike strangeness, was postulated to exist before it was discovered. A good review of charm has been given by Gaillard *et al.* (1975).

In the previous chapter, we pointed out that there was good evidence that baryons occur in multiplets of **1**, **8**, and **10** of $SU(3)$, and that mesons occur in mixed $\mathbf{1} + \mathbf{8}$ multiplets. If the group is enlarged to $SU(4)$, it is not *a priori* clear how the $SU(3)$ multiplets are to be incorporated into $SU(4)$ multiplets. Here we shall give the conventional way of doing this. Our enlargement is not based on the eightfold way of $SU(3)$, but rather on the quark model, which we shall discuss in Chapter 12.

For the baryons, we assume that the relevant Young tableaux are the same for both $SU(3)$ and $SU(4)$. With this hypothesis, the **1** of $SU(3)$ becomes a $\bar{\mathbf{4}}$ of $SU(4)$, and the **8** and **10** of $SU(3)$ become enlarged to two inequivalent **20** multiplets. We denote these multiplets by $\mathbf{20}_M$ and $\mathbf{20}_S$, respectively, where M refers to a Young tableau of mixed symmetry and S to a symmetric tableau. The baryon multiplets correspond to the Young tableaux

: **1** of $SU(3)$, $\bar{\mathbf{4}}$ of $SU(4)$,

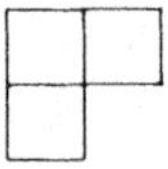
: **8** of $SU(3)$, $\mathbf{20}_M$ of $SU(4)$,

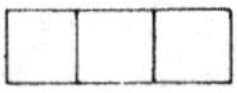
: **10** of $SU(3)$, $\mathbf{20}_S$ of $SU(4)$.

The meson **8** of $SU(3)$ belongs to the adjoint representation of this group. We assume that the meson **8** becomes a **15**, which belongs to the adjoint representation of $SU(4)$, and that the **1** of $SU(3)$ remains a **1** in $SU(4)$. Unlike the situation for the baryons, the meson multiplets correspond to different Young tableaux in $SU(3)$ and $SU(4)$, namely,

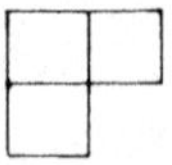 → 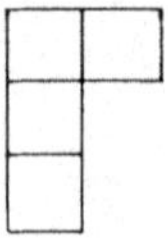 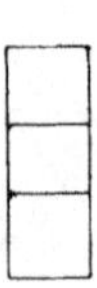 → 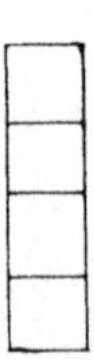

8 of $SU(3)$ **15** of $SU(4)$ **1** of $SU(3)$ **1** of $SU(4)$

In $SU(4)$, the Gell-Mann–Nishijima formula [see Eq. (6.11)] must be generalized to include charm C. The formula is[2]

$$Q = I_3 + \tfrac{1}{2}Y + C \tag{10.1}$$

or, defining

$$S = Y + C - B, \tag{10.2}$$

then

$$Q = I_3 + \tfrac{1}{2}(B + S + C). \tag{10.3}$$

If still other additive quantum numbers are discovered, these expressions will have to be generalized further if they are to apply to particles with the new quantum numbers.

For the first fundamental quartet **4**, the quantities I_3, B, Y, C, S, and Q can be written in terms of the diagonal $SU(4)$ generators of Eq. (6.27) and the unit matrix of Eq. (6.28). The formulas are

$$I_3 = \tfrac{1}{2}\lambda_3, \tag{10.4}$$

$$B = \tfrac{1}{3}\lambda_0, \tag{10.5}$$

$$Y = (1/\sqrt{3})\lambda_8 + (1/\sqrt{6})\lambda_{15} - \tfrac{1}{6}\lambda_0, \tag{10.6}$$

$$C = \tfrac{1}{4}(\lambda_0 - \sqrt{6}\,\lambda_{15}), \tag{10.7}$$

$$S = (1/\sqrt{3})\lambda_8 - (1/2\sqrt{6})\lambda_{15} - \tfrac{1}{4}\lambda_0, \tag{10.8}$$

$$Q = \tfrac{1}{2}\lambda_3 + (1/2\sqrt{3})\lambda_8 - (1/\sqrt{6})\lambda_{15} + \tfrac{1}{6}\lambda_0. \tag{10.9}$$

For the $\bar{\mathbf{4}}$ multiplet, all these quantities are replaced by their negatives. Since all representations can be constructed from the **4** and $\bar{\mathbf{4}}$, and since the quantum numbers of Eqs. (10.4)–(10.9) are additive, we can obtain the values of these quantum numbers for any multiplet.

Some physicists do not define all these additive quantum numbers as we have done. For example, hypercharge is sometimes defined in $SU(4)$ as $Y = \lambda_8/\sqrt{3}$, as it is in $SU(3)$. We do not use this definition because it causes charmed hadrons to have fractional hypercharge. In $SU(3)$, the charges of the members of any multiplet can be written in terms of traceless matrices. However, as we see from Eq. (10.9), this is not true in $SU(4)$. Thus, in $SU(4)$ the advantage of defining the hypercharge in terms of traceless matrices is lost, and we might as well use the definition (10.6).

[2] Different formulas, corresponding to different definitions of the hypercharge of charmed particles, can be found in the literature.

10.2 Heavy Mesons with and without Charm

In the $SU(4)$ symmetry scheme, mesons are assumed to belong to mixed **15** + **1** multiplets, or 16-plets. As of the time of this writing, the experimental evidence in favor of this assignment is not complete. Nevertheless, we expect that the $SU(4)$ scheme will be verified.

Using the methods of Chapter 7, we can obtain the $SU(3)$ content of any $SU(4)$ representation. For the **15** and **1** we obtain

$$\mathbf{15} \supset \mathbf{8} \oplus \mathbf{1} \oplus \mathbf{3} \oplus \bar{\mathbf{3}}, \tag{10.10}$$

$$\mathbf{1} \supset \mathbf{1}. \tag{10.11}$$

The **8** and the **1** of $SU(3)$ have charm $C = 0$, the **3** has $C = -1$, and the $\bar{\mathbf{3}}$ has $C = 1$. The **3** and $\bar{\mathbf{3}}$ of SU(3) have the Young tableaux

 and 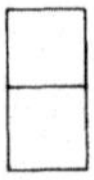

respectively. From Eqs. (7.42) and (7.45) we see that the **3** has triality 1 and the $\bar{\mathbf{3}}$ has triality -1. It is only the $SU(3)$ multiplets with $C = 0$ which are restricted to have triality 0.

In the $SU(4)$ scheme, the ψ meson, together with the vector meson nonet of Table 9.7, form an $SU(3)$-mixed decuplet with $C = 0$. The remaining

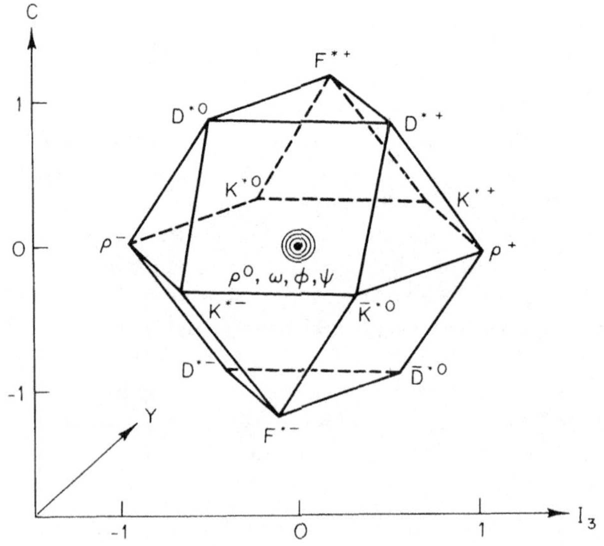

Fig. 10.1. $SU(4)$ weight diagram of the vector meson mixed **15** and **1** or 16-plet.

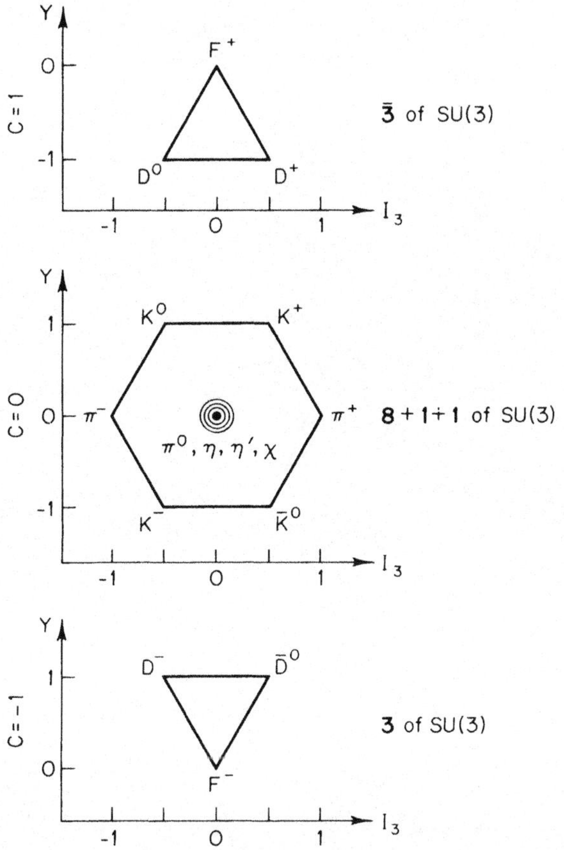

Fig. 10.2. *SU*(3) slices of the *SU*(4) weight diagram of the pseudoscalar meson 16-plet.

six vector mesons have charm. Because of the structure of the **15** of *SU*(4), the members of the **3** and $\bar{\mathbf{3}}$ must be antiparticles of each other.

The nomenclature for mesons belonging to the mixed 16-plet, but not included in the nonet of Table 9.5, is given in Table 10.1. The symbols for the charmed mesons are those of Gaillard *et al.* (1975).

As we have previously noted, the weight diagram of an *SU*(4) multiplet is three dimensional. We sketch in Fig. 10.1 the weight diagram of the mixed **15** and **1** or 16-plet of vector mesons expected if the broken symmetry is *SU*(4). For the pseudoscalar mesons, rather than give the *SU*(4) weight diagram, we give *SU*(3) slices corresponding to different values of charm. This is shown in Fig. 10.2.

In Table 10.2 we give the masses, mean lives, and principal decay modes of the family of ψ and χ mesons and of D, D^*, F, and F^* mesons. The data are adapted from the Particle Data Group (1976, 1977), Feldman (1977), and De Boer (1977).

TABLE 10.1

NOMENCLATURE FOR THE NEW MESONS[a]

Symbol	I	Y	S	C	$SU(3)$ multiplicity
ψ, χ[b]	0	0	0	0	1
D	$\frac{1}{2}$	-1	0	1	3
F	0	0	1	1	

[a] See Table 9.5 for nomenclature of other $C = 0$ mesons.

[b] The symbol ψ is used if C-parity is negative, χ if C-parity is positive.

TABLE 10.2

SOME PROPERTIES OF HEAVY MESONS

Symbol[a]	J^P	Mass (MeV)	Full width (MeV)	Principal decay modes[b]
χ?	0^+?	2850		$\gamma\gamma$
ψ	1^-	3095 ± 1	0.067 ± 0.12	e^+e^- $(7.3 \pm 0.5\%$ $\mu^+\mu^-$ $(7.4 \pm 0.5)\%$ hadrons
χ	0^+?	3414 ± 4		$\psi\gamma$ $(3 \pm 3)\%$ hadrons
χ	0^-?	3454 ± 7		
χ	1^+?	3508 ± 4		$\psi\gamma$ $(35 \pm 7)\%$ hadrons
χ	2^+?	3552 ± 6		$\psi\gamma$ $(14 \pm 6)\%$ hadrons
ψ	1^-	3684 ± 5	0.23 ± 0.06	e^+e^- $(0.9 \pm 0.2)\%$ $\mu^+\mu^-$ $(0.9 \pm 0.2)\%$ $\psi\pi\pi$ $(49 \pm 5)\%$ $\chi(3414)\gamma$ $(7 \pm 2)\%$ $\chi(3508)\gamma$ $(7 \pm 2)\%$ $\chi(3552)\gamma$ $(7 \pm 2)\%$
ψ	1^-	3772 ± 5	28 ± 7	$D\bar{D}$
ψ	1^-	~ 4100	(may be several states)	hadrons
ψ	1^-	4414 ± 7	33 ± 10	hadrons
D^0	0^-	1863 ± 1		$K^-\pi^+$ $K^-\pi^+\pi^+\pi^-$
D^+	0^-	1868 ± 1		$K^-\pi^+\pi^+$
D^{*0}	1^-	2006 ± 2		$D^0\gamma$
D^{*+}	1^-	2010 ± 3		$D^0\pi^0$
F^+	0^-	2030 ± 60		$\pi^+\eta$
F^{*+}	1^-	2140 ± 60		$F^+\gamma$

[a] A question mark after the symbol for a particle means that the experimental evidence for its existence needs confirmation. A question mark after the J^P value means that the quantum numbers need confirmation.

[b] Some of the decay modes are not yet known.

10.3 Charmed Baryons

As we have noted, the $SU(3)$ baryon octet becomes enlarged in $SU(4)$ to a $\mathbf{20}_M$, the $SU(3)$ decuplet becomes a $\mathbf{20}_S$, and the $SU(3)$ singlet becomes a $\bar{\mathbf{4}}$. The $SU(3)$ content of the $\mathbf{20}_M$, $\mathbf{20}_S$, and $\bar{\mathbf{4}}$ are

$$\mathbf{20}_M \supset \mathbf{8}_0 \oplus \mathbf{6}_1 \oplus \bar{\mathbf{3}}_1 \oplus \mathbf{3}_2, \tag{10.12}$$

$$\mathbf{20}_S \supset \mathbf{10}_0 \oplus \mathbf{6}_1 \oplus \mathbf{3}_2 \oplus \mathbf{1}_3, \tag{10.13}$$

$$\bar{\mathbf{4}} \supset \mathbf{1}_0 + \bar{\mathbf{3}}_1, \tag{10.14}$$

where the charm of each multiplet is written as a subscript.

Two different sets of symbols are in common use for the charmed baryons. The first, that of Gaillard *et al.* (1975), uses an entirely new set of symbols for these baryons. The second (Lichtenberg, 1975) keeps the usual notation for baryons of a given isospin and hypercharge (see Table 9.1), but uses a subscript to distinguish charmed from uncharmed baryons. In Table 10.3 we give both sets of symbols, but in the text, to avoid proliferation of symbols, we use only a notation with subscripts.

TABLE 10.3

NOMENCLATURE FOR CHARMED BARYONS

Symbol[a]		$SU(3)$ multiplicity	Isospin I	Hypercharge Y	Charm C
C_0	Λ_c	$\bar{3}$	0	0	1
C_1	Σ_c	6	1	0	1
S	Ξ_c	6	$\frac{1}{2}$	-1	1
A	Ξ_c^A	$\bar{3}$	$\frac{1}{2}$	-1	1
T	Ω_c	6	0	-2	1
X	Ξ_{cc}	3	$\frac{1}{2}$	-1	2
X_S	Ω_{cc}	3	0	-2	2
?	Ω_{ccc}	1	0	-2	3

[a] In the first column is the notation of Gaillard *et al.* (1975), in which new symbols are used for charmed baryons. In the second column is the notation of this book, in which the symbols of Table 9.1 are used with subscripts to denote charm. The superscript A on Ξ_c^A is necessary to distinguish this baryon from the Ξ_c, which belongs to a different $SU(3)$ multiplet.

In Fig. 10.3 we give a sketch of the $SU(4)$ weight diagram of the $\mathbf{20}_M$. This sketch is based on one given by Gaillard *et al.* In Fig. 10.4 we sketch the weight diagram of the $\mathbf{20}_S$. At present, very little is known experimentally about charmed baryons. There is, however, fair evidence for the Λ_c

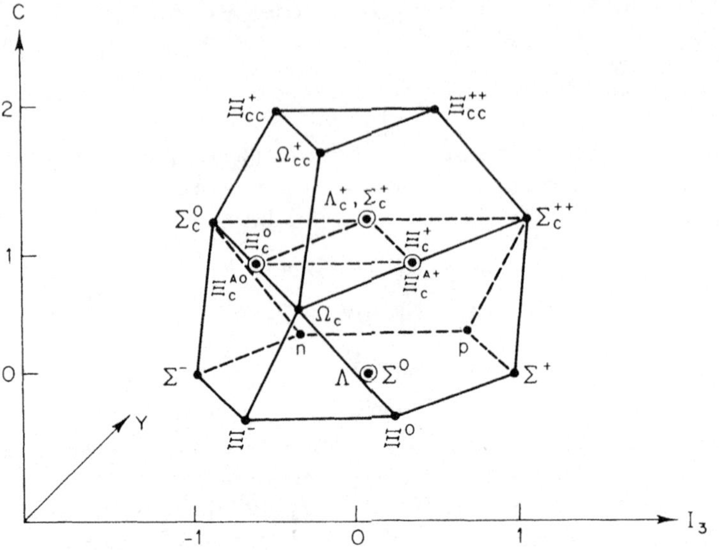

Fig. 10.3. $SU(4)$ weight diagram of the baryon $\mathbf{20}_M$. Of the charmed baryons, only the Λ_c^+ and possibly the Σ_c^+ have been seen.

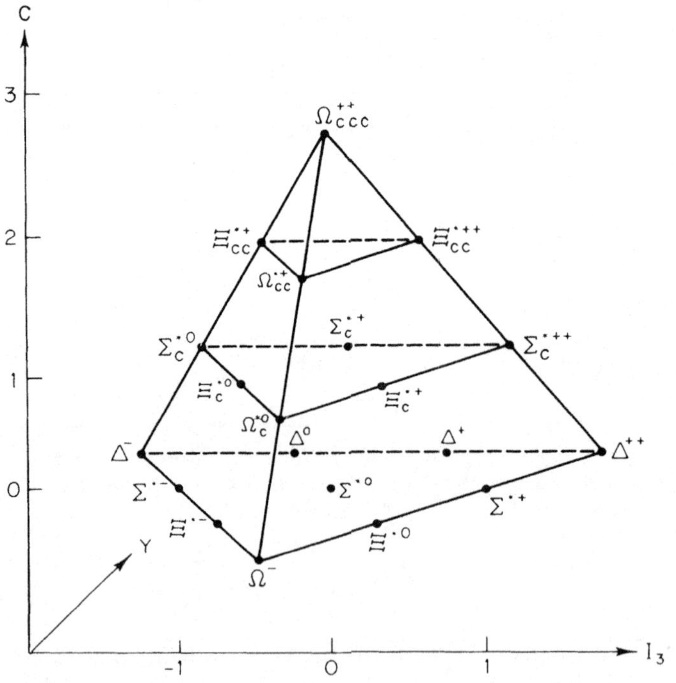

Fig. 10.4. $SU(4)$ weight diagram of the baryon $\mathbf{20}_S$. Of the charmed baryons, none has been observed except possibly the Σ_c^{*+}.

and lesser evidence for the Σ_c or Σ_c^* (Knapp *et al.*, 1976). See also Cazzoli *et al.* (1975). Some of the information from Knapp *et al.* is given in Table 10.4. It is amusing, but somewhat puzzling, that Knapp *et al.* claim to have seen the antiparticle $\bar{\Lambda}_c$, but not the Λ_c itself. If the evidence of Knapp *et al.* is confirmed, it will mark the first time an antibaryon has been seen before its baryon.

TABLE 10.4

SOME PROPERTIES OF CHARMED BARYONS

Symbol	Mass (MeV)	Width	Observed decay modes
Λ_c^+	2250	?	$\Lambda\pi^+\pi^+\pi^-$
Σ_c^+ or Σ_c^{*+}	2500	?	$\Lambda_c^+ + \pi^0$

10.4 $SU(4)$ Symmetry Breaking

If $SU(4)$ symmetry were exact, all members of a given $SU(4)$ multiplet would have the same mass. This is clearly not the case. For example, the π and D mesons, which are in the same $SU(4)$ multiplet, have masses of 139 MeV and 1870 MeV, respectively. Of course, as we have seen, the $SU(2)$ of isospin and the $SU(3)$ of isospin and hypercharge are also broken symmetries. But the symmetry breaking appears to get larger as we go to larger $SU(n)$ groups. Members of an $SU(2)$ isospin multiplet have mass differences less than 10 MeV, members of an $SU(3)$ multiplet have mass differences up to several hundred MeV, and members of an $SU(4)$ multiplet have mass differences of up to 1000 MeV or more.

Because of the large $SU(4)$ symmetry breaking, a simple $SU(4)$ mass formula, analogous to the Gell-Mann–Okubo formula of $SU(3)$, might not be a good approximation. Nevertheless, a number of authors, including Gaillard *et al.* (1975), Okubo (1975), and Moffat (1975) have written down $SU(4)$ mass formulas. As noted by Okubo (1975), an $SU(4)$ mass formula can be obtained from the hypothesis that the mass-breaking operator transforms like a component of a 15-dimensional representation of $SU(4)$. It is rather complicated to write down the resulting mass formula for a general $SU(4)$ representation, and we shall not do so. We shall also not write down here mass formulas for the special cases considered by Gaillard *et al.*, Okubo, and Moffat. Instead, we shall consider hadron mass formulas in Section 12.8 within the framework of the quark model.

CHAPTER 11

APPROXIMATE $SU(6)$ AND $SU(8)$

11.1 Dynamical Symmetry

A number of years ago Wigner (1937) considered the group $SU(4)$ as an approximate symmetry of nuclei. This symmetry can come about as follows. The nucleon is a doublet of the $SU(2)$ isospin group and also a doublet of the $SU(2)$ ordinary spin group. Thus, the nucleon has four degrees of freedom, and can be considered to belong to a quartet representation of the direct-product group $SU(2) \times SU(2)$. This direct product is contained in $SU(4)$. If the forces between nucleons are approximately independent of spin and isospin, then $SU(4)$ is an approximate symmetry of nuclei. Under such circumstances, the nucleon can be considered to belong to the first fundamental (four-dimensional) representation of $SU(4)$.

However, unlike the isospin $SU(2)$ group, which may be an approximate symmetry no matter how high the energy, $SU(4)$ is at best a dynamical symmetry which cannot be valid at high energy. We see this as follows. If $SU(4)$ is a valid symmetry, spin and isospin indices are indistinguishable and can be transformed into each other. Now we perform a Lorentz boost on a system with $SU(4)$ symmetry. Such a boost mixes spin and orbital angular momentum. However, the $SU(2)$ isospin group is an internal symmetry group and cannot be mixed with orbital angular momentum by a boost. Therefore a boost distinguishes between spin and isospin and destroys $SU(4)$ invariance.

If we have a free massive single particle, we can always consider it in its rest system, in which case $SU(4)$ might be a good symmetry. Likewise, if we have two particles moving very slowly with respect to each other, $SU(4)$ might hold to a good approximation. But if we have two particles moving rapidly with respect to each other, there is no frame of reference in which $SU(4)$ can hold. Thus, $SU(4)$ can at best be a dynamical symmetry which holds in the nonrelativistic limit.

Note that the $SU(4)$ we are discussing here differs greatly from the $SU(4)$ considered in the last chapter. The $SU(4)$ of isospin, strangeness, and charm is an internal symmetry group, while the present $SU(4)$ is a mixed group containing the internal isospin group and the group of ordinary spin.

Wigner's idea has been generalized to the case of elementary particles. Gürsey and Radicati (1964), Sakita (1964), and others considered $SU(6)$, and Moffat (1965) and Iwao (1965), among others, treated $SU(8)$.

The internal symmetry subgroup of elementary particles is apparently $SU(4)$ or an even larger group, so that the direct-product group of the internal symmetry group and the spin group is probably at least $SU(4) \times SU(2)$. Let us consider the case $SU(2n) \supset SU(n) \times SU(2)$, where $SU(n)$ is the internal symmetry group, and $SU(2)$ is the spin group. As we know $SU(n)$ is a simple group of rank $n - 1$ with $n - 1$ additive quantum numbers, so that the direct-product group $SU(n) \times SU(2)$ has n additive quantum numbers. However, the group $SU(2n)$ contains $2n - 1$ additive quantum numbers. The extra $n - 1$ of these (over and above the n quantum numbers of the direct-product group) do not seem particularly useful. Since spin is contained as a subgroup of $SU(2n)$, we expect to find multiplets of $SU(2n)$ containing states of different spin.

11.2 Classification of Hadrons in $SU(6)$

Let us first consider hadrons with charm zero, so that the underlying internal symmetry group is $SU(3)$, and the group including spin is $SU(6)$. In this case we wish to consider only those $SU(6)$ multiplets which include $SU(3)$ multiplets of zero triality. We ensure this in the following way.

Recall that in Chapter 7 we introduced an integer k defined in Eqs. (7.42) and (7.43) which gives the class of a representation of $SU(n)$ specified by a particular Young tableau. The integer k is just the number of boxes in the tableau, modulo n. Because the 0-triality multiplets correspond to $k = 0$, with $n = 3$, we must restrict ourselves to multiplets corresponding to Young tableaux with boxes which are a multiple of 3.

We distinguish baryon multiplets from meson multiplets in a similar way. The $k = 0$ representations have integral spin in $SU(2)$ and correspond to

Young tableaux with an even number of boxes, while the $k = 1$ representations have half-integral spin and correspond to tableaux with an odd number of boxes. Because mesons have integral spin they must belong to even tableaux, while baryons, with half-integral spin, must belong to odd tableaux.

Putting the $SU(3)$ and $SU(2)$ requirements together, we conclude that baryons must belong to multiplets corresponding to tableaux with 3, 9, 15, ... boxes, while mesons must belong to tableaux with 0, 6, 12, ... boxes. It follows in particular that, because baryons have half-integral spin, they cannot belong to $SU(6)$ singlets.

Let us first consider the baryons. It is simplest to consider $SU(6)$ multiplets corresponding to Young tableaux with only three boxes rather than to tableaux with 9, 15, or more boxes. There are only the following three possible arrangements of three boxes:

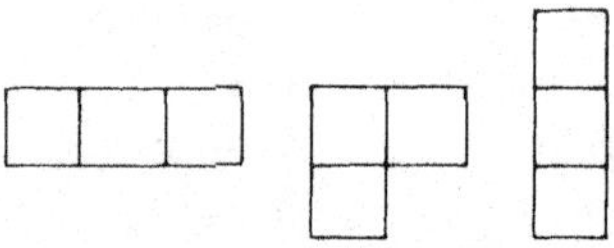

$SU(6)$: **56** **70** **20** (11.1)

where we have also given the dimensionality of the corresponding representations of $SU(6)$. We see that the 56-dimensional representation is completely symmetric, the 20-dimensional representation is antisymmetric, and the 70-dimensional representation has mixed symmetry.

In order to classify the baryons according to $SU(6)$, we need to know which submultiplets of $SU(3) \times SU(2)$ are contained in a given multiplet of $SU(6)$. We call this the $SU(3) \times SU(2)$ *content* of the multiplet. To obtain the content of the three-box tableaux of $SU(6)$ we look at the three-box tableaux of the $SU(3)$ and $SU(2)$ subgroups. The possible $SU(3)$ tableaux are

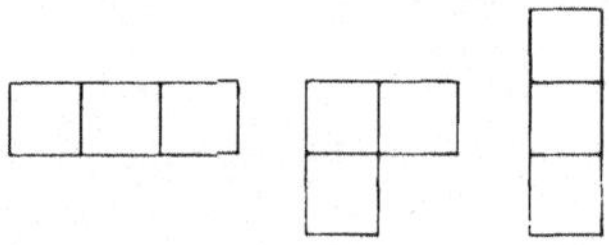

$SU(3)$: **10** **8** **1** (11.2)

and the possible $SU(2)$ tableaux are

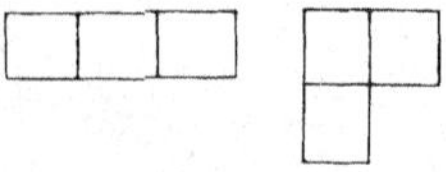

$SU(2)$: **4** **2** (11.3)

There is no antisymmetric $SU(2)$ tableau with three boxes, since $SU(2)$ has only two degrees of freedom.

Let us denote a multiplet of $SU(3) \times SU(2)$ by the symbol ${}^{2J+1}N$, where

$2J + 1$ is the $SU(2)$ multiplicity, and N is the $SU(3)$ multiplicity. We see from (11.2) and (11.3) that the only possible multiplets of $SU(3) \times SU(2)$ corresponding to three-box tableaux are the following:

$$^{4}10, \quad ^{2}10, \quad ^{4}8, \quad ^{2}8, \quad ^{4}1, \quad ^{2}1. \tag{11.4}$$

The multiplet $^{4}10$ is completely symmetric in both $SU(3)$ and $SU(2)$ indices, and therefore is totally symmetric (see Table 11.1). Thus, the $^{4}10$ multiplet

TABLE 11.1

SYMMETRY PROPERTIES OF PRODUCTS (OR SUMS OF PRODUCTS) OF TWO FACTORS WHICH ARE BASIS FUNCTIONS OF THE SYMMETRIC GROUP S_3

Factors	Product
SS or AA[a]	S
SA	A
SM or AM	M
MM	S, A, or M

[a] The symbols S, A, and M stand for symmetric, antisymmetric, and mixed symmetry, respectively.

must belong to the symmetric **56** of $SU(6)$. The $^{4}1$ is antisymmetric in $SU(3)$ indices and symmetric in $SU(2)$ indices, so that it is overall antisymmetric. Therefore it must belong to the **20** of $SU(6)$. Likewise, we see that the $^{2}10$, $^{2}1$, and $^{4}8$, each being the product of a symmetric or antisymmetric tensor with a mixed tensor, must have mixed overall symmetry and belong in the **70**. This leaves only the $^{2}8$, which is the product of two mixed tensors, unclassified. But the product of two mixed tensors can be overall symmetric, antisymmetric, or mixed. Therefore, a $^{2}8$ can go into a **56**, a **70**, or a **20** of $SU(6)$. We thus have

$$\mathbf{56} \supset {}^{4}10 + {}^{2}8, \tag{11.5}$$

$$\mathbf{70} \supset {}^{2}10 + {}^{2}1 + {}^{4}8 + {}^{2}8, \tag{11.6}$$

$$\mathbf{20} \supset {}^{4}1 + {}^{2}8. \tag{11.7}$$

We can check these assignments by counting states. For example, the $^{4}10$ contains 40 states and the $^{2}8$ contains 16 for a total of 56.

Since, experimentally, the baryon octet is an $SU(2)$ doublet and the baryon decuplet is an $SU(2)$ quartet, both these multiplets can be combined into the **56** multiplet of $SU(6)$. The justification for combining the octet and decuplet into the **56** is not compelling, since both the **70** and the **20** of $SU(6)$ contain an octet of spin $\frac{1}{2}$. However, if $SU(6)$ is an approximate symmetry, the decuplet must be accompanied by an octet, and the observed baryon octet is a good candidate.

Since parity is a quantum number outside $SU(6)$, it is essential that all members of a multiplet have the same parity. Since the baryon octet and decuplet are both positive-parity multiplets, their assignment to the **56** is allowed.

We next consider classification of the mesons. For simplicity we restrict ourselves to multiplets with tableaux of zero and six boxes, rather than considering multiplets with 12 or more boxes. The tableau of zero boxes is equivalent to a tableau with six boxes in a single column. Therefore, we consider the tableaux with six boxes. There are quite a few of these, and so we restrict ourselves to those which are self-conjugate. Our reason for doing so is that all known $SU(3)$ multiplets with zero charm contain both particles and antiparticles. It is only the self-conjugate diagrams which correspond to multiplets with this property.

The only self-conjugate tableaux with six boxes are

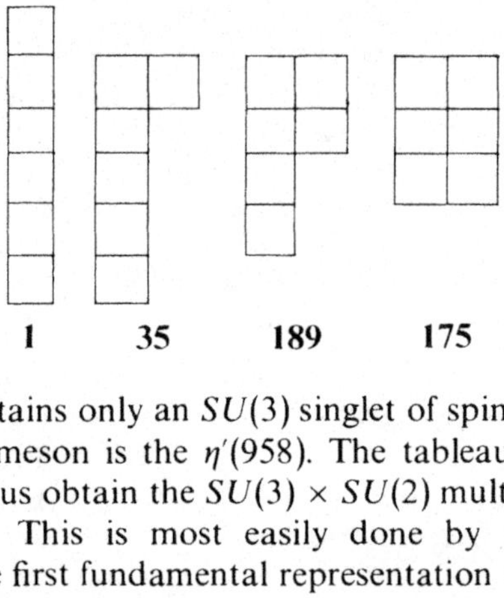

$SU(6)$: **1** **35** **189** **175**

The $SU(6)$ singlet contains only an $SU(3)$ singlet of spin zero. The most likely candidate for this meson is the $\eta'(958)$. The tableau of next lowest multiplicity is the **35**. Let us obtain the $SU(3) \times SU(2)$ multiplets contained in this $SU(6)$ multiplet. This is most easily done by considering the Kronecker product of the first fundamental representation of $SU(6)$ and its conjugate. Using Eq. (7.38), we have

$$\mathbf{6} \otimes \bar{\mathbf{6}} = \mathbf{35} \oplus \mathbf{1}. \qquad (11.8)$$

Making use of the fact that

$$\mathbf{6} \supset {}^{2}3,\ \bar{6} \supset {}^{2}\bar{3}, \qquad (11,9)$$

we obtain

$${}^{2}3 \times {}^{2}\bar{3} = {}^{(2\times 2)}(3 \times \bar{3}) = {}^{(3+1)}(8 + 1) = {}^{3}8 + {}^{1}8 + {}^{3}1 + {}^{1}1. \qquad (11.10)$$

Obviously the singlet of $SU(3) \times SU(2)$ must also go in the singlet of $SU(6)$, since the other representations of $SU(3) \times SU(2)$ have higher dimensions.

Thus the 24-dimensional, eight-dimensional, and three-dimensional representations of $SU(3) \times SU(2)$ must all belong in the 35 representation of $SU(6)$:

$$\mathbf{35} \supset {}^{3}8 + {}^{1}8 + {}^{3}1. \tag{11.11}$$

The arithmetic works out right: $24 + 8 + 3 = 35$.

If we look at the meson states of lowest mass, these belong to an octet and singlet with $J^P = 0^-$ and a mixed octet and singlet (or nonet) with $J^P = 1^-$. The pseudoscalar octet and the vector meson octet and singlet fill up the 35 multiplet nicely.

Note that these meson states all have negative parity. If the vector meson octet, the pseudoscalar octet, or the vector meson singlet had positive parity, $SU(6)$ would be violated but not $SU(2) \times SU(3)$. Thus, there is more predictive power in $SU(6)$ than in $SU(2) \times SU(3)$.

There is not convincing experimental evidence that any mesons belong to the **189** or **175** multiplets. Therefore we shall not consider them further. The interested reader may work out the $SU(3) \times SU(2)$ content of these multiplets.

If $SU(6)$ is an approximate dynamical symmetry, we perhaps have a clue toward understanding why the vector meson octet and singlet are more mixed than the pseudoscalar octet and singlet. The vector octet and singlet are both in the same multiplet of $SU(6)$, while the pseudoscalar octet and singlet are in different multiplets. Therefore, the vector mesons are more nearly degenerate in energy than the pseudoscalar mesons, and the $SU(3)$ mixing interaction has a chance to be more effective in the vector meson case.

Thus, $SU(6)$ provides a good framework for classifying the lowest mass baryons and mesons into multiplets. In classifying the states of higher mass, it is useful to consider the multiplets of a still larger group: $SU(6) \times O(3)$. However, it is most convenient to consider this group within the framework of the quark model, and so we postpone these considerations to the next chapter.

A number of other predictions may be made using $SU(6)$. One of the most successful is the prediction that the ratio of the magnetic moment of proton to neutron is given by (Bég *et al.*, 1964)

$$\mu(p)/\mu(n) = -3/2. \tag{11.12}$$

This compares to the experimental value

$$\mu(p)/\mu(n) = -1.46. \tag{11.13}$$

The prediction follows from $SU(6)$ and the assumption that the electric charge transforms like a component of an $SU(3)$ octet operator. We shall derive Eq. (11.12) in Chapter 12, using the quark model.

11.3 Classification of Hadrons in $SU(8)$

We now turn to $SU(8)$. In the case of baryons, as we mentioned in Chapter 10, we consider the same Young tableaux of three boxes that we treated for $SU(6)$. In the absence of experimental information about charmed baryons, this choice appears arbitrary, but the real motivation comes from the quark model, which we introduce in the next chapter.

The relevant Young tableaux with three boxes are exhibited in Eq. (11.1), but the dimensionalities are different in $SU(8)$. The symmetric tableau has dimension **120**, the tableau of mixed symmetry has dimension **168**, and the antisymmetric tableau has dimension **56**. These numbers are readily calculated using the formulas of Section 7.2. Note that a **56** is antisymmetric in $SU(8)$, but symmetric in $SU(6)$.

The decomposition of these multiplets into multiplets of $SU(4) \times SU(2)$ proceeds along the same lines as the decomposition of $SU(6)$ multiplets into multiplets of $SU(3) \times SU(2)$. We therefore give only the results, leaving the details as an exercise for the reader. The $SU(4) \times SU(2)$ content of these $SU(8)$ multiplets is

$$\mathbf{120} \supset {}^4 20_S + {}^2 20_M, \tag{11.14}$$

$$\mathbf{168} \supset {}^2 20_S + {}^2\bar{4} + {}^4 20_M + {}^2 20_M, \tag{11.15}$$

$$\mathbf{56} \supset {}^4\bar{4} + {}^2 20_M. \tag{11.16}$$

If we compare Eqs. (11.14)–(11.16) with Eqs. (11.5)–(11.7), we see that, in order to go from $SU(6)$ to $SU(8)$, we need merely replace the $SU(3)$ multiplicities by the $SU(4)$ multiplicities corresponding to the same Young tableaux. Thus, the procedure is easily generalized to any $SU(2n) \supset SU(n) \times SU(2)$.

In the case of the mesons, we do not keep the same Young tableaux as we enlarge the group, as we have already remarked in the last chapter. In classifying mesons according to $SU(3)$, we have seen that they occur in mixed $\mathbf{8} \oplus \mathbf{1}$ multiplets or nonets. Likewise in $SU(4)$, we believe that the mesons occur in mixed $\mathbf{15} \oplus \mathbf{1}$ or 16-plets. More generally, in $SU(n)$, we expect the mesons to occur in multiplets of $(\mathbf{n}^2 - \mathbf{1}) \oplus \mathbf{1}$. Generalizing from the case of $SU(6)$, we assume that this pattern also holds in the case of multiplets of

$SU(2n)$. Again, this choice will be justified in the next chapter. Then the relevant meson multiplets of $SU(8)$ correspond to the Young tableaux

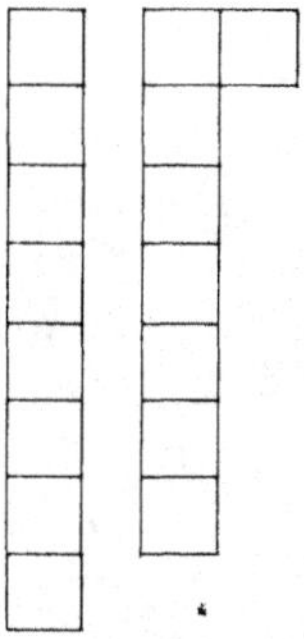

$$SU(8): \quad \mathbf{1} \quad \mathbf{63} \tag{11.17}$$

Likewise, Eq. (11.11) is readily generalized to $SU(8)$. It becomes

$$\mathbf{63} \supset {}^3 15 + {}^1 15 + {}^3 1. \tag{11.18}$$

For any $SU(2n)$, the result is

$$(\mathbf{2n})^2 - \mathbf{1} \supset {}^3(n^2 - 1) + {}^1(n^2 - 1) + {}^3 1. \tag{11.19}$$

The singlet for any $SU(2n)$ of course remains the same

$$\mathbf{1} \supset {}^1 1. \tag{11.20}$$

11.4 Troubles with $SU(6)$ and $SU(8)$

Using $SU(6)$, we can make a number of predictions about hadron–hadron scattering. For example, consider meson–baryon scattering

$$M + B \rightarrow M + B, \tag{11.21}$$

where M is a meson belonging to the **35**, and B is a baryon belonging to the **56**. To see how many independent amplitudes are necessary to describe this process, we decompose the Kronecker product by the method of Young tableaux. We obtain

$$\mathbf{56} \otimes \mathbf{35} = \mathbf{56} \oplus \mathbf{70} \oplus \mathbf{700} \oplus \mathbf{1134}. \tag{11.22}$$

We see that all scatterings of the form (11.21) (and there are many such processes accessible to experiment) are predicted to be given in terms of only four amplitudes. The relations between the various amplitudes can be obtained by using the $SU(6)$ Clebsch–Gordan coefficients.

Unfortunately, however, there is no reason to expect these $SU(6)$

predictions to agree with experiment. We can see this as follows. Since $SU(3)$ is a subgroup of $SU(6)$, whenever $SU(3)$ is badly broken, $SU(6)$ will also fail to give good predictions. Now at low energy, the mass splitting of the $SU(3)$ multiplets will lead to large violations of $SU(3)$ symmetry. An extreme example is the case in which the energy is sufficiently low so that some reactions are forbidden by energy conservation, while other reactions related by $SU(3)$ symmetry are allowed. So $SU(6)$ is expected to give poor results at low energy. But in Section 11.1 we argued that $SU(6)$ cannot be a good symmetry at high energy. Therefore, we cannot expect the predictions of $SU(6)$ for hadron–hadron scattering to be good at any energy.

The situation with respect to the strong decays of unstable hadrons is even worse. Since the $SU(2)$ subgroup of $SU(6)$ is the spin group, spin and orbital angular momentum are separately conserved in any interaction which is invariant under $SU(6)$.

Suppose we consider the decay of the ρ meson, which has spin 1. Experimentally, the ρ is observed to decay strongly into two pions with a decay width Γ of more than 100 MeV. However, pions have spin 0. Thus, in the decay, the spin of the ρ is converted into orbital angular momentum of the pions. But nonrelativistic $SU(6)$ has no mechanism to convert spin into orbital angular momentum, and the strong $\rho \to 2\pi$ decay is forbidden by the symmetry. Similarly the Δ baryon has spin $\frac{3}{2}$ and is observed to decay strongly into a nucleon and a pion with total spin $\frac{1}{2}$. Again the extra unit of spin is converted into orbital angular momentum, with the pion and nucleon emerging in a state of relative orbital angular momentum $L = 1$. But the nonrelativistic $SU(6)$ again forbids this decay. We conclude that although $SU(6)$ may be useful in the classification of states, it certainly does not lead to the correct selection rules for decays. One should regard $SU(6)$ as holding at best in the rest frame of a particle. A particle in its rest frame has no orbital angular momentum, and so for such states $SU(6)$ has a chance to be good.

Because of this difficulty, and others related to the nonrelativistic character of $SU(6)$, numerous attempts have been made to find a relativistic generalization. Most of these attempts have involved consideration of noncompact groups. Some of the defects of $SU(6)$ have in fact been overcome in this way, but only at the expense of other difficulties of principle. One possible exception is the group $SU(6)_W$, introduced by Lipkin and Meshkov (1965). This group allows the decays $\rho \to 2\pi$ and $\Delta \to N\pi$ to occur. Furthermore, it gives a number of moderately successful predictions concerning relative hadron decay rates and hadron–hadron scattering cross sections in the forward and backward directions. Pais (1966), in his excellent review, discusses the many attempts to make relativistic generalizations of $SU(6)$ and gives many references.

In Section 9.1, we assumed that the Hamiltonian H describing elementary particles consists of two parts H_0 and H', where H_0 is invariant under the direct product group $SU(3) \times P$ (where P is the Poincaré group) and H' breaks the symmetry. Unfortunately, we cannot extend this idea to $SU(6)$, because we do not have any physical interpretation of the direct-product group $SU(6) \times P$. If a group is to be a direct product, all the generators of one of the groups of the product must commute with all the generators of the other. But if the $SU(2)$ contained in $SU(6)$ is interpreted as the spin group, its generators do not commute with the angular momentum generators of P. This is just a reflection of the fact that the total angular momentum is conserved, but not spin and orbital angular momentum separately.

Since $SU(6) \times P$ does not make sense physically, a number of people have considered the possibility of embedding $SU(6)$ and P in a group G which is not a direct product. But as we have already remarked in Section 9.1, O'Raifeartaigh (1968) and others have shown that such a scheme does not work if G is a finite Lie group.

An approach which we favor is to recognize that $SU(6)$ is an approximate dynamical symmetry, and therefore cannot be a symmetry of the strong interaction Hamiltonian. Rather, we assume that the Hamiltonian is of such a form that when certain static properties of hadrons are calculated with it, the results are the same as those given by approximate $SU(6)$ invariance.

Since $SU(8)$ contains $SU(6)$ as a subgroup, all our remarks about the troubles with $SU(6)$ apply to $SU(8)$ as well. Furthermore, because $SU(4)$ is more badly broken than $SU(3)$, $SU(8)$ has even further troubles when regarded as a symmetry of nature.

We shall not dwell on these troubles further. Rather, in the next chapter, we shall discuss a model which does not incorporate $SU(6)$ or $SU(8)$ invariance, but nevertheless can be used to obtain some of the good predictions of these groups without giving the bad predictions. This is the quark model.

CHAPTER 12

THE QUARK MODEL

12.1 Predecessors to the Quark Model

Some time ago Fermi and Yang (1949) proposed a model in which a pion is a bound state of a nucleon and antinucleon. This model can be generalized to include a large number of other mesons with zero hypercharge. In this scheme, the nucleon isospin doublet and the antinucleon doublet are the fundamental building blocks out of which meson states of zero hypercharge are constructed.

We can fit this model into the framework of group theory. The state vectors of the nucleon

$$u_1 = p, \qquad u_2 = n, \tag{12.1}$$

can be considered as a fundamental doublet of $U(2)$, where the $SU(2)$ subgroup is the isospin, and the $U(1)$ subgroup is the baryon number. (In this case $I = \frac{1}{2}$, $B = 1$.) Likewise the antinucleon state vectors

$$v_1 = -\bar{n}, \qquad v_2 = \bar{p}, \tag{12.2}$$

can be considered to be a second fundamental doublet of $U(2)$, again with $I = \frac{1}{2}$, but with baryon number $B = -1$. The meson isospin multiplicities in this model are obtained from the Clebsch–Gordan series of two doublets:

$$\mathbf{2} \otimes \bar{\mathbf{2}} = \mathbf{3} \oplus \mathbf{1}. \tag{12.3}$$

We distinguish between **2** and $\bar{\mathbf{2}}$ because the group is $U(2)$, not $SU(2)$. However, the multiplicities of the Clebsch–Gordan series are the same for **2** and $\bar{\mathbf{2}}$.

We see from Eq. (12.3) that the model leads to a prediction that all mesons with $Y = 0$ should be either isospin triplets or singlets. This prediction is in agreement with the known facts up to the present except for the possible occurrence of relatively weakly coupled exotic mesons.

The model can be generalized to include excited baryon states with hypercharge $Y = 1$. Such excited baryons are considered as bound states of two nucleons and an antinucleon. The Clebsch–Gordan series is

$$\mathbf{2} \otimes \mathbf{2} \otimes \bar{\mathbf{2}} = \mathbf{2} \oplus \mathbf{2} \oplus \mathbf{4}. \tag{12.4}$$

Thus, the model predicts that all baryons of hypercharge $Y = 1$ should be isospin doublets and quartets. This prediction is also in accord with the presently known facts except possibly for weakly coupled exotic baryons.

Recall that the strangeness S of a hadron is defined by $S = Y - B$. Use of the quantum number S rather than Y lets us state the predictions of the model succinctly: The nucleon and antinucleon doublets are the building blocks of all hadrons of zero strangeness.

This scheme, while attractive in many ways, is insufficient to account for particles with strangeness different from zero. Sakata (1956) suggested the simplest extension of the Fermi–Yang model to include strangeness different from zero. Sakata's idea was to add one additional state to the fundamental building blocks: namely the Λ hyperon, which has the quantum numbers $I = 0$, $B = 1$, $S = -1$. Sakata postulated that all hadrons can be constructed from the triplet p, n, Λ, and their antiparticles.

Ikeda *et al.* (1959) took the next step by regarding the proton, neutron, and Λ to be a fundamental triplet of $U(3)$. As with the Fermi–Yang model, a meson is given by a bound state of a baryon and antibaryon. The relevant Clebsch–Gordan series is

$$\mathbf{3} \otimes \bar{\mathbf{3}} = \mathbf{8} \oplus \mathbf{1}. \tag{12.5}$$

Thus, the Sakata model leads to the successful prediction that mesons should occur in octets and singlets. Furthermore, if the interaction between baryon and antibaryon depends only weakly on F-spin (unitary spin), the octet and singlet will not differ appreciably in mass. With the assumption that this is the case, the model predicts that mesons should occur in multiplets of nine. Thus, within the framework of the model, we can obtain a dynamical explanation for the existence of nonets.

Although the Sakata model provides a good description of meson multiplicities, it leads to difficulties with the baryons. The reason is as follows: Since according to the model, the p, n, and Λ fill the fundamental triplet, the Σ and Ξ must belong to another multiplet. We can form other multiplets with

$B = 1$ from two Sakata triplets and an antitriplet. We can readily obtain the Clebsch–Gordan series with Young tableaux. We have

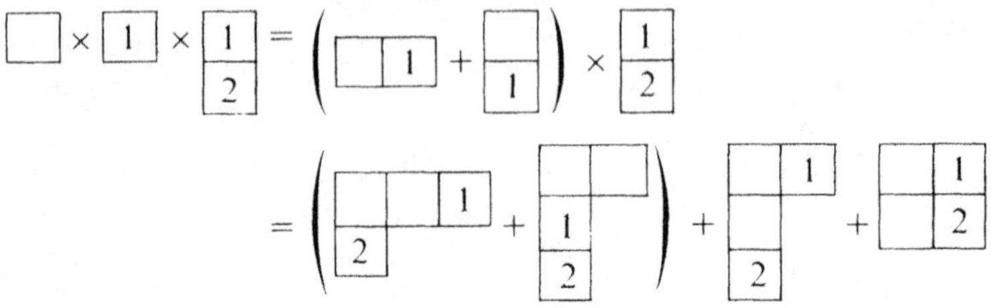

or, removing columns of three, we obtain

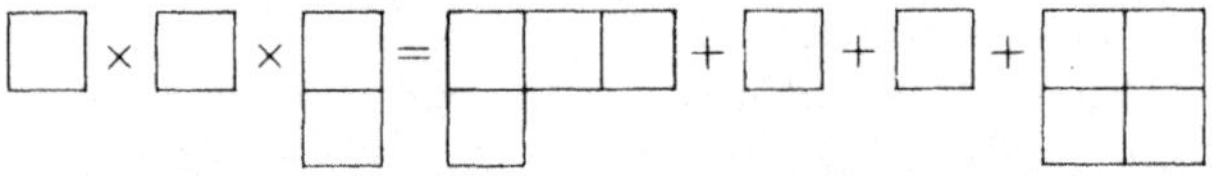

Then we have

$$\mathbf{3} \otimes \mathbf{3} \otimes \bar{\mathbf{3}} = \mathbf{15} \oplus \mathbf{3} \oplus \mathbf{3} \oplus \bar{\mathbf{6}}. \tag{12.6}$$

In the Sakata model, we must decompose the Kronecker product of $\mathbf{3} \otimes \mathbf{3} \otimes \bar{\mathbf{3}}$ because we want states with baryon number $B = 1$. In the model, each $\mathbf{3}$ has $B = 1$ and $\bar{\mathbf{3}}$ has $B = -1$. Then, since baryon number is an additive quantum number, from $\mathbf{3} \otimes \mathbf{3} \otimes \bar{\mathbf{3}}$ we get states with $B = 1$. From Eq. (12.6), we should expect to find baryon multiplets belonging to the $\mathbf{15}$, $\bar{\mathbf{6}}$, or $\mathbf{3}$ multiplets of $U(3)$. The Σ and Ξ must belong to a 15 with $J^P = \frac{1}{2}^+$ in this scheme, as the $\bar{\mathbf{6}}$ and $\mathbf{3}$ do not have the right quantum numbers. But the other members of this multiplet seem to be missing.

Similarly the Δ must belong to a 15 with $J^P = \frac{3}{2}^+$, since there is no $I = \frac{3}{2}$ state in the $\bar{\mathbf{6}}$ or $\mathbf{3}$ representations. But the $\mathbf{15}$ contains, in addition to these multiplets an $I = 1$ multiplet of strangeness $S = 1$ which has not been observed. Another difficulty is that there is no place for the Ω in the $\mathbf{15}$, $\bar{\mathbf{6}}$, or $\mathbf{3}$ representations. We shall not discuss the other difficulties with this baryon classification scheme except to note that there seems no way to obtain good agreement with experiment.

12.2 Quarks with Flavor, Including Charm

It was because of the failure of the Sakata model to give the correct baryon multiplicities that Gell-Mann and Ne'eman put the nucleon and Λ into an octet together with the Σ and Ξ. But if the p, n, and Λ do not belong to a fundamental triplet of $U(3)$, perhaps other, still undiscovered particles do. Gell-Mann (1964a) and Zweig (1964) independently proposed a model in which baryons and mesons are composites of a fundamental triplet of

$U(3)$.[1] Gell-Mann called these particles "quarks"; Zweig called them "aces." The name quark has stuck.[2]

One notation for the three quarks is u, d, and s [for up, down, and sideways (or strange)]. Other names, such as p, n, λ are frequently used in the literature. Often, it is convenient to use the symbols q_i ($i = 1, 2, \ldots, n$) for a model with n quarks.

Shortly after Gell-Mann and Zweig introduced the quark model with three quarks, a number of authors generalized the model to include a fourth quark.[3] Of these authors, we mention Bjorken and Glashow (1964), who called the fourth quark a quark with charm. We denote the charmed quark by the symbol c, although p' is sometimes used in the literature.

Originally, Bjorken and Glashow introduced a fourth quark because they thought that nature might have some kind of symmetry between quarks and leptons. Since at that time four leptons were known (the electron, the muon, and their respective neutrinos), according to the argument, there should be four quarks. Actually, this reasoning says that there should be at least four quarks, as we have no reason to believe that the known leptons constitute all the leptons that exist in nature.

Now that there is rather good evidence (Perl, 1977) for the existence of an additional charged lepton τ (and lesser evidence for its accompanying neutrino ν_τ), there should exist at least six quarks according to the argument of Bjorken and Glashow. The fifth and sixth quarks are often given the symbols b and t. Already, many authors have postulated the existence of more than four quarks. However, as of the time of this writing, there is no hard experimental evidence for more quarks coming from hadron spectroscopy. A good review of the subject of more than four quarks has been given by Harari (1976).

The different kinds of quarks are now often called quarks of different *flavors*. We do not have any principle which says how many flavors of quarks should exist. Therefore, even apart from any symmetry between quarks and leptons, we should not be surprised if in the future new hadrons are discovered which require additional quarks to account for them.

In the quark model in its simplest form, baryons are bound states of three quarks, and mesons are bound states of a quark–antiquark pair. The

[1] Actually, Goldberg and Ne'eman (1963) were the first to construct baryons out of a fundamental triplet with baryon number $B = \frac{1}{3}$. Gell-Mann and Zweig first raised the possibility that this triplet might consist of actual particles.

[2] Gell-Mann took the name from a quote from *Finnegan's Wake* by James Joyce: "Three quarks for Muster Mark." In this context, quark is archaic for croak. In German, the word quark has a number of meanings, including slime and cheese.

[3] Now that the model contains at least four quarks, the reference "three quarks" should be reinterpreted as referring to the three quark *colors*, discussed in the next section.

basic reason for this assumption is, as we shall see, that it leads to a classification scheme for the hadrons which is in agreement with experiment. Because a baryon consists of a state of three quarks, quarks are assumed to have baryon number $B = \frac{1}{3}$. Also, in order for baryons to have half-integral spin, it is necessary for quarks to have this property also. Following Gell-Mann and Zweig, we assume, for simplicity, that the quarks all have spin $\frac{1}{2}$ and the same parity (defined to be positive).

The u and d quarks are approximately degenerate in mass, and constitute an $SU(2)$ isospin doublet. The u and d quarks can, therefore, account for the additive quantum number I_3 in hadrons. The third quark s enlarges the group from $SU(2)$ to $SU(3)$, and carries the additive quantum number strangeness. The fourth quark c makes the group $SU(4)$, and carries the additive quantum number charm. A model with n quark flavors, therefore, can form the basis for an $SU(n)$ symmetry with $n - 1$ additive quantum numbers. This $SU(n)$ symmetry can be broken by giving the quarks different masses and by letting their interactions depend on flavor.

In Table 12.1 we list some of the properties of the four flavors of quarks which the model needs at present to account for almost all the presently known hadrons. We shall discuss later how the quark quantum numbers are arrived at and how the quark masses are estimated.

TABLE 12.1

QUANTUM NUMBERS OF THE QUARKS[a]

Symbols			Isospin I	I_3	Hypercharge Y	Strangeness S	Charm C	Charge Q	Mass[b] (MeV)
u	p	q_1	$\frac{1}{2}$	$\frac{1}{2}$	$\frac{1}{3}$	0	0	$\frac{2}{3}$	336
d	n	q_2	$\frac{1}{2}$	$-\frac{1}{2}$	$\frac{1}{3}$	0	0	$-\frac{1}{3}$	339
s	λ	q_3	0	0	$-\frac{2}{3}$	-1	0	$-\frac{1}{3}$	467
c	p'	q_4	0	0	$-\frac{2}{3}$	0	1	$\frac{2}{3}$	1550

[a] All quarks have baryon number $B = \frac{1}{3}$ and spin and parity $J^P = \frac{1}{2}^+$.
[b] These are estimated effective masses of bound quarks.

The symmetry of $SU(4)$ would be exact if all quarks had the same mass and the same interactions. However, in the model, the u and d quarks have effective masses of about 300 MeV, the s quark has an effective mass of about 500 MeV, and the c quark has an effective mass of about 1500 MeV. These effective masses are the apparent masses of quarks when bound in hadrons, and are not necessarily the masses of free quarks. The small mass difference between the u and d quarks (estimated between 2 and 5 MeV) gives rise to a small breaking of $SU(2)$ symmetry. Because the mass splitting of d and s

quarks is larger (about 100 to 200 MeV), $SU(3)$ is more badly broken. Likewise, because the c quark is about 1000 MeV heavier than the s quark, $SU(4)$ symmetry is even more badly broken than $SU(3)$. If additional quarks are needed in the model, their effective masses will determine in part how badly the higher symmetries are broken. In later sections of this chapter we shall discuss how the estimates of quark masses are arrived at.

In this book, although we have emphasized the groups $SU(3)$ and $SU(4)$, we have usually used methods which work for any $SU(n)$. Therefore, if more quarks are needed, the reader should find it easy to use the appropriate $SU(n)$. Of course, there remains the possibility that there are n quarks, but that the relevant broken internal symmetry group is not $SU(n)$. If so, the general methods of this book may not be needed for elementary particle physics. In any case, the existing experimental evidence points to the fact that $SU(3)$ at least, and probably also $SU(4)$, will remain subgroups of any larger broken internal symmetry group. (There is also the possibility that strongly interacting particles will some day be discovered which do not fit into the framework of the quark model at all.)

Let us first consider hadrons in a model with only three quark flavors, and then later include the charmed quark. It is convenient to consider the group $SU(3)$ rather than $U(3)$, and to treat the baryon number separately. Consider the decomposition of the Kronecker product $\mathbf{3} \otimes \mathbf{3} \otimes \mathbf{3}$. Using Young tableaux, we obtain

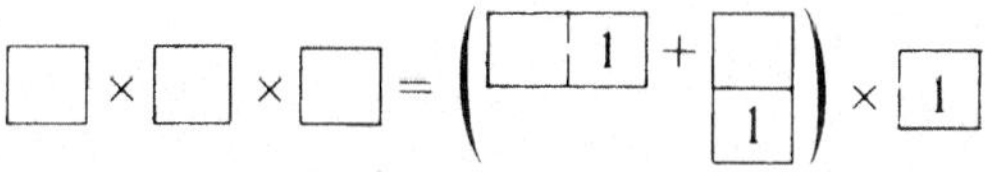

or

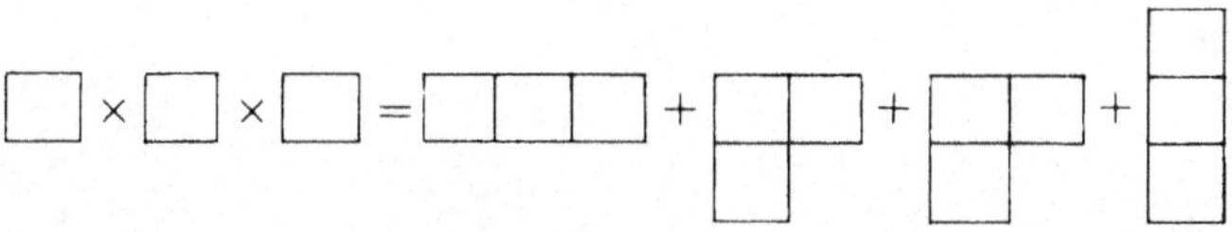

(12.7)

This reduction holds for any $SU(n)$ with $n > 2$. In particular, the $SU(3)$ multiplicities are

$$\mathbf{3} \otimes \mathbf{3} \otimes \mathbf{3} = \mathbf{10} \oplus \mathbf{8} \oplus \mathbf{8} \oplus \mathbf{1}. \tag{12.8}$$

We see from Eq. (12.8) that if we regard a baryon as made up of three triplets, rather than two triplets and an antitriplet, we obtain a baryon decuplet, two octets, and a singlet. Baryon multiplicities of 10, 8, and 1 have been observed experimentally. Because of the inequivalence of the $\mathbf{3}$ and $\bar{\mathbf{3}}$ representations of $SU(3)$, the decomposition of $\mathbf{3} \otimes \mathbf{3} \otimes \mathbf{3}$ is quite different from the decomposition of $\mathbf{3} \otimes \mathbf{3} \otimes \bar{\mathbf{3}}$. [Compare Eqs. (12.6) and (12.8).]

In the model a meson is made of a quark and antiquark. Thus, the Clebsch–Gordan series for mesons is given in the quark model by Eq. (12.5), just as in the Sakata model. As a consequence, mesons are predicted to occur in octets and singlets, or nonets.

In Section 9.1, we defined a hadron with $|B| \leq 1$ as exotic if it does not belong to a **1**, **8**, or **10** multiplet of $SU(3)$. We now give a definition of an exotic hadron within the framework of the quark model. A baryon is *exotic* if its quantum numbers are such that it cannot be a bound state of three quarks, and a meson is *exotic* if its quantum numbers are such that it cannot be a bound state of a quark and antiquark. With this definition all nonexotic baryons not containing any c quarks belong to **1**, **8**, or **10** multiplets, and all nonexotic mesons not containing c or $\bar{c}$ belong to **1** or **8**. However, our present definition is superior to our previous one, in that the definition in terms of quark composition is independent of the number of quark flavors. In other words, our previous definition applied only to $SU(3)$ without charm, whereas the present definition applies to any $SU(n)$. In $SU(4)$, for example, nonexotic charmed baryons (baryons containing one or more c quarks) can belong to $SU(3)$ submultiplets of **6**, **3**, and $\bar{\mathbf{3}}$. Likewise, nonexotic charmed mesons can belong to $SU(3)$ multiplets of **3** or $\bar{\mathbf{3}}$. We shall postpone discussion of exotic hadrons until Section 12.10.

Let us consider the baryon decuplet, which has the Young tableau $\begin{array}{|c|c|c|}\hline \ & \ & \ \\ \hline\end{array}$, and is therefore a symmetric state of three quarks. We can make the following identification between the baryons of the decuplet and the standard arrangements of this tableau:

$$\begin{array}{|c|c|c|}\hline 1&1&1\\ \hline\end{array} = \Delta^{++}, \quad \begin{array}{|c|c|c|}\hline 1&1&2\\ \hline\end{array} = \Delta^{+}, \quad \begin{array}{|c|c|c|}\hline 1&2&2\\ \hline\end{array} = \Delta, \quad \begin{array}{|c|c|c|}\hline 2&2&2\\ \hline\end{array} = \Delta^{-}$$

$$\begin{array}{|c|c|c|}\hline 1&1&3\\ \hline\end{array} = \Sigma^{*+}, \quad \begin{array}{|c|c|c|}\hline 1&2&3\\ \hline\end{array} = \Sigma^{*0}, \quad \begin{array}{|c|c|c|}\hline 2&2&3\\ \hline\end{array} = \Sigma^{*-}$$

$$\begin{array}{|c|c|c|}\hline 1&3&3\\ \hline\end{array} = \Xi^{*0}, \quad \begin{array}{|c|c|c|}\hline 2&3&3\\ \hline\end{array} = \Xi^{*-} \tag{12.9}$$

$$\begin{array}{|c|c|c|}\hline 3&3&3\\ \hline\end{array} = \Omega^{-}.$$

These basis functions are fully written out in Eq. (8.87) with $u_i \to q_i$. In particular, we see from Eq. (12.9) that the Δ^{++} is made from three identical quarks of type q_1, the Δ^{-} from three identical quarks of type q_2, and the Ω^{-} from three identical quarks of type q_3. Then, from the values of the additive quantum numbers of Δ^{++}, Δ^{-}, and Ω, we can deduce the corresponding quantum numbers of the quarks, which must have $\frac{1}{3}$ these values. Since The Δ^{++} has $I_3 = \frac{3}{2}$, $Y = 1$, $Q = 2$, the first quark q_1 has $I_3 = \frac{1}{2}$, $Y = \frac{1}{3}$, $Q = \frac{2}{3}$. Similarly, since Δ^{-} has $(I_3 Y Q) = (-\frac{3}{2}, 1, -1)$ the second quark q_2

has $(I_3YQ) = (-\frac{1}{2}, \frac{1}{3}, -\frac{1}{3})$. Likewise, since Ω has $(I_3YQ) = (0, -2, -1)$ the third quark q_3 has $(I_3YQ) = (0, -\frac{2}{3}, -\frac{1}{3})$. Since all baryons have baryon number $B = 1$, the quarks must have baryon number $B = \frac{1}{3}$. From the above arguments, one can see how the quantum numbers of Table 12.1 can be simply obtained for the *u*, *d*, and *s* quarks.

Of course, once the quantum numbers of the *u* and *d* quarks are determined, a knowledge of the quantum numbers of a single baryon carrying strangeness, such as the Λ baryon, suffices to fix the quantum numbers of the *s* quark. Likewise, once the quantum numbers of the *u*, *d*, and *s* are determined, we need only to know the quantum numbers of a single charmed baryon, such as the Λ_c baryon, to fix the quantum numbers of the *c* quark. (We note, however, that the quantum numbers of the *c* quark were assigned on other grounds, as the existence of this quark was postulated before the discovery of any charmed hadrons.)

Our notation of Table 9.1 gives a symbol to a baryon which depends only on its isospin and hypercharge. But we can see from Table 12.1 that the *s* and *c* quarks have the same values of these quantum numbers. It is therefore natural to give the same basic symbol to a baryon independently of how many *c* quarks replace *s* quarks in the baryon. The maximum in the simple model is of course three. We distinguish between baryons containing *c* and *s* quarks by using a subscript *c* for each charmed quark the baryon contains. Thus, for example, a Λ^0 contains the quarks *u d s*, while the Λ_c^+ contains *u d c*. As another example, the Ξ^0 contains *u s s*, the Ξ_c^+ contains *u s c*, and the Ξ_{cc}^{++} contains *u c c*. Note that the charge of a baryon (written as a superscript when we wish to denote the charge explicitly) is increased by unity for each *c* quark that replaces an *s* quark. The nomenclature of Table 10.3 for the charmed baryons conforms to the notation we have described here.

In the four-quark model, mesons should occur in multiplets of 16 states, corresponding to the different arrangements of four quarks and four antiquarks. The $SU(4)$ Clebsch–Gordan series is

$$\mathbf{4} \otimes \bar{\mathbf{4}} = \mathbf{15} \oplus \mathbf{1}. \tag{12.10}$$

In the *n*-quark model, this series is

$$\mathbf{n} \otimes \bar{\mathbf{n}} = \mathbf{n^2 - 1} \oplus \mathbf{1}. \tag{12.11}$$

But because the singlet mixes with the $n^2 - 1$ multiplet, we expect to see multiplets of n^2 mesons.

Turning to the baryons, in the four-quark model we have instead of Eq. (12.8):

$$\mathbf{4} \otimes \mathbf{4} \otimes \mathbf{4} = \mathbf{20}_S \oplus \mathbf{20}_M \oplus \mathbf{20}_M + \bar{\mathbf{4}}. \tag{12.12}$$

More generally, with n quarks ($n > 2$) we have

$$\mathbf{n} \otimes \mathbf{n} \otimes \mathbf{n} = \tfrac{1}{6}\mathbf{n}(\mathbf{n}+1)(\mathbf{n}+2) \oplus \tfrac{1}{3}\mathbf{n}(\mathbf{n}+1)(\mathbf{n}-1) \oplus \tfrac{1}{3}\mathbf{n}(\mathbf{n}+1)(\mathbf{n}-1) \oplus \tfrac{1}{6}\mathbf{n}(\mathbf{n}-1)(\mathbf{n}-2). \quad (12.13)$$

Using Young tableaux, it is straightforward to obtain the $SU(m)$ submultiplets of $SU(n)$, in particular, the $SU(3)$ submultiplets. We shall omit the details.

12.3 Colored Quarks

As far as is known, all particles with integral spin obey Bose statistics (are bosons), and all particles with half-integral spin obey Fermi statistics (are fermions). Thus, it is natural to assume that quarks, having spin $\frac{1}{2}$, are fermions. There is one difficulty with this assumption, however, as we shall now describe. The wave function of a collection of identical fermions must be antisymmetric under the interchange of all the coordinates of any two of them. But this does not seem to be the case for the wave functions of identical quarks in baryons.

Consider as an example the Δ^{++} baryon, which in the model is a bound state of three identical u quarks. According to the conventional description of the Δ^{++}, the three quarks have their spins pointing in the same direction, leading to a value of $J = \frac{3}{2}$ for the spin of the Δ^{++}, provided the orbital angular momentum of the quarks is 0.

If the quarks in the Δ^{++} are bound in an attractive potential, the state of lowest energy is an S state (0 orbital angular momentum). But such a state is ordinarily symmetric under the interchange of the spatial coordinates of any two quarks. Likewise, the spin $\frac{3}{2}$ wave function is symmetric under the interchange of the spin coordinates of any two quarks. Thus the Δ^{++} baryon appears to be symmetric, rather than antisymmetric under the combined interchange of the space and spin coordinates of the quarks. In other words, the quarks are apparently acting like bosons rather than fermions.

We can give similar arguments that the quarks in the proton and neutron apparently behave like bosons. In this case we must include the interchanges of isospin coordinates. Furthermore, the ratio of the magnetic moment of the proton and neutron is most easily explained if the quark wave functions are symmetric, as we shall show later. A more detailed discussion of quark wave functions is given by Kokkedee (1969).

We now consider possible ways out of this difficulty. It is certainly possible to construct an antisymmetric spatial wave function with zero orbital angular momentum. One such choice is

$$\psi(\mathbf{r}_1\mathbf{r}_2\mathbf{r}_3) = (\mathbf{r}_1 \cdot \mathbf{r}_2 \times \mathbf{r}_3 + \mathbf{r}_2 \cdot \mathbf{r}_3 \times \mathbf{r}_1 + \mathbf{r}_3 \cdot \mathbf{r}_1 \times \mathbf{r}_2)\phi(r_1 r_2 r_3) \quad (12.14)$$

where $\mathbf{r}_1$, $\mathbf{r}_2$, and $\mathbf{r}_3$ are the spatial coordinates of the quarks, and ϕ is totally symmetric in r_1, r_2, and r_3. We assume ψ is normalized to unity, i.e., $(\psi, \psi) = 1$. The wave function of Eq. (12.14) is more complicated than a symmetric spatial wave function. Although the total orbital angular momentum of $\psi(\mathbf{r}_1\mathbf{r}_2\mathbf{r}_3)$ is zero, the relative angular momentum of any two quarks is greater than zero. The parity of this wave function is positive. The problem is that quark dynamics would have to be very complicated to give such an antisymmetric state as the ground state.

A more intriguing possibility has been suggested by Greenberg (1964). According to Greenberg, quarks are not fermions, but obey parastatistics; specifically they are parafermions of order three. The wave functions of identical particles which obey parastatistics are basis functions of multidimensional representations of the symmetric group S_n. In particular, identical parafermions of order three satisfy the rule that at most three of them can be put in a given state. Thus three paraquarks could be bound in a symmetric state, but not a fourth. This assumption does not by itself explain why antisymmetric combinations of three quarks (or combinations with mixed symmetry) are not also observed. It must be postulated that the forces are such that the symmetric states be lowest in energy. This *symmetric quark model* has been used in most of the nonrelativistic treatments of quarks bound in hadrons. See, for example, Greenberg (1964), Dalitz (1967), and Morpurgo (1968).

Greenberg's parafermion model is essentially equivalent to a model in which quarks have an additional degree of freedom which can take on three values. In baryons, quarks are assumed to be antisymmetric in this new degree of freedom, now called color.

We suggested in 1970, in the first edition of this book, that the three quark colors be called red, white, and blue. However, it may be preferable to adopt the notation of some other authors who subsequently named the colors red, green, and blue. These are primary colors, and, when mixed, lead to white. The known hadrons, then, formed of quarks colored in primary colors, have no color themselves. We leave it up to the reader to decide whether this far-fetched analogy is useful. Greenberg and Nelson (1977) have written a good review of colored quarks.

Shortly after Greenberg's paper on parafermion quarks appeared, Han and Nambu (1965) suggested a model in which each type (now called flavor) of quark comes in three varieties. Han and Nambu noted that with the multiplication of the number of quarks by 3, there was no need to give the quarks fractional charge. (See also Cabibbo *et al.*, 1967.)

If quarks are integrally charged, not all the quarks of a given flavor have the same charge. However, the *average* charge of the three differently colored quarks of a given flavor must be fractional in order for the model to agree with experiment.

Models with integrally charged quarks are often said to have Han–Nambu color, while models with fractionally charged quarks are often said to have Gell-Mann–Zweig color. (Perhaps the latter should be Greenberg color). In either case, the underlying symmetry of the color degree of freedom is usually taken to be $SU(3)$. In a model with fractionally charged quarks, this symmetry may be exact. However, in a model with integrally charged quarks, color symmetry is necessarily broken by the electromagnetic interaction. This is because quarks of the same flavor and different colors do not all have the same value of the charge.

12.4 Quarks and Local Gauge Field Theory

If the quark model is to be taken seriously, the dynamics of interacting quarks must be considered. Because this is a book primarily on symmetry rather than dynamics, we here only sketch a few of the ideas about quark dynamics, referring the reader elsewhere for more detailed treatments.

Weinberg (1967) and Salam (1968) proposed that quarks and leptons interact both electromagnetically and weakly via the exchange of vector gauge fields. According to their idea, the electromagnetic and weak interactions are both governed by a single coupling constant, the fine-structure constant α. In a theory which exhibits local gauge invariance, the gauge field, like the electromagnetic field, is massless, and the interaction is long range. However, it is well known that the weak interaction has a very short range. In the Weinberg–Salam model, the short-range character of the weak interaction is accomplished by spontaneous symmetry breaking. This means that although the interaction is invariant under the symmetry, the eigenstates of the system are not.

Glashow *et al.* (1970) showed that the predictions of the Weinberg–Salam model would exhibit improved agreement with experiment if a charmed quark were introduced into the model. Before the discovery of charmed particles in 1976, the paper of Glashow *et al.* was regarded as the best indirect argument for the existence of a charmed quark. A few of the main features of this argument follow.

We first introduce the idea of the coupling of charged and neutral currents to gauge fields. The ordinary electromagnetic current is neutral, since it couples to neutral photons, which are the quanta of the electromagnetic field. An example of such a neutral current is provided by a positron–electron pair. On the other hand, for a long time the weak current was thought to be only charged, coupling to a charged weak boson (as yet undiscovered) called the W. An example of such a charged weak current is a positron–neutrino pair. However, in the Weinberg–Salam model, the constraints of gauge-invariance also require the existence of a weak neutral current coupling to a weak neutral boson (as yet undiscovered) called Z.

At the time the Weinberg–Salam model was proposed, there was no evidence for the existence of weak neutral currents. On the contrary, in strangeness-changing processes, the experimental evidence was that weak neutral currents did not exist. For example, the strangeness-changing decay $\Lambda \to p + e^- + \bar{\nu}$ proceeds via a charged current and is experimentally seen. On the other hand, the hypothetical decay $\Lambda \to n + e^- + e^+$, which would require a neutral current, has not been observed.

Glashow *et al.* (1970) noted that if a charmed quark is introduced, the unwanted strangeness-changing neutral current could be canceled out. They also noted that a strangeness-conserving neutral current, about which there was no experimental evidence either way, is allowed. Thus, the Weinberg–Salam model, as modified by Glashow *et al.*, predicted the existence both of strangeness-conserving weak neutral currents and charmed particles. Subsequently, both have been observed.

In the model, it is quarks and leptons which interact with the gauge fields. The charged lepton currents are (e, ν_e), (μ, ν_μ) and presumably, but with less evidence, also (τ, ν_τ). The charged quark currents are (u, d'), (c, s'), and perhaps others. In the model, the u and c quarks are those of Table 12.1. However, the d' and s' are orthogonal linear combinations of the d and s quarks, namely,

$$d' = d \cos \theta_C + s \sin \theta_C, \qquad s' = -d \sin \theta_C + s \cos \theta_C \tag{12.15}$$

where the angle θ_C is Cabibbo's (1963) angle. Once the charged currents are given, the neutral currents are specified by the gauge-invariance of the theory. Good discussions of these ideas have been given by Weinberg (1974) and Wolfenstein (1977).

It has been postulated by a number of authors (see, e.g., Weinberg, 1973; Gross and Wilczek, 1973; De Rújula *et al.*, 1975) that the strong interactions between quarks are also governed by gauge fields, in this case associated with the color degree of freedom. The quanta of the gauge fields are assumed to be an octet of vector bosons called gluons. These gluons carry the quantum numbers of color (color charge), but are electrically neutral.

In some theories of this kind, the effective gauge coupling constant is small and decreasing at high momentum or short distance. This property, discovered by Politzer (1973) and Gross and Wilczek (1973), is called *asymptotic freedom*.

According to the theory, if the color symmetry is spontaneously broken, the color gluons can acquire masses. On the other hand, if the symmetry is unbroken, the gluons remain massless. But it is known that the strong interaction has a short range rather than the long range which is expected from the exchange of massless particles. It has, therefore, been speculated (see, e.g., De Rújula *et al.*, 1975) that the theory has properties so as to confine particles having color, such as quarks and gluons. This confinement property has been called *infrared slavery*.

The confinement of quarks in the theory might not be absolute. In this case, the quarks might be confined by a finite potential barrier. According to quantum mechanics, given enough time, the quarks would tunnel through this barrier, and given enough energy, they would overcome it. Of course, in a model, the barrier could be made sufficiently high so as to lead to effective confinement.

There have also been attempts (e.g., Pati and Salam, 1973, 1974; Georgi and Glashow, 1974) to unify strong, as well as electromagnetic and weak interactions within the framework of the quark model. We shall not go into the details of such models, but shall merely point out that in some of them baryon number and lepton number are not strictly conserved.

12.5 Baryon and Meson Wave Functions

It is useful for many applications to be able to write hadron wave functions in terms of the quark wave functions. In this section, we shall do so, assuming that the underlying unitary symmetry is exact (except for the vector mesons), although, of course, we know that it is actually broken. We shall treat the cases of $SU(3)$ and $SU(4)$ explicitly, but the generalization to any $SU(n)$ is straightforward.

In the quark model, baryon wave functions are bound-state wave functions of three quarks, and meson wave functions are bound-state wave functions of a quark–antiquark pair. (Exceptions to this rule will be discussed in Section 12.10.) These wave functions include a spatial part, a spin part, a flavor unitary-symmetry part, and a color unitary-symmetry part.

The color unitary-symmetry part of the wave function is assumed to be an $SU(3)$ singlet. This color wave function can be written down explicitly using the methods of Chapter 8. Let us do so for a baryon composed of three quarks and for a meson composed of a quark and antiquark. We use the symbols r, g, b (for red, green, and blue) to indicate the different color degrees of freedom. Then the baryon color wave function is

$$B_{\text{color}} = (rgb - rbg - grb + gbr - bgr + brg)/\sqrt{6}. \qquad (12.16)$$

Likewise, the meson color wave function is

$$M_{\text{color}} = (r\bar{r} + g\bar{g} + b\bar{b})/\sqrt{3}. \qquad (12.17)$$

In the case of baryons, the wave function is antisymmetric under the interchange of any two colors. The remainder of the baryon wave function is therefore totally symmetric under any interchange.

The spatial part of a hadron wave function cannot be written down explicitly, as its form depends on the unknown details of quark dynamics. However, if we assume that the forces between quarks are basically attractive, the lowest-energy states will have symmetric spatial wave functions.

We next turn to the spin part of the wave function. Again we assume that the lowest-lying states have zero orbital angular momentum. Then the spin wave functions $\chi_m^{(3/2)}$ (for $m = \frac{3}{2}$ and $\frac{1}{2}$) of the $\mathbf{20}_S$ of $SU(4)$ or the decuplet of $SU(3)$ are

$$\chi_{3/2}^{(3/2)} = \alpha\alpha\alpha, \qquad \chi_{1/2}^{(3/2)} = (aa\beta + \alpha\beta\alpha + \beta\alpha\alpha)/\sqrt{3}, \tag{12.18}$$

where α and β denote quark spin functions with the third component $\frac{1}{2}$ and $-\frac{1}{2}$, respectively. The members of the baryon $\mathbf{20}_M$ of $SU(4)$ or the octet of $SU(3)$ have spin $\frac{1}{2}$. There are two ways to construct a spin-$\frac{1}{2}$ wave function from three doublets. The Clebsch–Gordan series is

$$\mathbf{2} \otimes \mathbf{2} \otimes \mathbf{2} = \mathbf{4} \oplus \mathbf{2} \oplus \mathbf{2}.$$

Because of the equivalence of $\mathbf{2}$ and $\bar{\mathbf{2}}$ of $SU(2)$, we obtain the same spin multiplicities as in the Fermi–Yang model [see Eq. (12.4)]. We choose as the two possibilities wave functions χ_m and χ'_m which are symmetric and antisymmetric, respectively, under the interchange of the spin coordinates of the first two quarks. These functions are given by (for $m = \frac{1}{2}$)

$$\chi_{1/2} = (2\alpha\alpha\beta - \alpha\beta\alpha - \beta\alpha\alpha)/\sqrt{6}, \qquad \chi'_{1/2} = (\alpha\beta\alpha - \beta\alpha\alpha)/\sqrt{2}. \tag{12.19}$$

TABLE 12.2

BARYON DECUPLET WAVE FUNCTIONS CONSTRUCTED FROM THE BASIS VECTORS OF THREE QUARKS

Baryon	Wave function
Δ^{++}	uuu
Δ^{+}	$(uud + udu + duu)/\sqrt{3}$
Δ^{0}	$(udd + dud + ddu)\sqrt{3}$
Δ^{-}	ddd
Σ^{*+}	$(uus + usu + suu)/\sqrt{3}$
Σ^{*0}	$(uds + usd + dus + dsu + sud + sdu)/\sqrt{6}$
Σ^{*-}	$(dds + dsd + sdd)/\sqrt{3}$
Ξ^{*0}	$(uss + sus + ssu)/\sqrt{3}$
Ξ^{*-}	$(dss + sds + ssd)/\sqrt{3}$
Ω^{-}	sss

TABLE 12.3

WAVE FUNCTIONS OF CHARMED BARYONS BELONGING TO THE $\mathbf{20}_S$ OF $SU(4)$

Baryon	$SU(3)$ multiplicity	Wave function
Σ_c^{*++}	6	$(uuc + ucu + cuu)/\sqrt{3}$
Σ_c^{*+}	6	$(udc + ucd + duc + dcu + cud + cdu)/\sqrt{6}$
Σ_c^{*0}	6	$(ddc + dcd + cdd)/\sqrt{3}$
Ξ_c^{*+}	6	$(usc + ucs + suc + cus + scu + csu)/\sqrt{6}$
Ξ_c^{*0}	6	$(dsc + dcs + sdc + cds + scd + csd)/\sqrt{6}$
Ω_c^{*0}	6	$(ssc + scs + css)/\sqrt{3}$
Ξ_{cc}^{*++}	3	$(ucc + cuc + ccu)/\sqrt{3}$
Ξ_{cc}^{*+}	3	$(dcc + cdc + ccd)/\sqrt{3}$
Ω_{cc}^{*+}	3	$(scc + csc + ccs)/\sqrt{3}$
Ω_{ccc}^{++}	1	ccc

TABLE 12.4

BARYON OCTET WAVE FUNCTIONS CONSTRUCTED FROM THE BASIS VECTORS OF THREE QUARKS[a]

Baryon	Wave function
	Octet 1
p	$(2uud - udu - duu)/\sqrt{6}$
n	$(udd + dud - 2ddu)/\sqrt{6}$
Λ^0	$\frac{1}{2}(usd + sud - dsu - sdu)$
Σ^+	$(2uus - usu - suu)/\sqrt{6}$
Σ^0	$(2uds + 2dus - usd - dsu - sud - sdu)/\sqrt{12}$
Σ^-	$(2dds - dsd - sdd)/\sqrt{6}$
Ξ^0	$(uss + sus - 2ssu)/\sqrt{6}$
Ξ^-	$(dss + sds - 2ssd)/\sqrt{6}$
	Octet 2
p	$(udu - duu)/\sqrt{2}$
n	$(udd - dud)/\sqrt{2}$
Λ^0	$(2uds - 2dus + sdu - dsu + usd - sud)/\sqrt{12}$
Σ^+	$(usu - suu)/\sqrt{2}$
Σ^0	$\frac{1}{2}(usd + dsu - sud - sdu)$
Σ^-	$(dsd - sdd)/\sqrt{2}$
Ξ^0	$(uss - sus)/\sqrt{2}$
Ξ^-	$(dss - sds)/\sqrt{2}$

[a] There are two different linearly independent octets, the first being symmetric under the interchange of the flavor indices of the first two quarks, and the second being antisymmetric under this interchange.

Next we consider the flavor part of the wave functions. From the quark and antiquark basis vectors, we can construct the wave functions of any hadron multiplet. For the $SU(3)$ decuplet these are easily obtained from Eqs. (12.9) and (8.87). To avoid writing subscripts, we use the notation u, d, s, and c instead of the u_i of Eq. (8.87). In Table 12.2 we give the decuplet wave functions with the notation that the symbol for a particle stands for its flavor wave function. In Table 12.3 we generalize to $SU(4)$, giving the wave functions of remaining members of the $\mathbf{20}_S$. All baryons of Table 12.3 have charm. These wave functions of Tables 12.2 and 12.3 can be constructed from the lowering operators $E_{i+1,i}$ operating on the quark basis vectors. The method applies to any $SU(n)$, as we have noted in Chapter 8.

TABLE 12.5

WAVE FUNCTIONS OF CHARMED BARYONS BELONGING TO THE $\mathbf{20}_M$ OF $SU(4)$

Baryon	$SU(3)$ multiplicity	Wave function
		First $\mathbf{20}_M$
Σ_c^{++}	6	$(2uuc - ucu - cuu)/\sqrt{6}$
Σ_c^{+}	6	$(2udc + 2duc - ucd - dcu - cud - cdu)/\sqrt{12}$
Σ_c^{0}	6	$(2ddc - dcd - cdd)/\sqrt{6}$
Ξ_c^{+}	6	$(2usc + 2suc - ucs - scu - cus - csu)/\sqrt{12}$
Ξ_c^{0}	6	$(2dsc + 2sdc - dcs - scd - cds - csd)/\sqrt{12}$
Ω_c^{0}	6	$(2ssc - scs - css)/\sqrt{6}$
Λ_c^{+}	$\bar{3}$	$\frac{1}{2}(ucd + cud - dcu - cdu)$
Ξ_c^{A+}	$\bar{3}$	$\frac{1}{2}(ucs + cus - scu - csu)$
Ξ_c^{A0}	$\bar{3}$	$\frac{1}{2}(dcs + cds - scd - csd)$
Ξ_{cc}^{++}	3	$(ucc + cuc - 2ccu)/\sqrt{6}$
Ξ_{cc}^{+}	3	$(dcc + cdc - 2ccd)/\sqrt{6}$
Ω_{cc}^{+}	3	$(scc + csc - 2ccs)/\sqrt{6}$
		Second $\mathbf{20}_M$
Σ_c^{++}	6	$(ucu - cuu)/\sqrt{2}$
Σ_c^{+}	6	$\frac{1}{2}(ucd + dcu - cud - cdu)$
Σ_c^{0}	6	$(dcd - cdd)/\sqrt{2}$
Ξ_c^{+}	6	$\frac{1}{2}(ucs + scu - cus - csu)$
Ξ_c^{0}	6	$\frac{1}{2}(dcs + scd - cds - csd)$
Ω_c^{0}	6	$(scs - css)/\sqrt{2}$
Λ_c^{+}	$\bar{3}$	$(2udc - 2duc + cdu - dcu + ucd - cud)/\sqrt{12}$
Ξ_c^{A+}	$\bar{3}$	$(2usc - 2suc + csu - scu + ucs - cus)/\sqrt{12}$
Ξ_c^{A0}	$\bar{3}$	$(2dsc - 2sdc + csd - scd + dcs - cds)/\sqrt{12}$
Ξ_{cc}^{++}	3	$(ucc - cuc)/\sqrt{2}$
Ξ_{cc}^{+}	3	$(dcc - cdc)/\sqrt{2}$
Ω_{cc}^{+}	3	$(scc - csc)/\sqrt{2}$

We next construct the baryon octet $SU(3)$ wave functions. We see from Eq. (12.8) that two octets can be constructed from the basis vectors of three quarks. Because of this, the Clebsch–Gordan coefficients are not determined from the group properties alone. We shall choose the first octet to be symmetric in the flavor indices of the first two quarks and the second octet to be antisymmetric in these indices. With this choice, the wave functions of the two baryon octets are given in Table 12.4. In Table 12.5 we generalize to $SU(4)$, giving the wave functions of the remaining members

TABLE 12.6

WAVE FUNCTIONS OF BARYONS BELONGING TO THE $\bar{\mathbf{4}}$ OF $SU(4)$

Baryon	$SU(3)$ multiplicity	Wave function
Λ^{*0}	1	$(uds - dus + dsu - usd + sud - sdu)/\sqrt{6}$
Λ_c^{*+}	$\bar{3}$	$(udc - duc + dcu - ucd + cud - cdu)/\sqrt{6}$
Ξ_c^{*+}	$\bar{3}$	$(usc - suc + scu - ucs + cus - csu)/\sqrt{6}$
Ξ_c^{*0}	$\bar{3}$	$(dsc - sdc + scd - dcs + cds - csd)/\sqrt{6}$

TABLE 12.7

WAVE FUNCTIONS OF THE PSEUDOSCALAR MESON 15-PLET AND SINGLET OF $SU(4)$, ASSUMING NO $SU(3)$ OR $SU(4)$ MIXING

	Meson	$SU(3)$ multiplicity	Wave function
15	K^+	8	$u\bar{s}$
	K^0	8	$d\bar{s}$
	π^+	8	$-u\bar{d}$
	π^0	8	$(u\bar{u} - d\bar{d})/\sqrt{2}$
	π^-	8	$d\bar{u}$
	η	8	$(u\bar{u} + d\bar{d} - 2s\bar{s})/\sqrt{6}$
	$\bar{K}^0$	8	$-s\bar{d}$
	K^-	8	$s\bar{u}$
	F^+	$\bar{3}$	$c\bar{s}$
	D^+	$\bar{3}$	$-c\bar{d}$
	D^0	$\bar{3}$	$c\bar{u}$
	$\bar{D}^0$	3	$-u\bar{c}$
	D^-	3	$-d\bar{c}$
	F^-	3	$-s\bar{c}$
	χ	1	$(u\bar{u} + d\bar{d} + s\bar{s} - 3c\bar{c})/\sqrt{12}$
1	η'	1	$\frac{1}{2}(u\bar{u} + d\bar{d} + s\bar{s} + c\bar{c})$

of the $\mathbf{20}_{\mathrm{M}}$. The baryons of Table 12.5 all have charm. The wave functions Ξ_c and Ξ_c^A are likely to be mixed because of $SU(3)$ symmetry breaking. We have not included this effect in Table 12.5.

Orthogonal to both the $\mathbf{20}_{\mathrm{S}}$ and $\mathbf{20}_{\mathrm{M}}$ are the baryon wave functions of the $\bar{\mathbf{4}}$ of $SU(4)$. These are shown in Table 12.6.

We now turn to the meson wave functions, which are constructed from q and $\bar{q}$ states. Here the situation is complicated because of $SU(3)$ and $SU(4)$ mixing. In Table 12.7 we give the wave functions of the $SU(4)$ 15-plet and singlet pseudoscalar mesons, assuming no mixing. The actual meson wave functions are surely different from those of Table 12.7, as we shall discuss shortly. The sign conventions of Table 12.7 are dictated by Eq. (8.79) and its generalization to $SU(4)$.

At least in part because the quarks have different masses, $SU(3)$ and especially $SU(4)$ are badly broken. This fact should lead to mixing among states with the same values of spin, parity, isospin, strangeness, and charm. Thus, for example, among the pseudoscalar mesons, the wave functions of the η, χ, and η' ought to be linear combinations of the wave functions of Table 12.7. The nature of these linear combinations will, however, depend upon the dynamics.

TABLE 12.8

WAVE FUNCTIONS OF THE 16 VECTOR MESONS, ASSUMING IDEAL $SU(3)$ AND $SU(4)$ MIXING

Meson	$SU(3)$ multiplicity	Wave function
K^{*+}	8	$u\bar{s}$
K^{*0}	8	$d\bar{s}$
ρ^+	8	$-u\bar{d}$
ρ^0	8	$(u\bar{u} - d\bar{d})/\sqrt{2}$
ρ^-	8	$d\bar{u}$
$\bar{K}^{*0}$	8	$-s\bar{d}$
K^{*-}	8	$s\bar{u}$
ω	mixed	$(u\bar{u} + d\bar{d})/\sqrt{2}$
φ	mixed	$s\bar{s}$
ψ	mixed	$c\bar{c}$
F^{*+}	$\bar{3}$	$c\bar{s}$
D^{*+}	$\bar{3}$	$-c\bar{d}$
D^{*0}	$\bar{3}$	$c\bar{u}$
$\bar{D}^{*0}$	3	$-u\bar{c}$
D^{*-}	3	$-d\bar{c}$
F^{*-}	3	$-s\bar{c}$

For the vector mesons, rather than give the same wave functions as for the pseudoscalars, we shall give wave functions for so-called *ideal mixing*. In order to obtain ideal mixing, we assume that, except for isospin invariance, the interaction between quark and antiquark plays a negligible role in determining the physical states. Rather, the masses of the quarks primarily determine the meson masses and wave functions. In Table 12.8 we give the wave functions of the ideally mixed $SU(4)$ 16-plet of vector mesons.

For simplicity, we discuss ideal mixing for $SU(3)$ instead of $SU(4)$. We can define an $SU(3)$ mixing angle for the vector mesons as follows. The ω and φ wave functions are linear combinations of the $SU(3)$ octet V_8 and singlet V_1 whose wave functions are

$$V_8 = (u\bar{u} + d\bar{d} - 2s\bar{s})/\sqrt{6}, \qquad V_1 = (u\bar{u} + d\bar{d} + s\bar{s})/\sqrt{3}, \qquad (12.20)$$

where we are assuming that the $c\bar{c}$ contribution is negligible. Then we define the mixing angle θ by writing the physical φ and ω in terms of V_8 and V_1 as

$$\varphi = -V_8 \cos\theta + V_1 \sin\theta, \qquad \omega = V_8 \sin\theta + V_1 \cos\theta. \qquad (12.21)$$

We can see from this equation that we have ideal mixing, if $\sin\theta = 1/\sqrt{3}$, or $\theta \approx 35°$. But this angle is very close to the empirical $SU(3)$ mixing angle we deduced in Chapter 9 from the Gell-Mann–Okubo mass formula (see Table 9.11). We conclude that the vector mesons are not far from being ideally mixed. From Table 9.11 we also see that the tensor mesons are also close to being ideally mixed, but that the pseudoscalars are not.

Because the vector mesons are nearly ideally mixed, perhaps it is not a bad approximation to neglect the contribution to the vector meson masses from the quark–antiquark interaction. Then we can use the observed meson masses to estimate the quark masses. This is done as follows. The ρ and ω mesons are made of u and d quarks, whose mass difference we neglect. Then, if our picture is correct, the ρ and ω ought to have the same mass, and this mass should be twice that of the u or d quark. We see from Table 9.7 that the mean mass of the ω and ρ is 776 MeV, and that their mass difference is only 13 MeV. We therefore estimate that the u and d quarks have masses around 390 MeV. Likewise, from the known φ meson mass of 1020 MeV, we estimate that the s quark has a mass around 510 MeV, and from the ψ meson mass of 3095 MeV (Table 10.2), we estimate that the c quark has a mass of 1550 MeV. According to this picture, the mass of the K^* should be the sum of the masses of a u and s quark, or around 900 MeV, and the mass of a D^* should be the sum of the masses of a c and u quark, or around 1940 MeV. The actual mass of the K^* is 894 MeV, in good qualitative agreement with the prediction. On the other hand, the mass of the D^* is 2010 MeV, somewhat higher than predicted. Thus, the masses estimated by this argument, namely

$$m_u \approx m_d \approx 390 \text{ MeV}, \qquad m_s \approx 510 \text{ MeV}, \qquad m_c \approx 1550 \text{ MeV} \quad (12.22)$$

are probably not very accurate.

The spins and parities of the meson 16-plets are accounted for quite naturally by the model. It is convenient to use spectroscopic notation to describe the quantum numbers of a bound state of a quark and antiquark. In this notation the orbital angular momentum is denoted by $S, P, D, F, \ldots$ for $l = 0, 1, 2, 3, \ldots$. The spin multiplicity (either singlet or triplet) is denoted by a left superscript and the total angular momentum is denoted by a right subscript. With this notation the pseudoscalar 16-plet is most simply classified as a ${}^1S_0$ state of the quark–antiquark system, the vector 16-plet as a ${}^3S_1$ state, and the tensor 16-plet as a ${}^3P_2$ state. The parities of these mesons are then correctly given, since the parity P of a fermion–antifermion pair with orbital angular momentum l is given by

$$P = (-1)^{l+1}. \quad (12.23)$$

We next consider the charge-conjugation properties of the $I_3 = Y = 0$ members of the meson 16-plets within the model. A particle–antiparticle pair with $I_3 = Y = 0$ has charge-conjugation parity C given by

$$C = (-1)^{l+s}, \quad (12.24)$$

where s is the spin of the pair. The proof of Eq. (12.24), which we shall not give, makes use of the usual relation between spin and statistics.

Nonexotic mesons with $I = Y = 0$ must satisfy Eqs. (12.23) and (12.24). Thus, any meson with odd spin and parity and $C = +$ would be considered to be exotic. This can be seen as follows: Since P is odd, from Eq. (12.23), l is even. Now, since a quark has spin $\frac{1}{2}$, the only two possibilities for s are $s = 0$ or 1. But, by conservation of angular momentum, if the meson spin is odd and l is even, s must be 1. Therefore by Eq. (12.24), C must be negative, in contradiction to our assumption. Therefore, the meson cannot be made from a quark and antiquark and so must be exotic.

In the model, pseudoscalar mesons are assigned $(ls) = (00)$, the vectors (01), scalars and tensors (11). Thus, the model predicts that the scalars, pseudoscalars and tensors which are eigenstates of C should have eigenvalue +, while the vectors should have eigenvalue −. These predictions are in agreement with experiment. On the other hand, axial vector mesons can have $(ls) = (11)$ or (10), and so can have eigenvalue either + or −. Both values have been observed for the axial vectors.

12.6 $SU(6)$, $SU(8)$, and the Quark Model

As we have discussed in Chapter 11, the dynamical groups $SU(6)$ and $SU(8)$ have both attractive and unattractive features as approximate symmetries of hadrons. Here, we shall treat these groups within the framework of the quark model and show that some of the unpleasant aspects can be avoided. For simplicity, we shall restrict most of the discussion to $SU(6)$, as the generalization to $SU(8)$ and to any $SU(2n) \supset SU(n) \times SU(2)$ is straightforward.

If a quark has spin $\frac{1}{2}$, it is a doublet of the spin group $SU(2)$ as well as a triplet of $SU(3)$ and thus is a fundamental sextet of $SU(3) \times SU(2)$. But the quark basis functions can equally well be regarded as a fundamental sextet of $SU(6)$. We choose as basis functions of $SU(6)$, the following six quark states u_i:

$$\begin{aligned} u_1 &= q_1\alpha, & u_2 &= q_2\alpha, & u_3 &= q_3\alpha, \\ u_4 &= q_1\beta, & u_5 &= q_2\beta, & u_6 &= q_3\beta, \end{aligned} \tag{12.25}$$

where here the q_i are the quark basis functions of $SU(3)$ and α and β are states with spin "up" and "down," respectively.

When we consider the interactions of two or more quarks, we find there is a difference between $SU(6)$ and $SU(3) \times SU(2)$, since one multiplet of $SU(6)$ may contain several multiplets of $SU(3) \times SU(2)$. The quark model provides a natural way for us to incorporate $SU(6)$ as an approximate dynamical symmetry. If the interactions of quarks are taken to be Lorentz invariant (as we believe they should be), then $SU(6)$ will automatically be broken in high energy processes. This is because a Lorentz transformation acts on the states of the $SU(2)$ spin subgroup of $SU(6)$, but does not affect the internal symmetry subgroup $SU(3)$. Then, any symmetry transformation such as $SU(6)$, which allows transformations which mix spin and internal symmetry coordinates, must be broken. However, the interaction could be such as to preserve $SU(6)$ invariance in a low-energy approximation in which the quark velocities are negligibly small.

In the model, a baryon is supposed to consist of a bound state of three quarks. If the bound quarks are moving at relativistic velocities inside the baryon, it is difficult to see how $SU(6)$ can be an approximate dynamical symmetry of baryons. Therefore relativistic effects which break $SU(6)$ symmetry must be small. Morpurgo (1965) has discussed under what conditions such an approximation might be valid. Briefly, Morpurgo argues that if quarks are bound by a potential of range a, then a typical momentum k of a bound quark is $k \approx 1/a$. Now if the range of the potential is much greater than the inverse of a quark mass m, i.e., $a \gg 1/m$, then $k \ll m$, and a non-relativistic approximation might be a good one. This argument does not hold for a singular potential, for example, a Yukawa potential $\sim (a/r)e^{-r/a}$. It is not known whether the argument is good for any system with a binding energy comparable to the masses of the constituents, as has been conjectured to be the case for quarks.

We have remarked that the reduction of the Kronecker product of three quark state vectors given in Eq. (12.7) holds for any $SU(n)$ with $n > 2$. In particular for $SU(6)$, using the methods described in Chapter 7, we obtain the following multiplicities [or see Eq. (12.13)]

$$\mathbf{6} \otimes \mathbf{6} \otimes \mathbf{6} = \mathbf{56} \oplus \mathbf{70} \oplus \mathbf{70} \oplus \mathbf{20}. \tag{12.26}$$

Regarding the quark also as a fundamental multiplet $^2 3$ of $SU(3) \times SU(2)$, Eq. (12.26) becomes

$$\begin{aligned} {}^2 3 \times {}^2 3 \times {}^2 3 &= {}^{(2 \times 2 \times 2)}(3 \times 3 \times 3) \\ &= {}^{(4+2+2)}(10 + 8 + 8 + 1) \\ &= {}^4 10 + {}^4 8 + {}^4 8 + {}^4 1 + {}^2 10 + {}^2 8 \\ &\quad + {}^2 8 + {}^2 1 + {}^2 10 + {}^2 8 + {}^2 8 + {}^2 1. \end{aligned} \tag{12.27}$$

We have already used symmetry arguments in Section 11.2 to show how these multiplets of $SU(3) \times SU(2)$ are classified into multiplets of $SU(6)$. See Eqs. (11.5), (11.6), and (11.7).

The $SU(6)$ states of the **56** are particularly easy to construct from the basis vectors of three quarks, since the **56** belongs to the totally symmetric tableau $\boxed{}\boxed{}\boxed{}$. We shall not give all 56 of these baryon wave functions. However, it is instructive to write down several of these functions as examples. The function denoted by the Young tableau $\boxed{1}\boxed{1}\boxed{4}$ is given by

$$\boxed{1}\boxed{1}\boxed{4} = (u_1 u_1 u_4 + u_1 u_4 u_1 + u_4 u_1 u_1)/\sqrt{3}.$$

Using Eq. (12.25), this becomes

$$\boxed{1}\boxed{1}\boxed{4} = q_1 q_1 q_1 (\alpha\alpha\beta + \alpha\beta\alpha + \beta\alpha\alpha)/\sqrt{3}. \tag{12.28}$$

Thus, this function factors into an $SU(3)$ decuplet function Δ^{++} times the spin $\frac{3}{2}$ function of Eq. (11.10). Thus we have

$$\boxed{1}\boxed{1}\boxed{4} = \Delta^{++}\chi_{3/2}^{(3/2)}. \tag{12.29}$$

Two examples which are somewhat more interesting are given by the Young tableaux $\begin{array}{|c|c|c|}\hline 1&1&5\\\hline\end{array}$ and $\begin{array}{|c|c|c|}\hline 1&2&4\\\hline\end{array}$. These functions are

$$\begin{array}{|c|c|c|}\hline 1&1&5\\\hline\end{array} = (u_1u_1u_5 + u_1u_5u_1 + u_5u_1u_1)/\sqrt{3}$$

$$= (q_1q_1q_2\alpha\alpha\beta + q_1q_2q_1\alpha\beta\alpha + q_2q_1q_1\beta\alpha\alpha)/\sqrt{3},$$

$$\begin{array}{|c|c|c|}\hline 1&2&4\\\hline\end{array} = (u_1u_2u_4 + u_1u_4u_2 + u_2u_1u_4 + u_2u_4u_1 + u_4u_1u_2 \tag{12.30}$$

$$+ u_4u_2u_1)/\sqrt{6}$$

$$= (q_1q_2q_1\alpha\alpha\beta + q_1q_1q_2\alpha\beta\alpha + q_2q_1q_1\alpha\alpha\beta$$

$$+ q_2q_1q_1\alpha\beta\alpha + q_1q_1q_2\beta\alpha\alpha + q_1q_2q_1\beta\alpha\alpha)/\sqrt{6}.$$

These $SU(6)$ functions can be written in terms of the $SU(3)$ and spin functions defined in Section 12.5. We obtain

$$\begin{aligned} \begin{array}{|c|c|c|}\hline 1&1&5\\\hline\end{array} &= (\Delta^+\chi_{1/2}^{(3/2)} + p_1\chi_{1/2} + p_2\chi'_{1/2})/\sqrt{3}, \\ \begin{array}{|c|c|c|}\hline 1&2&4\\\hline\end{array} &= (2\Delta^+\chi_{1/2}^{(3/2)} - p_1\chi_{1/2} - p_2\chi'_{1/2})/\sqrt{6}, \end{aligned} \tag{12.31}$$

where p_1 and p_2 are the wave functions of the first and second octets of Table 12.4. The spin wave functions $\chi_{1/2}$ and $\chi'_{1/2}$ are given in Eq. (12.19). From Eq. (12.31) we see that the proton $SU(6)$ wave function is

$$p = (p_1\chi_{1/2} + p_2\chi'_{1/2})/\sqrt{2}. \tag{12.32}$$

Note from Eq. (12.31) that an alternative way of writing the proton $SU(6)$ wave function is

$$p = \sqrt{\tfrac{2}{3}}\begin{array}{|c|c|c|}\hline 1&1&5\\\hline\end{array} - \sqrt{\tfrac{1}{3}}\begin{array}{|c|c|c|}\hline 1&2&4\\\hline\end{array} \tag{12.33}$$

Since this proton function is a linear combination of the symmetric functions $\begin{array}{|c|c|c|}\hline 1&1&5\\\hline\end{array}$ and $\begin{array}{|c|c|c|}\hline 1&2&4\\\hline\end{array}$, it is obviously totally symmetric. However, we wish to reemphasize that we can assume that the proton wave function is symmetric without assuming all the consequences of $SU(6)$ invariance. To take an extreme example, the function of Eq. (12.32) might hold even if strange particles did not exist. More practically, Eq. (12.32) can hold even though the Λ, Σ, and Ξ hyperons are not degenerate in mass with the proton. This kind of $SU(6)$-breaking is a consequence of $SU(3)$-breaking rather than a consequence of the difference between internal and spin degrees of freedom. This means that the proton function of Eq. (12.32) follows if we make the lesser assumption that the proton is a basis function of a symmetric representation of $SU(4)$, where $SU(4)$ contains the spin and isospin $SU(2)$ groups. If we apply Eq. (12.7) to $SU(4)$, we obtain

$$\mathbf{4} \otimes \mathbf{4} \otimes \mathbf{4} = \mathbf{20}_S \oplus \mathbf{20}_M \oplus \mathbf{20}_M \oplus \bar{\mathbf{4}}, \tag{12.34}$$

where we have distinguished the symmetric multiplet from the multiplets of mixed symmetry by the letters S and M, since they have the same multiplicities. Regarded as states of $SU(2) \times SU(2)$, Eq. (12.34) becomes

$$^{2}2 \times {}^{2}2 \times {}^{2}2 = {}^{(4+2+2)}(4 + 2 + 2). \tag{12.35}$$

Using arguments analogous to those used in Section 11.2, we find

$$\begin{aligned} \mathbf{20}_{\mathrm{S}} &\supset {}^{4}4 + {}^{2}2, \\ \mathbf{20}_{\mathrm{M}} &\supset {}^{4}2 + {}^{2}4 + {}^{2}2, \\ \bar{\mathbf{4}} &\supset {}^{2}2. \end{aligned} \tag{12.36}$$

These multiplicities apply to the baryons of hypercharge $Y = 1$ (or nonstrange baryons). The spin-$\frac{3}{2}$ baryon Δ and the spin-$\frac{1}{2}$ nucleon N fit nicely into the $\mathbf{20}_{\mathrm{S}}$. Thus, we see that a symmetric nucleon wave function does not depend on $SU(6)$. However, the $SU(4)$ nucleon wave function is identical to the one predicted from $SU(6)$.

We now point out that we can have a symmetric nucleon wave function even if $SU(4)$ is broken. For example, it is very possible for an $SU(4)$ symmetry-breaking interaction to split the mass degeneracy of the Δ and N without disturbing the symmetry of either the Δ or N wave functions under interchange of quark indices. Therefore, the assumption that the nucleon wave function is symmetric is less restrictive than the assumption of either $SU(4)$ or $SU(6)$.

We can generalize from $SU(6)$ to $SU(8)$ by defining, instead of Eq. (12.25), the following eight quark states u_i:

$$\begin{aligned} &u_1 = q_1\alpha, \quad u_2 = q_2\alpha, \quad u_3 = q_3\alpha, \quad u_4 = q_4\alpha, \\ &u_5 = q_1\beta, \quad u_6 = q_2\beta, \quad u_7 = q_3\beta, \quad u_8 = q_4\beta \end{aligned} \tag{12.37}$$

and proceeding as before. The $SU(4) \times SU(2)$ content of the relevant $SU(8)$ multiplets is given in Eqs. (11.14), (11.15), and (11.16).

The relevant baryon multiplicities are obtained in the reduction of the direct product

$$\mathbf{8} \otimes \mathbf{8} \otimes \mathbf{8} = \mathbf{120} \oplus \mathbf{168} \oplus \mathbf{168} \oplus \mathbf{56},$$

where the **120** is symmetric under the combined flavor and spin indices of the quarks, the **56** is antisymmetric, and the **168** has mixed symmetry. The lowest-lying baryon multiplet belongs to the symmetric **120**. For the mesons we have

$$\mathbf{8} \otimes \bar{\mathbf{8}} = \mathbf{63} \oplus \mathbf{1}.$$

Meson multiplets belong to a mixed $\mathbf{63} \oplus \mathbf{1}$. These Clebsch–Gordan series for the baryons and mesons are special cases of the series Eqs. (12.13) and (12.11), respectively.

TABLE 12.9

SIMPLIFIED WAVE FUNCTIONS OF THE SPIN-$\frac{1}{2}$ BARYONS

Baryon	Wave function[a]	Baryon	Wave function[a]
p	$uud\chi_m$	Σ_c^{++}	$uuc\chi_m$
n	$ddu\chi_m$	Σ_c^{+}	$udc\chi_m$
Λ^0	$uds\chi'_m$	Σ_c^{0}	$ddc\chi_m$
Σ^+	$uus\chi_m$	Ξ_c^{+}	$usc\chi_m$
Σ^0	$uds\chi_m$	Ξ_c^{0}	$dsc\chi_m$
Σ^-	$dds\chi_m$	Ω_c^{0}	$ssc\chi_m$
Ξ^0	$ssu\chi_m$	Λ_c^{+}	$udc\chi'_m$
Ξ'	$ssd\chi_m$	Ξ_c^{A+}	$usc\chi'_m$
		Ξ_c^{A0}	$dsc\chi'_m$
		Ξ_{cc}^{++}	$ccu\chi_m$
		Ξ_{cc}^{+}	$ccd\chi_m$
		Ω_{cc}^{+}	$ccs\chi_m$

[a] The spin wave functions χ_m and χ'_m (for $m = \frac{1}{2}$) are given in Eq. (12.19).

The baryon wave functions which we have written down in this section and in the previous one are rather long and cumbersome to use. In many calculations we can obtain identical results using simplified wave functions.

TABLE 12.10

SIMPLIFIED WAVE FUNCTIONS OF THE SPIN-$\frac{3}{2}$ BARYONS

Baryon	Wave function[a]	Baryon	Wave function[a]
Δ^{++}	$uuu\chi_m^{(3/2)}$	Σ_c^{*++}	$uuc\chi_m^{(3/2)}$
Δ^{+}	$uud\chi_m^{(3/2)}$	Σ_c^{*+}	$udc\chi_m^{(3/2)}$
Δ^{0}	$ddu\chi_m^{(3/2)}$	Σ_c^{*0}	$ddc\chi_m^{(3/2)}$
Δ^{-}	$ddd\chi_m^{(3/2)}$	Ξ_c^{*+}	$usc\chi_m^{(3/2)}$
Σ^{*+}	$uus\chi_m^{(3/2)}$	Ξ_c^{*0}	$dsc\chi_m^{(3/2)}$
Σ^{*0}	$uds\chi_m^{(3/2)}$	Ω^{*0}	$ssc\chi_m^{(3/2)}$
Σ^{*-}	$dds\chi_m^{(3/2)}$	Ξ_{cc}^{*++}	$ccu\chi_m^{(3/2)}$
Ξ^{*0}	$ssu\chi_m^{(3/2)}$	Ξ_{cc}^{*+}	$ccd\chi_m^{(3/2)}$
Ξ^{*-}	$ssd\chi_m^{(3/2)}$	Ω_{cc}^{*+}	$ccs\chi_m^{(3/2)}$
Ω^{-}	$sss\chi_m^{(3/2)}$	Ω_{ccc}^{++}	$ccc\chi_m^{(3/2)}$

[a] The spin wave functions $\chi_m^{(3/2)}$ are given in Eq. (12.18) for $m = \frac{1}{2}$ and $\frac{3}{2}$. Because $\chi_m^{(3/2)}$ is completely symmetric, the order of writing the flavor wave functions is immaterial; for example, it is just as good to write $\Delta^+ = duu\chi_m^{(3/2)}$ as the way it appears in the table.

In writing these simplified wave functions we shall omit the color variable and treat quarks of different flavor as distinguishable particles. We confine ourselves to the 120 multiplet of $SU(8)$ and assume that the spatial part of the wave function is symmetric under the interchange of any two quarks. In order that the overall wave function have the proper symmetry, the spin wave function must be symmetric under the interchange of two quarks of the same flavor. This means that we force identically flavored quarks to obey effective Bose statistics in all but the color variables.

The simplified space-spin wave functions of members of the 120 multiplet are given in Tables 12.9 and 12.10. Similar wave functions have been used by Franklin (1968) among others. In subsequent sections we shall make use of these wave functions to calculate properties of the baryons, such as their magnetic moments.

12.7 Baryon Magnetic Moments

In order to calculate the baryon magnetic moments, we must assume something about the quark magnetic moments and about the baryon wave functions. To a first approximation, the baryon magnetic moments can be expressed as a vector sum of the quark moments plus a contribution from any orbital angular momentum of the quarks.

Let us write the magnetic moment operator $\boldsymbol{\mu}_q$ of a quark as

$$\boldsymbol{\mu}_q = \mu_q \boldsymbol{\sigma} \tag{12.38}$$

where $\boldsymbol{\sigma}$ is the Pauli spin operator and μ_q is a constant which depends on the flavor of the quark. Then, assuming that the orbital angular momenta of the quarks is zero, the magnetic moment operator $\boldsymbol{\mu}$ of a baryon is given by

$$\boldsymbol{\mu} = \sum_{i=1}^{3} \mu_q(i)\boldsymbol{\sigma}(i) \tag{12.39}$$

where the sum is over the three quarks in a baryon. In writing down this expression, we are neglecting any contribution to $\boldsymbol{\mu}$ from exchange currents. The value of the magnetic moment μ_B of any baryon is the expectation value of μ_3 (the third or z component of $\boldsymbol{\mu}$) with respect to a baryon wave function B which is maximally polarized along the z axis, that is,

$$\mu_B = \left(B, \sum_i \mu_q(i)\sigma_3(i)B\right). \tag{12.40}$$

It is possible to evaluate μ_B in terms of the μ_q for any baryon once the flavor and spin wave functions of the baryon are specified. We shall see that the baryon magnetic moment calculation favors quark wave functions which are symmetric in the combined flavor and spin variables.

We shall now obtain the magnetic moments of the spin-$\frac{1}{2}$ baryons of the $\mathbf{20}_\mathrm{M}$ in terms of the quark moments using the simplified baryon wave functions of Table 12.9. The interested reader can convince himself by direct calculation that use of the simplified wave functions leads to the same results as use of the full $SU(8)$ wave functions of the **120**.

First we evaluate the proton magnetic moment in some detail to illustrate how the calculations go. We substitute the proton wave function of Table 12.9 into Eq. (12.40). We obtain

$$\mu_B = \left(uud\chi_{1/2}, \sum_i \mu_q(i)\sigma_3(i)uud\chi_{1/2}\right). \tag{12.41}$$

Note that the order of the factors in the wave function is important. For example uud is short for $u(1)\,u(2)\,d(3)$. Let us evaluate the first term in Eq. (12.41) using Eq. (12.19). We get

$$\begin{aligned}
&(uud\chi_{1/2}, \mu_q(1)\sigma_3(1)uud\chi_{1/2})\\
&= \tfrac{1}{6}(uud[2\alpha\alpha\beta - \alpha\beta\alpha - \beta\alpha\alpha], \mu_q(1)\sigma_3(1)uud[2\alpha\alpha\beta - \alpha\beta\alpha - \beta\alpha\alpha])\\
&= \tfrac{1}{6}(uud[2\alpha\alpha\beta - \alpha\beta\alpha - \beta\alpha\alpha], \mu_u\, uud[2\alpha\alpha\beta - \alpha\beta\alpha + \beta\alpha\alpha])\\
&= \tfrac{2}{3}\mu_u.
\end{aligned}$$

We have used the following properties of the operators μ_q and σ_3:

$$\begin{aligned}
&\mu_q u = \mu_u u, \quad \mu_q d = \mu_d d, \quad \mu_q s = \mu_s s, \quad \mu_q c = \mu_c c,\\
&\sigma_3 \alpha = \alpha, \quad \sigma_3 \beta = -\beta
\end{aligned} \tag{12.42}$$

and the fact that the wave functions are orthonormal. The second term in Eq. (12.41) must be the same as the first, because the simplified proton wave function is symmetric under the interchange of the first two quarks. Evaluating the third term by the same method we get

$$(uud\chi_{1/2}, \mu_q(3)\sigma_3(3)uud\chi_{1/2}) = -\tfrac{1}{3}\mu_d.$$

Therefore, we finally obtain the result

$$\mu_p = (4\mu_u - \mu_d)/3. \tag{12.43}$$

Because the neutron wave function of Table 12.9 is obtained from the proton wave function by the permutation $u \leftrightarrow d$, we can easily obtain the expression for the neutron magnetic moment by the same permutation in Eq. (12.43). We thus have

$$\mu_n = (4\mu_d - \mu_u)/3. \tag{12.44}$$

In the same way as we evaluated the proton magnetic moment we can obtain the magnetic moments of the Λ^0 and Σ^0. They are

$$\mu_{\Lambda^0} = \mu_s, \tag{12.45}$$

$$\mu_{\Sigma^0} = (2\mu_u + 2\mu_d - \mu_s)/3. \tag{12.46}$$

The magnetic moments of all the other spin-$\frac{1}{2}$ baryons of the $\mathbf{20}_\mathrm{M}$ can be obtained from Eqs. (12.43), (12.45), and (12.46) by permutation of quark flavors, making use of Table 12.9. For example, from Table 12.9 we see that the Ξ^- wave function can be obtained from the proton wave function by the permutation $u \leftrightarrow s$. Therefore, if we make the permutation $u \leftrightarrow s$ in Eq. (12.43), we obtain the Ξ^- magnetic moment.

In a similar manner, if we compute the magnetic moments of, for example, the Δ^{++}, the Δ^+, and the Σ^{*0}, then by permutations of quark flavors, we can obtain the magnetic moments of all the spin-$\frac{3}{2}$ baryons belonging to the $\mathbf{20}_\mathrm{S}$. Because these are less amenable to experimental measurement, we shall not write them down.

There are also some nonvanishing transition magnetic moments μ'_{BB} where B and B' are two different baryons. The definition of μ'_{BB} is

$$\mu'_{BB} = \left(B, \sum_i \mu_q(i)\sigma_3(i)B'\right), \tag{12.47}$$

where, if B and B' have different spins, the baryon with the smaller spin is maximally polarized along the z axis, and both baryons have the same value of J_z. These transition moments are important in calculating decays from one baryon to another via photon emission.

Using the quark model, we have shown how to obtain the magnetic moments of 40 baryons (belonging to the $\mathbf{20}_\mathrm{M}$ and $\mathbf{20}_\mathrm{S}$) as well as a number of transition moments, in terms of only four parameters: the magnetic moments of the four quarks. This shows the excellent predictive power of the quark model. Expressions for the baryon magnetic moments in terms of the quark moments have been written down for uncharmed baryons by Franklin (1968) and for charmed baryons by Johnson and Shah-Jahan (1977). Franklin (1975) pointed out that the results for charmed baryons could be obtained by permutation of quark flavors.

We can turn the problem around, and express the quark magnetic moments in terms of the magnetic moments of only four baryons. Choosing p, n, Λ, and Σ_c^{++} as the four baryons, we obtain

$$\mu_u = (\mu_n + 4\mu_p)/5, \qquad \mu_d = (\mu_p + 4\mu_n)/5, \tag{12.48}$$

$$\mu_s = \mu_\Lambda, \qquad \mu_c = 4(\mu_n + 4\mu_p)/5 - 3\mu_{\Sigma_c^{++}}. \tag{12.49}$$

Unfortunately, no magnetic moment of a charmed baryon has yet been measured, and therefore we cannot obtain the magnetic moment of the charmed quark. Using the experimentally measured proton, neutron, and Λ moments taken from the Particle Data Group (1976), we obtain

$$\mu_u = 1.85, \qquad \mu_d = -0.97, \qquad \mu_s = -0.67 \pm 0.06 \tag{12.50}$$

in units of the proton magneton $e\hbar/(2m_p c)$. We can use these values to obtain an estimate of the quark masses by assuming that the quarks have Dirac magnetic moments, that is

$$\mu_q = Q_q e\hbar/(2m_q c). \tag{12.51}$$

We obtain

$$m_u = 338 \text{ MeV}, \qquad m_d = 322 \text{ MeV}, \qquad m_s = 467 \text{ MeV}. \tag{12.52}$$

These mass values are similar to those we estimated from the vector meson masses [see Eq. (12.22)]. This fact gives us confidence, not only that the light quark masses are in the few hundred MeV range, but that, to a good approximation, the quarks have Dirac magnetic moments. We emphasize again that these are the effective masses of bound quarks. We can say nothing about masses of free quarks at present, and are not sure whether free quarks exist.

There is one disturbing aspect to the masses of Eq. (12.52). It is that the mass of the d quark appears to be smaller than the mass of the u quark. There is other evidence, to be discussed in Section 12.9, that the d quark has a slightly heavier mass than the u quark. We attribute the discrepancy to our neglect of a small amount of orbital angular momentum in the wave functions plus our neglect of relativistic effects such as exchange currents.

It is customary to neglect the mass difference between the u and d quarks in calculations of baryon magnetic moments. If we do so, then

$$\mu_d = -\tfrac{1}{2}\mu_u,$$

and Eqs. (12.43) and (12.44) become

$$\mu_p = \tfrac{3}{2}\mu_u, \qquad \mu_n = -\mu_u. \tag{12.53}$$

Thus the predicted ratio of the proton-to-neutron magnetic moments is

$$\mu_p/\mu_n = -1.50 \qquad \text{(quark model)}.$$

The experimental value of this ratio is

$$\mu_p/\mu_n = -1.46 \qquad \text{(experiment)}.$$

The close agreement between the simple model and experiment is considered a triumph for the so-called symmetric quark model (the quark model with symmetric baryon wave functions). If the wave functions are assumed to be antisymmetric, the calculated value of the ratio is in serious disagreement with experiment.

If we neglect the mass difference $m_d - m_u$ and assume the quarks have Dirac magnetic moments, then we obtain predictions for all the baryon magnetic moments in terms of the proton and Λ moments. The values of the magnetic moments of those baryons which have been measured experimentally, together with the calculated values, are given in Table 12.11.

TABLE 12.11

MAGNETIC MOMENTS OF THE BARYONS IN PROTON MAGNETONS

Name	Measured moment[a]	Calculated moment
p	2.79	input
n	-1.91	-1.86
Λ	-0.67 ± 0.06	input
Σ^+	2.62 ± 0.41	2.70 ± 0.03
Σ^-	-1.48 ± 0.37	-1.02 ± 0.03
Ξ^-	-1.85 ± 0.75	-0.6 ± 0.1

[a] Measured values from the Particle Data Group (1976).

12.8 Hadron Mass Splittings

We shall use the quark model within the framework of $SU(8)$ to obtain expressions for hadron masses in terms of a small number of parameters which refer to properties of quarks and their interactions. We can eliminate these parameters to obtain sum rules relating the masses of various hadrons belonging to a given multiplet of $SU(8)$. Our method is a generalization to $SU(8)$ (Hendry and Lichtenberg, 1975; Franklin, 1975) of a method used by Federman *et al.* (1966) within the framework of $SU(6)$. Johnson (1976) has further generalized to the case of six quarks and $SU(12)$. The procedure works independently of the number of quark flavors.

In the model, the mass of a hadron belonging to a given multiplet of $SU(8)$ arises from a contribution which is common to all members of a multiplet, a contribution equal to the sum of the masses of the quarks the hadron contains, plus a contribution from the sum of two-body interaction energies. These interaction energies depend on the flavor and spin configuration of the quarks, as well as on whether the interaction is between two quarks or between a quark and an antiquark. We write the interaction energy between the ith and jth quark (or ith quark and jth antiquark) as $V^S_{q_iq_j}$, where q_i and q_j are flavor indices and S is a spin index. In the case of baryons, it is an approximation to assume, as we shall, that the interaction energy between any two quarks is independent of the flavor and spin configuration of the third quark.

With these assumptions, the mass operator m for a meson is

$$m = m_0 + m_{q_1} + m_{q_2} + V^S_{q_1q_2} \tag{12.54}$$

where m_0 is the same for all members of a multiplet. For a baryon, the mass operator is

$$m = m_0 + \sum_i m_{q_i} + \sum_{i<j} V^S_{q_iq_j}. \tag{12.55}$$

Then, according to the model, the mass of any meson m_M is

$$m_M = (M, mM), \tag{12.56}$$

where M stands for the meson wave function. Likewise, the mass of any baryon is

$$m_B = (B, mB), \tag{12.57}$$

where B is the baryon wave function.

Let us evaluate (12.57) for the proton, using the simplified proton wave function of Table 12.9. We get

$$m_p = \left(uud\chi_{1/2},\ \left[m_0 + \sum_i m_{q_i} + \sum_{i<j} V^S_{q_iq_j}\right] uud\chi_{1/2}\right). \tag{12.58}$$

The matrix element of $\sum_i m_{q_i}$ is easy to evaluate. We get

$$\left(p, \sum_i m_{q_i} p\right) = 2m_u + m_d.$$

Likewise the matrix element of $V^S_{q_1q_2}$ is simply

$$(p, V^S_{12}\, p) = V^1_{uu}$$

Next let us evaluate the matrix element of $V^S_{q_1q_3}$. It is easy to carry out the flavor part of the matrix element. We get

$$(uud\chi_{1/2},\ V^S_{q_1q_3}\, uud\chi_{1/2}) = (\chi_{1/2}\, V^S_{ud}\, \chi_{1/2}).$$

Our notation is a little abstract: we must remember that V^S_{ud} operates on the spin wave functions of the first and third quarks. It is not trivial to evaluate this spin matrix element because $\chi_{1/2}$ is not a spin eigenfunction of the first and third quarks. We therefore expand $\chi_{1/2}$ in terms of these eigenfunctions, obtaining

$$\chi_{1/2} = [\tfrac{1}{2}(\alpha\alpha\beta + \beta\alpha\alpha) + \tfrac{3}{2}(\alpha\alpha\beta - \beta\alpha\alpha) - \alpha\beta\alpha]/\sqrt{6}.$$

We then operate on $\chi_{1/2}$ with V^S_{ud}, obtaining

$$V^S_{ud}\chi_{1/2} = [\tfrac{1}{2}(\alpha\alpha\beta + \beta\alpha\alpha)V^1_{ud} + \tfrac{3}{2}(\alpha\alpha\beta - \beta\alpha\alpha)V^0_{ud} - \alpha\beta\alpha V^1_{ud}]/\sqrt{6}.$$

Then, carrying out the spin matrix element, we get

$$(\chi_{1/2}, V^S_{ud}\chi_{1/2}) = \tfrac{1}{4}V^1_{ud} + \tfrac{3}{4}V^0_{ud}.$$

By symmetry, the expectation value of $V^S_{q_2q_3}$ is the same as that of $V^S_{q_1q_3}$. Substituting all these results in Eq. (12.58), we get

$$m_p = m_0 + 2m_u + m_d + V^1_{uu} + \tfrac{1}{2}V^1_{ud} + \tfrac{3}{2}V^0_{ud}. \tag{12.59}$$

If we neglect the mass difference of the u and d quark and assume by isospin conservation that $V^1_{uu} = V^1_{ud}$, we get

$$m_N = m_0 + 3m_u + \tfrac{3}{2}V^1_{uu} + \tfrac{3}{2}V^0_{ud}, \tag{12.60}$$

where N stands for a nucleon (either a proton or neutron). In similar fashion we can obtain expressions for the masses of all members of the $\mathbf{20}_M$ and $\mathbf{20}_S$. Of course it is not necessary to repeat the calculation many times. We need only evaluate the expressions for a few baryons, for example, for the Λ, Σ^0, and Σ^{*0}, and obtain results for the remainder by permutation of quark flavors. There are 39 other expressions, analogous to that of Eq. (12.59), for the masses of baryons belonging to $\mathbf{20}_M$ and $\mathbf{20}_S$. We shall not write them down. Using those expressions, however, we can eliminate the unknown quantities m_0, m_i, and V^S_{ij} to obtain a number of relations among the baryon masses. Let us neglect the small mass differences within isospin multiplets. Then, using the notation that the symbol for a hadron denotes its mass, we obtain the following relations among the baryon masses (Hendry and Lichtenberg, 1975):

$$\Omega - \Delta = 3(\Xi^* - \Sigma^*), \tag{12.61a}$$

$$\Omega_{ccc} - \Delta = 3(\Xi^*_{cc} - \Sigma_c^*), \tag{12.61b}$$

$$\Omega_{ccc} - \Omega = 3(\Omega^*_{cc} - \Omega_c^*), \tag{12.61c}$$

$$\Omega_c^* - 2\Xi_c^* + \Sigma_c^* = \Omega - 2\Xi^* + \Sigma^*, \tag{12.61d}$$

$$\Xi^* - \Sigma^* = \Xi - \Sigma, \tag{12.62a}$$

$$\Xi^*_{cc} - \Sigma_c^* = \Xi_{cc} - \Sigma_c, \tag{12.62b}$$

$$\Omega^*_{cc} - \Omega_c^* = \Omega_{cc} - \Omega_c, \tag{12.62c}$$

$$\Xi^* - \Delta = \Xi - N + \tfrac{3}{2}(\Lambda - \Sigma), \qquad (12.62d)$$

$$\Xi_{cc}^* - \Delta = \Xi_{cc} - N + \tfrac{3}{2}(\Lambda_c - \Sigma_c), \qquad (12.62e)$$

$$\Omega - 2\Xi^* + \Sigma^* = \Omega_c - 2\Xi_c + \Sigma_c, \qquad (12.62f)$$

$$\Omega_c^* - 6\Xi_c^* + \Sigma_c^* + 4\Sigma^* = \Omega_c - 6\Xi_c^A + \Sigma_c + 4\Sigma. \qquad (12.62g)$$

To the extent that these relations can be compared to experiment, they are in fair, but not complete, agreement.

The first four of these relations are among baryons of the $\mathbf{20}_S$. The remainder connect baryons of the $\mathbf{20}_S$ and $\mathbf{20}_M$. This method, with no further assumptions, does not enable us to obtain any relations involving only baryons belonging to the $\mathbf{20}_M$. In particular, we do not obtain the well-known Gell-Mann–Okubo octet mass formula [Eq. (9.47)]. We also do not get the decuplet equal-spacing rule [Eq. (9.42)]. We do obtain these two formulas, however, if we assume that there is a relation among the quark–quark interaction energies, namely

$$V_{us}^1 = \tfrac{1}{2}(V_{uu}^1 + V_{ss}^1). \qquad (12.63)$$

This says that (if quarks have their spins aligned) the interaction energy between a u and an s quark is the average of the interaction energy of two u quarks and two s quarks. If we include the charmed quark, we can hypothesize that the quark–quark interaction energies satisfy the following conditions, analogous to that of Eq. (12.63):

$$\begin{aligned} V_{uc}^1 &= \tfrac{1}{2}(V_{uu}^1 + V_{cc}^1), \\ V_{sc}^1 &= \tfrac{1}{2}(V_{ss}^1 + V_{cc}^1). \end{aligned} \qquad (12.64)$$

Using Eqs. (12.63) and (12.64), we obtain the following additional mass relations (including the Gell-Mann–Okubo octet and decuplet formulas):

$$\Xi_c^* - \Sigma_c^* = \Omega - \Xi^* = \Xi^* - \Sigma^* = \Sigma^* - \Delta, \qquad (12.65a)$$

$$\Omega_{ccc} - \Xi_{cc}^* = \Xi_{cc}^* - \Sigma_c^* = \Sigma_c^* - \Delta, \qquad (12.65b)$$

$$\Omega_{cc}^* - \Omega_c^* = \Omega_c^* - \Omega = \Sigma_c^* - \Sigma^*, \qquad (12.65c)$$

$$2(N + \Xi) = 3\Lambda + \Sigma, \qquad (12.66a)$$

$$2(N + \Xi_{cc}) = 3\Lambda_c + \Sigma_c, \qquad (12.66b)$$

$$\Omega_c - \Xi_c = \Xi_c - \Sigma_c, \qquad (12.66c)$$

$$2(\Sigma_c - \Sigma) = 2\Xi_{cc} + \Xi_c - 3\Xi_c^A, \qquad (12.66d)$$

$$\Omega_{cc} - \Xi_{cc} + \Xi - \Sigma = 2(\Xi_c - \Sigma_c). \qquad (12.66e)$$

None of the mass relations involving charmed baryons has yet been tested experimentally. Because $SU(8)$ is more badly broken than $SU(6)$, these relations may not be in very good agreement with experiment.

We next consider the mass splitting within isospin multiplets. We can obtain relations for these mass differences by the simple expedient of not neglecting the mass difference $m_d - m_u$ or the differences between the interactions V^S_{uu}, V^S_{ud}, and V^S_{dd}. Some of these relations are (Franklin, 1975)

$$n - p = \Sigma^- - \Sigma^+ + \Xi^0 - \Xi^-, \quad (12.67a)$$

$$n - p = \Delta^0 - \Delta^+ = \tfrac{1}{3}(\Delta^- - \Delta^{++}), \quad (12.67b)$$

$$n - p = \Sigma^{*-} - \Sigma^{*+} + \Xi^{*0} - \Xi^{*-}, \quad (12.67c)$$

$$n - p = \Sigma_c^{\ 0} - \Sigma_c^{++} + \Xi_{cc}^{++} - \Xi_{cc}^{+}, \quad (12.67d)$$

$$n - p = \Sigma_c^{*0} - \Sigma_c^{*++} + \Xi_c^{*++} - \Xi_{cc}^{*+}, \quad (12.67e)$$

$$\Sigma^+ + \Sigma^- - 2\Sigma^0 = \Sigma_c^{++} + \Sigma_c^{\ 0} - 2\Sigma_c^{\ +}, \quad (12.67f)$$

$$\Sigma^+ + \Sigma^- - 2\Sigma^0 = \Sigma^{*+} + \Sigma^{*-} - 2\Sigma^{*0}. \quad (12.67g)$$

The first of these formulas is the Coleman–Glashow (1961) relation, which has been derived in a variety of ways, (see, for example, Section 9.5). The other formulas have also been derived in different ways. There are no serious disagreements with experiment, but few have been tested.

We now turn to the mesons. We shall confine our attention to the vector mesons because they are nearly ideally mixed, and therefore their wave functions are relatively simple (see Table 12.8). We cannot obtain any mass relations just from the assumption that the quark–antiquark interaction is a two-body interaction, because a meson contains just two particles—a quark–antiquark pair. However, if we assume that relations like those of Eqs. (12.63) and (12.64) hold for the quark–antiquark interaction energies, then, using the same method we illustrated for the proton, we obtain the following mass relations (Hendry and Lichtenberg, 1975):

$$2K^* = \rho + \varphi, \quad (12.68a)$$

$$2D^* = \rho + \psi, \quad (12.68b)$$

$$2F^* = \varphi + \psi. \quad (12.68c)$$

We have neglected the splitting of isospin multiplets. From Tables 9.7 and 10.2, we can see that the first of these relations agrees quite well with experiment, while the second disagrees with experiment by about 60 MeV, and the third by about 80 MeV.

We next consider the mass splitting within meson isospin multiplets. Again, just the assumption of two-body forces is not useful. Therefore we adopt a simple model, namely that the electromagnetic interaction between quarks is the cause of the splitting of isospin multiplets. In particular we assume that, aside from the mass difference $m_d - m_u$, the isospin splitting is caused by a sum of a Coulomb and a magnetic-moment interaction between quarks. This interaction is

$$V^S_{q_i q_j} = \alpha Q_{q_i} Q_{q_j}/r_{ij} - (8\pi/3)\boldsymbol{\mu}_{q_i} \cdot \boldsymbol{\mu}_{q_j} \delta(\mathbf{r}_{ij}), \tag{12.69}$$

where $\alpha = 1/137$ is the fine-structure constant. For a discussion of the magnetic-moment term, see, for example, Abragam (1961, p. 170). This interaction has been assumed to apply to quarks by a number of authors, the early papers including those by Kuo and Yao (1965), Miyamoto (1966), and Gal and Scheck (1967). However, Celmaster (1976) has recently argued that more terms are needed. Despite Celmaster's work, let us use Eq. (12.69) to illustrate how electromagnetic mass splittings can be calculated. We simplify Eq. (12.69) by assuming that quarks have Dirac magnetic moments [Eq. (12.51)]. We get

$$V^S_{q_i q_j} = \alpha Q_{q_i} Q_{q_j}[1/r_{ij} - 2\pi \boldsymbol{\sigma}_i \cdot \boldsymbol{\sigma}_j \delta(\mathbf{r}_{ij})/(3m_{q_i} m_{q_j})]. \tag{12.70}$$

Using methods we illustrated when we evaluated magnetic moments, it is straightforward (Lichtenberg, 1975b) to evaluate the matrix elements of $V^S_{q_i q_j}$ in terms of the quantities C^S_{ij} and M^S_{ij} defined by

$$C^S_{ij} = \alpha\langle 1/r_{ij}\rangle, \qquad M^S_{ij} = 2\pi\alpha\langle\delta(\mathbf{r}_{ij})\rangle/(3m_{q_i} m_{q_j}), \tag{12.71}$$

where $\langle\ \rangle$ denotes an expectation value with respect to a spatial wave function. Because we do not know the spatial wave functions, we regard the C^S_{ij} and M^S_{ij} as parameters. We can see from their definitions that they are positive definite quantities. In terms of these parameters, the meson electromagnetic mass differences can be written:

$$\pi^+ - \pi^0 = \tfrac{1}{2}C^0_{uu} + \tfrac{3}{2}M^0_{uu}, \tag{12.72a}$$

$$K^0 - K^+ = m_d - m_u - \tfrac{1}{3}C^0_{us} - M^0_{us}, \tag{12.72b}$$

$$D^+ - D^0 = m_d - m_u + \tfrac{2}{3}C^0_{uc} + 2M^0_{cu}, \tag{12.72c}$$

$$\rho^+ - \rho^0 = \tfrac{1}{2}C^1_{uu} - \tfrac{1}{2}M^1_{uu}, \tag{12.72d}$$

$$K^{*0} - K^{*+} = m_d - m_u - \tfrac{1}{3}C^1_{us} + \tfrac{1}{3}M^1_{us}, \tag{12.72e}$$

$$D^{*+} - D^{*0} = m_d - m_u + \tfrac{2}{3}C^1_{uc} - \tfrac{2}{3}M^1_{uc}, \tag{12.72f}$$

where we have neglected the $m_d - m_u$ in M^S_{ij}, as this is a small correction to a term which is already small.

From Eq. (12.72a) we see that

$$\pi^+ > \pi^0, \tag{12.73}$$

in agreement with experiment. From Eq. (12.72b), and the experimental fact that $K^0 - K^+ = 4$ MeV, we see that

$$m_d - m_u > 4 \text{ MeV} \tag{12.74}$$

because the contribution from the interaction terms is negative. It then follows from Eq. (12.72c) that

$$D^+ - D^0 > 4 \text{ MeV}, \tag{12.75}$$

a value consistent with experiment.

Expressions for baryon electromagnetic mass differences analogous to those of Eq. (12.72) can be written down (Lichtenberg, 1976, 1977), but we shall not do so here. Only a little new information is obtained unless one makes further assumptions, for example that the expectation values $\langle 1/r_{ij} \rangle$ and $\langle \delta(\mathbf{r}_{ij}) \rangle$ are related by a symmetry such as $SU(6)$ or $SU(8)$.

12.9 Hadron Decays and Zweig's Rule

Let us consider the ω and φ mesons. Both of these mesons have the same quantum numbers, namely $J^{PC} = 1^{--}$, both belong to the same mixed 16-plet of $SU(4)$, and both are unstable, decaying principally into psuedoscalar mesons via strong interactions. However, as can be seen from Table 9.7, the ω has a larger decay width than the φ, although naively we would expect the φ to have the larger width because it is more massive. The puzzle becomes greater when we look at the partial decay width of the ω and φ into three pions. This partial decay width is more than an order of magnitude larger for the ω than for the φ.

We conclude that there is something in the dynamics that inhibits the decay of the φ into three pions. We do not fully understand the dynamical mechanism, but it can be characterized by a rule, known as Zweig's rule or the Okubo–Zweig–Iizuka rule (Okubo, 1963; Zweig, 1964; Iizuka, 1966), which applies to other processes as well.

Zweig's rule is most readily explained in terms of quark line diagrams. A quark line diagram of a hadron is a composite picture of the hadron in terms of its quark constituents, illustrated as lines in the picture. In

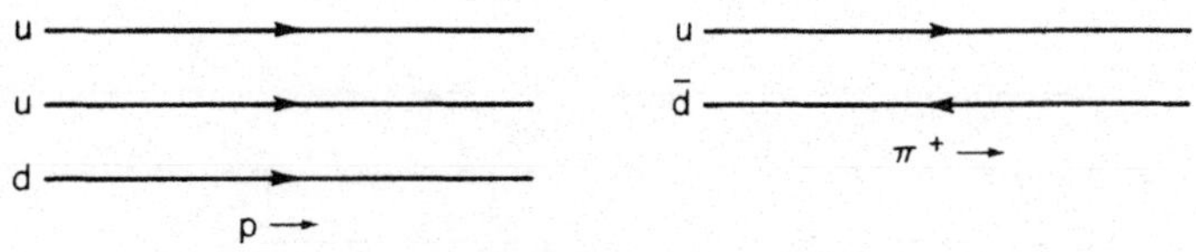

Fig. 12.1. Quark line diagrams of a proton and π^+ meson.

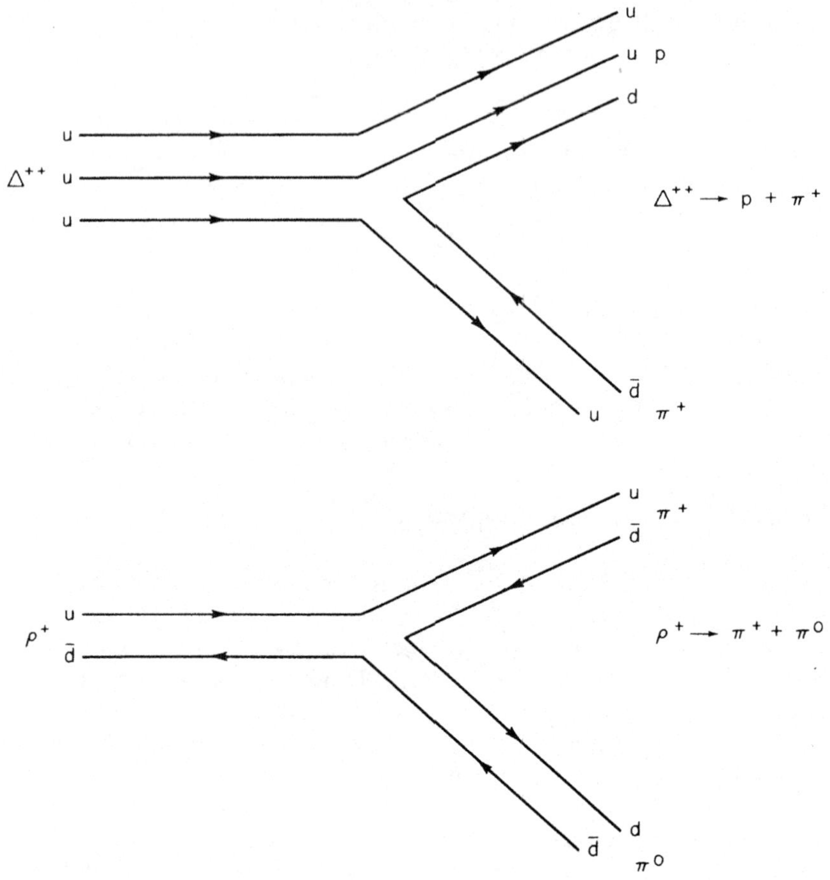

Fig. 12.2. Quark line diagrams illustrating the decays $\Delta^{++} \to p + \pi^+$ and $\rho^+ \to \pi^+ + \pi^0$. The π^0 wave function is a linear combination of $u\bar{u}$ and $d\bar{d}$ so another quark line diagram must also contribute to ρ decay.

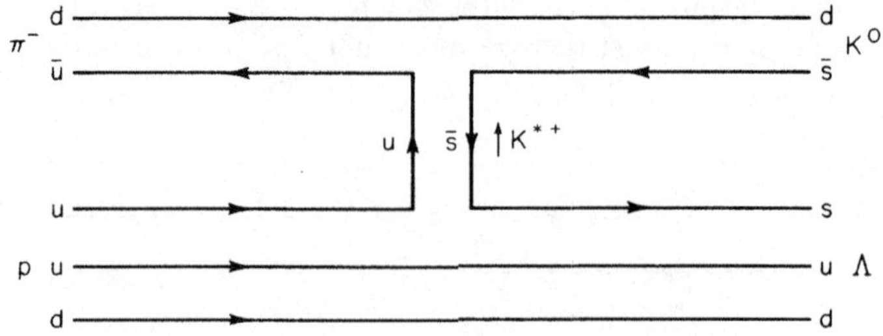

Fig. 12.3. Quark line diagram illustrating the reaction $\pi^- + p \to K^0 + \Lambda$.

Fig. 12.1 we illustrate quark line diagrams of a free proton and a free π^+ meson. Quark line diagrams illustrating the decays

$$\Delta^{++} \to p + \pi^+ \qquad \text{and} \qquad \rho^+ \to \pi^+ + \pi^0$$

are given in Fig. 12.2. In both these cases a quark–antiquark pair must be created. In Fig. 12.3 we show a quark line diagram illustrating the reaction

$$\pi^- + p \to K^0 + \Lambda.$$

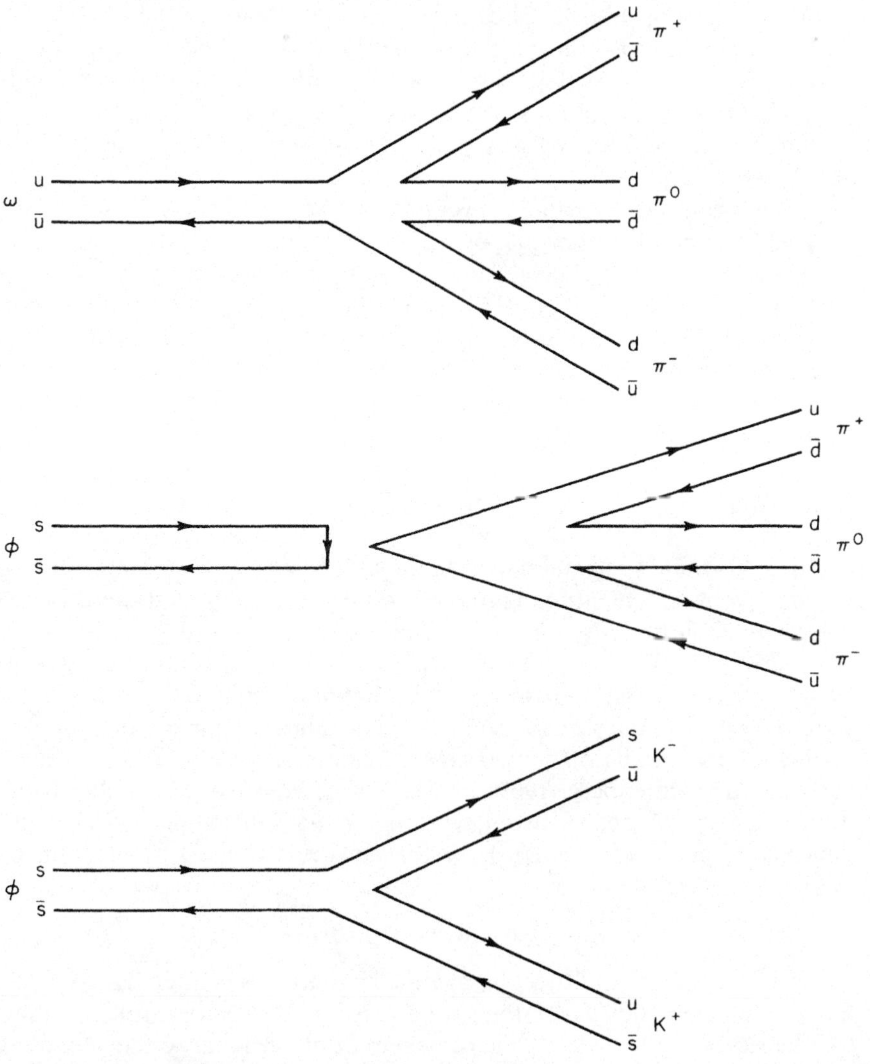

Fig. 12.4. Quark line diagrams showing that $\omega \to \pi^+ + \pi^0 + \pi^-$ and $\varphi \to K^+ + K^-$ are allowed by Zweig's rule, but that $\varphi \to \pi^+ + \pi^0 + \pi^-$ is inhibited.

The particular dynamical mechanism illustrated is the exchange of a K^* meson. (A K meson cannot be exchanged except weakly, because parity is violated if one pseudoscalar meson changes into two.)

A simplified statement of Zweig's rule is that processes which correspond to disconnected quark line diagrams are inhibited. A diagram is disconnected if one or more hadrons can be isolated by a line which does not cut any quark lines. Okubo postulated an essentially equivalent rule in terms of $SU(3)$ algebraic properties before the quark model was invented.

We illustrate in Fig. 12.4 how the decay $\omega \to 3\pi$ is allowed by Zweig's rule, but the decay $\varphi \to 3\pi$ is inhibited. This picture assumes that the ω and φ are ideally mixed, so that the ω contains only $u\bar{u}$ and $d\bar{d}$ quarks, while the φ contains only $s\bar{s}$ quarks. We have seen earlier in this chapter that the assumption of ideal mixing is a good one. In Fig. 12.4 we also illustrate the allowed decay $\varphi \to K^+ + K^-$.

Zweig's rule also explains the very narrow width of the ψ meson. The ψ is envisioned as a composite of a charmed quark–antiquark pair, and therefore, according to Zweig's rule, the ψ would like to decay into mesons containing charmed quarks. But because the ψ has a mass less than twice the mass of a charmed meson, energy conservation forbids such a decay. Thus, the only energetically possible decays of the ψ are into particles not containing charmed quarks. Hence, Zweig's rule is violated, and all the ψ decays are inhibited.

12.10 Diquarks and Exotic Hadrons

In Gell-Mann's original paper on the quark model (Gell-Mann, 1964a), he has a footnote mentioning the possibility that a bound state of two quarks, or diquark, might be stable. Since then, a number of authors (e.g., Ida and Kobayashi, 1966; Lichtenberg and Tassie, 1967; Ono, 1974; Capps, 1974) have considered in detail models containing diquarks. In particular, a model of excited baryon states in which one quark acquires orbital angular momentum, while the other two quarks remain as an S-wave diquark has been useful (Lichtenberg, 1969, 1975a). Let us consider this model within the framework of $SU(8)$. From two quarks, each belonging to an eight-dimensional multiplet of $SU(8)$, we can construct diquark states with the following multiplicities:

$$\mathbf{8} \otimes \mathbf{8} = \mathbf{36} \oplus \mathbf{28}. \tag{12.76}$$

The **36** diquark multiplet is symmetric under the interchange of the flavor and spin indices of the quarks, while the **28** multiplet is anti-symmetric. Now the lowest-lying baryon multiplet belongs to the **120**, which is symmetric. This means that in constructing the lowest-lying baryons from a quark and diquark, we need the symmetric **36** diquark rather than the antisymmetric **28**. Because in the model, excited baryon states arise from quark excitations, we keep the diquark in a **36** multiplet. Of course, any of the three quarks may be excited to a higher energy state. But, according to the model, if one quark is excited, the other two remain unexcited relative to each other. Also, the diquark may release a quark to be captured by the third quark, which then becomes an unexcited diquark. The $SU(4) \times SU(2)$ content of the **36** diquark is

$$\mathbf{36} \supset {}^{3}\mathbf{10} + {}^{1}\mathbf{6}. \tag{12.77}$$

The symmetric diquark is of course antisymmetric in color.

The relevant Clebsch–Gordan series for baryons in the model is

$$\mathbf{36} \otimes \mathbf{8} = \mathbf{120} \oplus \mathbf{168}. \tag{12.78}$$

Thus, according to the model, baryons should occur in $SU(8)$ multiplets of **120** and **168**, but not in multiplets of **56**, as the latter would arise from an antisymmetric diquark.

For classification of the known hadrons, it is much more convenient to consider $SU(6)$ rather than $SU(8)$, because so few charmed hadrons are known. The **120** and **168** of $SU(8)$ reduce to the **56** and **70**, respectively, while the antisymmetric **56** of $SU(8)$ reduces to the **20** of $SU(6)$. The $SU(6)$ Clebsch–Gordan series is

$$\mathbf{21} \otimes \mathbf{6} = \mathbf{56} \oplus \mathbf{70}. \tag{12.79}$$

Thus far, there is good evidence for the existence of baryons belonging to the **56** and **70** of $SU(6)$, but there is no strong evidence for any baryons belonging to the **20**, in agreement with the quark–diquark picture of baryons.

It is possible in principle to form a meson as a bound state of a diquark–antiquark pair (Carroll *et al.*, 1968). The relevant Clebsch–Gordan series in $SU(8)$ is

$$\mathbf{36} \otimes \overline{\mathbf{36}} = \mathbf{1} \oplus \mathbf{63} \oplus \mathbf{1232}. \tag{12.80}$$

Because in the ordinary quark model, mesons must be in a mixed **64** multiplet of $SU(8)$, the **1232** multiplet contains manifestly exotic mesons. However, even the **1** and **63** multiplets may be exotic in the following sense: If a meson is composed of a quark and antiquark, its parity and C-parity are given by Eqs. (12.23) and (12.24). If the meson has quantum numbers such that these relationships are not satisfied, the meson cannot be made of a $q\bar{q}$, and so is exotic. We have already seen in Section 12.5 that a meson with odd spin and parity but even C-parity must be exotic. Such a meson may be composed of a diquark–antiquark pair, however. For further discussion of exotic mesons made of diquark–antiquark pairs, see, for example, Rosenzweig (1976).

But mesons do not need to contain S-wave diquarks to be exotic. Another possibility is for a meson to contain two quark–antiquark pairs, with these pairs being relatively loosely bound to each other (Bander *et al.*, 1976). Likewise if an extra quark–antiquark pair is bound to the three quarks in a baryon, the baryon may be exotic.

It is here convenient to generalize the simple model of a baryon composed of three quarks and a meson composed of a quark–antiquark pair. In a relativistic model, quark–antiquark pairs can be created from the vacuum. This means that in a more realistic model, a baryon is composed of three quarks plus a "sea" of quark–antiquark pairs. However, only if this sea carries quantum numbers different from the vacuum is the baryon exotic. Similarly a meson contains a sea of quark–antiquark pairs. This expanded picture of baryons and mesons in the quark model has proved quite useful in a variety of applications, including high energy scattering. But, for the purposes of symmetry, unless hadrons are exotic, we may ignore the sea.

We shall not review the evidence for the existence of exotic hadrons. So far this evidence is not conclusive. But even if exotic hadrons exist, the simple quark model will remain valuable to describe a wide variety of properties of ordinary mesons and baryons.

12.11 Orbital Excitations

We have dwelt in considerable detail on the spin-$\frac{1}{2}$ and spin-$\frac{3}{2}$ baryons and on the pseudoscalar and vector mesons. Within the framework of the quark model, these are all states with total orbital angular momentum zero. Many of the mesons and baryons of higher mass have angular momentum which is sufficiently high that the quarks must be in states with orbital angular momentum greater than zero. We list in Table 12.12 the possible values of the total orbital angular momentum of the quarks in mesons and baryons with a given spin and parity. As we have noted, the group $SU(2n) \supset SU(n) \times SU(2)$ contains spin, but not orbital angular momentum. Therefore, to the extent that $SU(2n)$ is a good symmetry, the interactions between quarks should not couple spin and orbital angular momentum. This suggests that as a first approximation we should consider states of the direct-product group $SU(2n) \times O(3)$, where $O(3)$ is the orthogonal group in three dimensions and includes both proper rotations and reflections. The multiplets belonging to this group will be distinguished by quantum numbers of flavor and spin [$SU(2n)$] and by orbital angular momentum and parity [$O(3)$].

We denote such a multiplet by (N, L^P), where N is the $SU(2n)$ multiplicity, L is the orbital angular momentum, and P is the parity.

TABLE 12.12

POSSIBLE VALUES OF THE TOTAL ORBITAL ANGULAR MOMENTUM L OF QUARKS BOUND IN BARYONS OR QUARK–ANTIQUARK PAIRS BOUND IN MESONS[a]

Mesons		Baryons		Mesons		Baryons	
J^P	L	J^P	L	J^P	L	J^P	L
0^-	0	$\frac{1}{2}^+$	0, 1, 2,	2^-	2	$\frac{5}{2}^+$	1, 2, 3, 4
0^+	1	$\frac{1}{2}^-$	1, 2,	2^+	1, 3	$\frac{5}{2}^-$	1, 2, 3, 4
1^-	0, 2	$\frac{3}{2}^+$	0, 1, 2, 3	3^-	2, 4	$\frac{7}{2}^+$	2, 3, 4, 5
1^+	1	$\frac{3}{2}^-$	1, 2, 3	3^+	3	$\frac{7}{2}^-$	2, 3, 4, 5

[a] The spin and parity of a hadron is denoted by J^P.

Of course $SU(2n) \times O(3)$ is a badly broken symmetry. If the symmetry-breaking interaction between quarks includes a spin–orbit term, then a multiplet of $SU(2n) \times O(3)$ will split into a number of $SU(n)$ multiplets of different mass, each with a different value of total angular momentum J. These multiplets in turn will be split by an $SU(n)$-breaking interaction.

Because very few states containing charmed quarks have been seen, let us consider the case $n = 3$ (Greenberg, 1964; Dalitz, 1967, 1968, 1969).

For simplicity, we shall let that part of the quark–quark or quark–antiquark interaction which is invariant under $SU(6) \times O(3)$ be a harmonic oscillator potential. The energy levels of a harmonic oscillator are of course equally spaced and increase according to the number v of excitation quanta. In the meson case the relative motion of the quark–antiquark pair is described by a single three-dimensional harmonic oscillator. The ordering of the energy levels according to the number v of excitation quanta is

$$\begin{aligned} v &= 0, & L^P &= 0^-, \\ v &= 1, & L^P &= 1^+, \\ v &= 2, & L^P &= 2^-, \quad 0^-, \\ v &= 3, & L^P &= 3^+, \quad 1^+. \end{aligned} \tag{12.81}$$

Each of these levels should correspond to two multiplets of $SU(6) \times O(3)$, a $(35, L^P)$ and a $(1, L^P)$. We have already considered the ground state $L = 0^-$ multiplets $(35, 0^-)$ and $(1, 0^-)$. Since the quark–antiquark system can have spin 0 or 1, the spin–orbit interaction splits these levels into nonets with $J^{PC} = 0^{-+}$, 1^{--}. The charge-conjugation quantum number applies only to the $Y = I_3 = 0$ members of the nonets. Next consider the $v = 1$ multiplets with $L^P = 1^+$ which are split into the following nonets

$$J^{PC} = 0^{++}, \quad 1^{++}, \quad 2^{++}, \quad 1^{+-}. \tag{12.82}$$

The experimental evidence for the existence of these nonets is given in Tables 9.8 and 9.9. As a last example we consider the $v = 2$ multiplets with $L^P = 2^-, 0^-$. These split into the following nonets

$$J^{PC} = 1^{--}, \quad 2^{--}, \quad 3^{--}, \quad 2^{-+}, \quad 0^{-+}, \quad 1^{--}. \tag{12.83}$$

There is only fragmentary experimental evidence concerning the existence of these nonets.

In the case of the baryons, we consider three quarks interacting by means of harmonic oscillator potentials. The level ordering appears in a number of places, for example, in Dalitz (1969). We shall make the usual assumption that the three-quark wave function is totally symmetric under the interchange of all the coordinates except color of any two quarks. With this restriction on the wave functions, the ordering of the multiplets is

$$\begin{aligned}
v = 0&: \quad (56, 0^+), \\
v = 1&: \quad (70, 1^-), \\
v = 2&: \quad (56, 2^+), \quad (56, 0^+), \quad (70, 2^+), \quad (70, 0^+), \quad (20, 1^+).
\end{aligned} \tag{12.84}$$

A spin–orbit interaction breaks these multiplets into singlets, octets, and decuplets. In addition, $SU(3)$ mixing can occur. We shall not give any detailed comparison with experiment except to note that there is substantial evidence for the existence of a $(56, 0^+)$ multiplet with $v = 0$, a $(70, 1^-)$ with $v = 1$, and a $(56, 2^+)$ and $(56, 0^+)$ with $v = 2$. There is at present little evidence for the existence of the other multiplets of Eq. (12.84) with $v = 2$.

A number of authors have calculated decay rates of baryons within the framework of the harmonic-oscillator quark model. Agreement with experiment for decay via both pion and photon emission is rather good. See, for example, Faiman and Hendry (1969).

The quark–diquark model and the three-quark model give quite different predictions for the baryon excited states classified according to the representations of $SU(6) \times O(3)$. For example, according to the quark–diquark model, baryons should belong either to the **56** or **70** multiplets of $SU(6)$, whereas in the quark model **20** multiplets should exist as well. Also, in a quark–diquark model, the orbital angular momentum L and parity P are related by $P = (-1)^L$, whereas in a three-quark model $P = (-1)^{l+l'}$, where l and l' are two internal angular momenta. Thus, in the three-quark model one cannot predict the parity from the total orbital angular momentum.

However, Mitra (1967) has proposed a three-quark model with special restrictions on the forces. It then turns out that spin and parity are related by $P = (-1)^L$, and also that baryons must belong to 56- or 70-dimensional multiplets. It is not clear how closely Mitra's model is related to the quark–diquark model.

We have already noted that the multiplets of $SU(6) \times O(3)$ for which there is the best experimental evidence are the $(56, 0^+)$, $(70, 1^-)$, and $(56, 2^+)$. All these well-established multiplets are consistent with the predictions of the quark–diquark model.

Recently, a number of physicists (e.g., Appelquist and Politzer, 1975; Eichten *et al.*, 1975; De Rújula *et al.*, 1975; Wills *et al.*, 1977) have considered hadron spectra arising from nonabelian gauge field theory plus a phenomenological term to confine the quarks. If the theory is reduced to nonrelativistic form, a potential between two quarks (or a quark and an antiquark) is obtained which (except for the confining term) resembles in form the potential between an electron and a positron. Such a potential contains, in addition to a term which goes as $1/r$, a spin-dependent interaction, a spin–orbit interaction, and a tensor interaction. The confining term can be a harmonic oscillator potential, but is usually assumed to be linear in the quark–quark separation. The confining interaction is necessary, not only to give qualitative agreement with the observed hadron spectra, but also to prevent the copious appearance of free quarks (see, e.g., Jackson, 1977).

Potentials of this form are usually applied to mesons to avoid the complications of the three-body problem. However, baryons can be treated if the confining term is a harmonic oscillator potential and the remaining terms are treated in perturbation theory. Another approach is to use a quark–diquark model and treat the baryon spectra as a two-body problem.

12.12 High Energy Scattering

Lipkin and Scheck (1966) have shown that some features of hadron–hadron collisions at high energy can be correctly predicted in terms of the quark model with one important assumption. Briefly, these authors assumed that the hadron–hadron scattering amplitude in the forward direction is the sum of all possible two-body quark–quark or quark–antiquark scattering amplitudes. There is a theorem in quantum mechanics, known as the optical theorem, which states that the imaginary part of the forward scattering amplitude is proportional to the total cross section. Therefore, the assumption of Lipkin and Scheck applies to the total cross sections.

Since there are many more hadrons than quarks, it is possible to obtain sum rules relating various hadron–hadron total cross sections. For simplicity we restrict ourselves to hadrons not containing any c quarks. We let the total cross section for the collision of particles a and b be $\sigma(ab)$. If the quark–quark and quark–antiquark amplitudes are $SU(3)$-invariant, then, following Lipkin and Scheck, we obtain the following sum rules (among others):

$$\begin{aligned}
\sigma(pp) - \sigma(np) &= \sigma(K^+p) - \sigma(K^+n),\\
\sigma(pp) + \sigma(\bar{p}p) &= \tfrac{3}{2}[\sigma(\pi^+p) + \sigma(\pi^-p)]\\
&\quad + \tfrac{1}{2}[\sigma(K^+p) + \sigma(K^-p) - \sigma(K^+n) - \sigma(K^-n)],\\
\sigma(pp) + \sigma(\bar{p}p) &= 2[\sigma(\pi^+p) + \sigma(\pi^-p)] - \tfrac{1}{2}[\sigma(K^+p) + \sigma(K^-p)],\\
\sigma(\bar{p}p) - \sigma(\bar{p}n) &= \sigma(K^-p) - \sigma(K^-n),\\
\sigma(\Lambda p) - \sigma(pp) &= \sigma(K^-n) - \sigma(\pi^+p).
\end{aligned} \tag{12.85}$$

What is remarkable about these rules is that they relate meson–baryon to baryon–baryon cross sections, and thus cannot be obtained from considerations of symmetry (except for supersymmetry in which mesons and baryons belong in the same multiplet). Other sum rules involving only meson–baryon cross sections are derivable from the model, but they can be obtained from $SU(6)$-invariance without quarks. The agreement with experiment at high energy $E \gtrsim 10$ GeV is rather good. See Lipkin and Scheck (1966) for further details. If, in the limit of very high energy, all quark–quark and quark–antiquark cross sections are equal, then we obtain a very simple relation between the baryon–baryon and meson–baryon total cross sections:

$$\sigma(BB) = \tfrac{3}{2}\sigma(MB). \tag{12.86}$$

These ideas have been extended by many authors. For example, Franco (1967) considered the corrections that would arise from double scattering of quarks in hadrons, using the theory of Glauber (1959). Applying double-scattering corrections, De Souza *et al.* (1968) obtained sum rules which are quadratic and cubic rather than linear in the hadron–hadron cross sections. We do not have the space to discuss other ramifications of these ideas.

The simple model of Lipkin and Scheck ignores the "sea" of quark–antiquark pairs. The fact that the model gives qualitative agreement with experiment shows that the sea is not necessary to account for the gross features of high-energy hadron–hadron scattering.

In recent years, much emphasis has been placed on lepton–hadron scattering. The leptons, which are apparently structureless particles, are used as probes of hadron structure, especially in very inelastic collisions at high momentum transfer. Such highly inelastic collisions are often referred to as *deep-inelastic scattering*.

We shall not discuss these experiments and their interpretation in any detail. But we feel it is important to point out that the experiments can be interpreted as providing evidence that the nucleon (which is the usual target) contains small point-like constituents called *partons* (Feynman, 1969). Furthermore, there is evidence that at least some of these partons have half-integral spin and so might be quarks. At present, it seems simplest to regard the deep-inelastic scattering experiments as providing indirect evidence for the existence of quarks within the nucleon. Further information on deep-inelastic lepton scattering, and references to the original literature, are contained in reviews by a number of authors, of which we mention only the ones by Roy (1975) and Yan (1976).

Still another type of experiment that has yielded much indirect information about quarks is a high-energy electron–positron colliding beam. Electrons and positrons are coupled to all charged particles, including hadrons and quarks, through the electromagnetic interaction. Therefore colliding-beam experiments yield much direct information about hadrons and indirect information about quarks. Most of the information about the ψ family of mesons and about charmed mesons has come from colliding beams.

Of particular importance is the ratio R of hadron production to $\mu^+\mu^-$ pair production in e^+e^- annihilation. According to quantum electrodynamics, the cross section $\sigma_{\mu\mu}$ for the reaction

$$e^+ + e^- \to \mu^+ + \mu^-$$

at high energy is

$$\sigma_{\mu\mu} = 4\pi\alpha^2/(3s), \tag{12.87}$$

where α is the fine-structure constant and s is the square of the total energy in the center-of-mass system. Likewise, the cross section σ_{qq} for producing quark–antiquark pairs is

$$\sigma_{qq} = 4\pi\alpha^2 \sum_i Q_{q_i}^2/(3s), \tag{12.88}$$

where the sum is over all the quarks of masses appreciably smaller than half the beam energy. Although, according to this picture, quarks are produced in e^+e^- annihilation, they are not seen. It is assumed that they do not escape the interaction region as free particles, but instead convert into hadrons by means of the creation of additional quark–antiquark pairs. If this is the case, the cross section for hadron production σ_h is given simply by $\sigma_h = \sigma_{qq}$, and the ratio R is

$$R = \sum_i Q_{q_i}^2. \tag{12.89}$$

The importance of this equation is that if it is valid it can give an answer to the question: How many quarks are there below a given energy?

At an energy well below threshold for producing charmed quark–antiquark pairs, the ratio R is given by

$$R = 3(Q_u^2 + Q_d^2 + Q_s^2) = 3[(\tfrac{2}{3})^2 + (-\tfrac{1}{3})^2 + (-\tfrac{1}{3})^2] = 2. \qquad (12.90)$$

The factor 3 multiplying the squares of the quark charges arises from the fact that each quark comes in three colors. Experimentally, at energies below 2 GeV, the average value of R is about 2. This fact constitutes dynamical evidence that quarks have a color degree of freedom. Above the charm threshold, using $Q_c = \frac{2}{3}$, we obtain for R

$$R = \tfrac{10}{3}. \qquad (12.91)$$

Experimentally, however, R is about 5. One possible explanation of the fact that the experimental value of R is so high is that one or more heavy leptons are produced, which subsequently decay chiefly to hadrons.

We shall not pursue this topic further, but shall call attention to reviews by Feldman and Perl (1975, 1977).

12.13 A Fifth Quark?

In a recent experiment, Herb *et al.* (1977) and Innes *et al.* (1977) have observed two or three peaks in the mass spectrum of $\mu^+\mu^-$ pairs produced in proton–nucleus collisions. It is plausible that these peaks come from the decays of hitherto undiscovered hadrons. If so, these new hadrons, called Υ,Υ', and (if it exists) Υ'', constitute some evidence for the existence of a fifth quark. The reason for suspecting that we are seeing a new phenomenon is that the masses of the members of the Υ family are much larger than the masses of the members of the ψ family of particles. The ψ's have masses around 3 to 4 GeV, whereas the Υ masses are

$$\Upsilon = 9.40 \pm 0.013 \text{ GeV}. \quad \Upsilon' = 10.01 \pm 0.04 \text{ GeV}, \quad \Upsilon'' = 10.40 \pm 0.12 \text{ GeV},$$

respectively.

Just as the ψ provided indirect evidence for the existence of a c quark (although the ψ does not have charm itself), the Υ (although presumably not having any new quantum number itself) constitutes indirect evidence for the existence of a new b quark. Presumably, just as the ψ is ideally mixed, containing a $c\bar{c}$ pair, the Υ is also ideally mixed and is composed of a $b\bar{b}$ pair. The new quantum number is sometimes referred to as *beauty*.

According to our picture, the reason that both the ψ and Υ have prominent decay modes into $\mu^+\mu^-$ pairs is that their decays into hadrons are inhibited by Zweig's rule. Consequently the leptonic decays have a chance to compete. Thus, the ψ can be said to have an anomalously small decay width because it has *hidden* charm, and likewise the Υ width is small because it has hidden beauty. If this view is correct, we should expect to discover in the not-too-distant future hadrons with manifest beauty, analogous to the hadrons like the D and F mesons which have manifest charm.

We shall not discuss here the enlarged multiplets of mesons and baryons resulting from enlarging the symmetry from $SU(4)$ to $SU(5)$. Using the methods of this book, the reader should be able to do this for himself. Furthermore, we have no more reason to think that there are only five quark flavors than we had to think that there were only four. Fortunately, the special unitary groups have sufficient regularities that it is easy to generalize from $SU(n)$ to $SU(n + 1)$.

12.14 Where Are the Quarks?

Physicists have searched for quarks on land, in the sea, and in the air. Jones (1977) has given descriptions, with references, of many of the unsuccessful attempts to find quarks at accelerators, in seawater, in rocks, and in cosmic rays. Goldhaber and Smith (1975) have also reviewed the search for quarks and other hypothetical particles.

Recently, LaRue *et al.* (1977) claimed to have found some evidence for quarks. They observed residual charges of $\pm\frac{1}{3}$ (modulo unity in units of the charge on the proton) on two out of three super conducting niobium balls which had been heat-treated on a tungsten substrate. The superconducting balls were levitated in a magnetic field, and their residual electric charges were measured by means of forces on the balls arising from an alternating electric field.

The method of LaRue *et al.* (1977) is basically a variant of the oil-drop experiment in which Millikan first measured the charge on the electron. It is interesting that in Millikan's notes, he remarks on evidence for an oil drop with fractional charge. However, Millikan was unable to repeat this result and wisely chose to publish the correct value of the charge on the electron.

There is a puzzling feature of the experiment of LaRue *et al.* Five niobium balls which had been heat-treated on a niobium substrate showed no evidence for fractional charge, in contrast to the tungsten-treated balls. If we take both results at face value, we conclude that quarks somehow have an affinity for tungsten. If so, this would help explain why many previous searches for quarks turned up negative results. In particular, in a very sensitive experiment using iron levitated in a magnetic field, Gallinaro *et al.* (1977) found no evidence for fractionally charged matter.

But it is difficult to think of a mechanism by which quarks, created, say, by cosmic rays in various kinds of matter, could somehow migrate to tungsten atoms. In view of the puzzle about how quarks get on the tungsten, and in view of possible sources of error in the experiment, we regard the experiment of LaRue *et al.* as not conclusive. However, if quarks have indeed been seen, it should not be very long before physicists learn much about their properties.

If quarks have not been seen, why not? The number of correct predictions appears to be too large to be explained as a coincidence. And the incorrect predictions can all be explained away as resulting from oversimplifications rather than from the failure of the quark concept.

Here we briefly point out three possible explanations for the failure to see quarks.

1. Quarks might be very massive. Although they seem light when bound in hadrons, these light masses are merely effective masses bearing an unknown relation to the masses of free quarks.

2. There might be a large repulsive barrier in the interaction between quarks. If the height of this barrier goes to infinity, the quarks will be absolutely confined. As we have noted, it has been conjectured that certain nonabelian gauge field theories have properties such that quarks are absolutely confined. On the other hand, the quark confining interaction might not be absolute, but strong enough to make free quarks exceedingly rare.

3. There might be an unusual mechanism to confine quarks, as in the dyon model of Schwinger (1969). In this model, quarks have magnetic charge as well as electric charge. Dirac (1931, 1948) has shown that if magnetic charge exists, electric charge is quantized. If quark charges violate the quantization condition, presumably free quarks cannot exist.

4. Quarks might have integral charge (Han and Nambu, 1965). Once quarks have a color degree of freedom, as we have noted in Section 12.3, there is no need for them to have fractional charge. It is very possible that integrally charged quarks may have escaped detection, especially if they are unstable, as in a model of Pati and Salam (1973).

Independently of whether quarks are discovered, the quark model provides a very useful way to enable us to understand many properties of hadrons (which are still called elementary particles in spite of their complexity) in terms of simpler entities. But the number of quarks necessary to do the job keeps growing. Already there have been suggestions that the quarks themselves are composite structures of still more elementary particles. The capacity of physicists to speculate is very large. But despite all our speculations, no doubt nature will continue to surprise us.

REFERENCES

Abragam, A. (1961). "Principles of Nuclear Magnetism." Oxford Univ. Press (Clarendon), London and New York.

Adler, S. L., and Dashen, R. F. (1968). "Current Algebras." Benjamin, New York.

Aharonov, Y., and Susskind, L. (1967). *Phys. Rev.* **155**, 1428.

Amati, D., Bacry, H., Nuyts, J., and Prentki, J. (1964). *Phys. Lett.* **11**, 190.

Appelquist, T., and Politzer, H. D. (1975). *Phys. Rev. Lett.* **34**, 43.

Aubert, J. J., *et al.* (1974). *Phys. Rev. Lett.* **33**, 1404.

Augustin, J. E., *et al.* (1974). *Phys. Rev. Lett.* **33**, 1406.

Baird, G. E., and Biedenharn, L. C. (1963). *J. Math. Phys.* **4**, 1449.

Baird, G. E., and Biedenharn, L. C. (1964). "Proc. 1st Coral Gables Conf. on Symmetry Principles at High Energy," p. 58. Freeman, San Francisco.

Bander, M., Shaw, G. L., Thomas, P., and Meshkov, S. (1976). *Phys. Rev. Lett.* **36**, 695.

Bargmann, V., (1964). *J. Math. Phys.* **5**, 682.

Bég, M. A. B., Lee, B. W., and Pais, A. (1964). *Phys. Rev. Lett.* **13**, 514.

Behrends, R. E., Dreitlein, J., Fronsdal, C., and Lee, B. W. (1962). *Rev. Mod. Phys.* **34**, 1.

Bernstein, J. (1968). "Elementary Particles and Their Currents." Freeman, San Francisco.

Bernstein, J., Feinberg, G., and Lee, T. D. (1965). *Phys. Rev.* **139**, B1650.

Bjorken, J. D., and Glashow, S. L. (1964). *Phys. Lett.* **11**, 255.

Cabibbo, N. (1963). *Phys. Rev. Lett.* **10**, 531 (reprinted in Gell-Mann and Ne'eman, 1964).

Cabibbo, N., Maiani, L., and Preparata, G. (1967). *Phys. Lett.* **25B**, 132.

Capps, R. H. (1974). *Phys. Rev. Lett.* **33**, 1637.

Carroll, J., Lichtenberg, D. B., and Franklin, J. (1968). *Phys. Rev.* **174**, 1681.

Cartan, E. (1933). "Sur la Structure des Groupes de Transformations Finis et Continue." Thèse, Paris, 1894; 2nd ed., 1933. Vuibert, Paris.

Casella, R. C. (1969). *Phys. Rev. Lett.* **22**, 554.

Cazzoli, E. G., *et al.* (1975). *Phys. Rev. Lett.* **34**, 1125.

Celmaster, W. (1976). *Phys. Rev. Lett.* **37**, 1042.
Chew, G. F., Gell-Mann, M., and Rosenfeld, A. H. (1964). *Sci. Am.* **210**, No. 2, 74.
Christenson, J., Cronin, J., Fitch, V., and Turlay, R. (1964). *Phys. Rev. Lett.* **13**, 138.
Coleman, S., and Glashow, S. L. (1961). *Phys. Rev. Lett.* **6**, 423 (reprinted in Gell-Mann and Ne'eman, 1964).
Condon, E. U., and Shortley, G. H. (1935). "The Theory of Atomic Spectra." Cambridge Univ. Press, London and New York.
Cook, C. L., and Murtaza, G. (1965). *Nuovo Cimento* **39**, 531.
Dalitz, R. H. (1967). "Proc. 13th Intern. Conf. on High Energy Phys." Univ. of California Press, Berkeley.
Dalitz, R. H. (1968). *In* "Meson Spectroscopy" (C. Baltay and A. Rosenfeld, eds.), p. 497. Benjamin, New York.
Dalitz, R. H. (1969). *In* "Pion Nucleon Scattering" (G. Shaw and D. Wong, eds.). Wiley, New York.
De Boer, W. (1977). Summer Institute on Particle Physics, SLAC, Stanford University, Stanford, California.
De Rújula, A., Georgi, H., and Glashow, S. L. (1975). *Phys. Rev.* **12**, 147.
De Rújula, A., Georgi, H., and Glashow, S. L. (1977). *Phys. Rev. Lett.* **38**, 317.
De Souza, P. D., and Lichtenberg, D. B. (1967). *Phys. Rev.* **161**, 1513.
De Souza, P. D., Heinz, R. M., and Lichtenberg, D. B. (1968). *Phys. Rev.* **169**, 1185.
de Swart, J. J. (1963). *Rev. Mod. Phys.* **35**, 916 (reprinted in Gell-Mann and Ne'eman, 1964).
Dirac, P. A. M. (1931). *Proc. Roy. Soc. (London) Ser. A* **133**, 60.
Dirac, P. A. M. (1948). *Phys. Rev.* **74**, 817.
Edmonds, A. (1957). "Angular Momentum in Quantum Mechanics." Princeton Univ. Press, Princeton, New Jersey.
Edmonds, A. (1962). *Proc. Roy. Soc. (London) Ser. A* **268**, 567.
Eichten, E., *et al.* (1975). *Phys. Rev. Lett.* **34**, 369.
Faiman, D., and Hendry, A. W. (1969). *Phys. Rev.* **180**, 1572, 1609.
Federman, P., Rubinstein, H. R., and Talmi, I. (1966). *Phys. Lett.* **22**, 208.
Feld, B. T. (1969). "Models of Elementary Particles." Blaisdell, Waltham, Massachusetts.
Feldman, G. J. (1977). "Proc. Fifth Intern. Conf. on Meson Spectroscopy." Northeastern Univ., Boston, to be published.
Feldman, G. J., and Perl, M. L. (1975). *Phys. Reports* **C 19**, 233.
Feldman, G. J., and Perl, M. L. (1977). *Phys. Reports*, to be published.
Fermi, E., and Yang, C. N. (1949). *Phys. Rev.* **76**, 1739.
Ferrara, S. (1976). *Riv. Nuovo Cimento* **6**, 105.
Feynman, R. P. (1969). *Phys. Rev. Lett.* **23**, 1415.
Feynman, R. P., and Gell-Mann, M. (1958). *Phys. Rev.* **109**, 193.
Franco, V. (1967). *Phys. Rev. Lett.* **18**, 1159.
Franklin, J. (1968). *Phys. Rev.* **172**, 1807.
Franklin, J. (1969). *Phys. Rev.* **180**, 1583; **181**, 1984.
Franklin, J. (1975). *Phys. Rev. D* **12**, 2077.
Gaillard, M. K., Lee, B. W., and Rosner, J. L. (1975). *Rev. Mod. Phys.* **47**, 277.
Gal, A., and Scheck, F. (1967). *Nucl. Phys. B* **2**, 110.
Gallinaro, G., Marinelli, M., and Morpurgo, G. (1977). *Phys. Rev. Lett.* **38**, 1225.
Gasiorowicz, S. (1966). "Elementary Particle Physics." Wiley, New York.
Gell-Mann, M. (1953). *Phys. Rev.* **92**, 833.
Gell-Mann, M. (1961). Cal. Tech. Rept. CTSL-20 (reprinted in Gell-Mann and Ne'eman, 1964).
Gell-Mann, M. (1962). *Phys. Rev.* **125**, 1067.
Gell-Mann, M. (1964). *Physics* **1**, 63 (reprinted in Gell-Mann and Ne'eman, 1964).

Gell-Mann, M. (1964a). *Phys. Lett.* **8**, 214 (reprinted in Gell-Mann and Ne'eman, 1964).
Gell-Mann, M. and Ne'eman, Y. (1964). "The Eightfold Way." Benjamin, New York.
Georgi, H., and Glashow, S. L. (1974). *Phys. Rev. Lett.* **32**, 438.
Glashow, S. L., and Rosenfeld, A. H. (1963), *Phys. Rev. Lett.* **10**, 192 (reprinted in Gell-Mann and Ne'eman, 1964).
Glashow, S. L., and Socolow, R. H. (1966). *Phys. Rev. Lett.* **15**, 329.
Glashow, S. L., Iliopoulos, J., and Maiani, L. (1970). *Phys. Rev. D* **2**, 1285.
Glauber, R. J. (1959). *In* "Lectures in Theoretical Physics," Vol. 1 (W. E. Brittin *et al.* eds.), p. 315. Interscience, New York.
Goldberg, H., and Ne'eman, Y. (1963). *Nuovo Cimento* **27**, 1.
Goldberger, M. L., and Treiman, S. B. (1958). *Phys. Rev.* **110**, 1478.
Goldhaber, A. S., and Smith, J. (1975). *Rep. Prog. Phys.* **38**, 731.
Goldhaber, G., *et al.* (1976). *Phys. Rev. Lett.* **37**, 255.
Gottfried, K. (1966). "Quantum Mechanics," Vol. I. Benjamin, New York.
Greenberg, O. W. (1964). *Phys. Rev. Lett.* **13**, 598.
Greenberg, O. W., and Nelson, C. A. (1977). *Phys. Reports* **C 32**, 69.
Gross, D. J., and Wilczek, F. (1973). *Phys. Rev. Lett.* **30**, 1343.
Gürsey, F., and Radicati, L. A. (1964). *Phys. Rev. Lett.* **13**, 173.
Gürsey, F., Lee, T. D., and Nauenberg, M. (1964). *Phys. Rev.* **135**, B467.
Haacke, E. M., Moffat, J. W., and Savaria, P. (1976). *J. Math. Phys.* **17**, 2041.
Hamermesh, G., (1962). "Group Theory." Addison-Wesley, Reading, Massachusetts.
Han, M. Y., and Biedenharn, L. C. (1970). *Phys. Rev. Lett.* **24**, 118.
Han, M. Y., and Nambu, Y. (1965). *Phys. Rev.* **139**, B1006.
Hara, Y. (1964). *Phys. Rev.* **134**, B701.
Harari, H. (1976). "Proc. Les Houches Summer School," to be published.
Heisenberg, W. (1932). *Z. Phys.* **77**, 1.
Hendry, A. W., and Lichtenberg, D. B. (1975). *Phys. Rev. D* **12**, 2756.
Herb, S. W., *et al.* (1977). *Phys. Rev. Lett.* **39**, 252.
Herglotz, G. (1911). *Ann. Physik* **36**, 493.
Ida, M., and Kobayashi, R. (1966). *Prog. Theoret. Phys.* **36**, 846.
Iizuka, J. (1966). *Prog. Theoret. Phys.* Suppl. No. **37–38**, 21.
Ikeda, M., Ogawa, S., and Ohnuki, Y. (1959), *Prog. Theoret. Phys.* **22**, 715.
Innes, W. R., *et al.* (1977). *Phys. Rev. Lett.* **39**, 1240.
Iwao, S. (1965). *Ann. Physics* **35**, 1.
Jackson, J. D. (1977). "European Conf. on Particle Physics." Budapest, proceedings to be published.
Jacobi, C. G. J. (1884). "Vorlesungen Uber Dynamik," Werke, Supplementband. Reimer, Berlin.
Johnson, R. J. (1976). *Phys. Rev. D* **14**, 3220.
Johnson, R. J., and Shah-Jahan, M. (1977). *Phys. Rev. D* **15**, 1400.
Jones, L. W. (1977). *Rev. Mod. Phys.* **49**, 717.
Katayama, Y., *et al.* (1962) *Prog. Theoret. Phys.* **28**, 675.
Knapp, B., *et al.* (1976). *Phys. Rev. Lett.* **37**, 882.
Kokkedee, J. J. J. (1969). "The Quark Model." Benjamin, New York.
Konopinski, E. J. (1966). "The Theory of Beta Radioactivity." Clarendon Press, Oxford.
Kuo, T. K., and Yao, T. (1965). *Phys. Rev. Lett.* **14**, 79.
LaRue, G., Fairbank, W. M., and Hebard, A. F. (1977). *Phys. Rev. Lett.* **38**, 1011.
Lee, T. D., and Wu, C. S. (1966). *Ann. Rev. Nucl. Sci.* **16**, 511.
Lee, T. D., and Yang, C. N. (1956). *Phys. Rev.* **104**, 254.
Lichtenberg, D. B. (1965). "Meson and Baryon Spectroscopy." Springer, New York.
Lichtenberg, D. B. (1969). *Phys. Rev.* **178**, 2197.

Lichtenberg, D. B. (1975). *Nuovo Cimento Lett.* **13**, 346.
Lichtenberg, D. B. (1975a). *Nuovo Cimento* **28** A, 563.
Lichtenberg, D. B. (1975b). *Phys. Rev. D* **12**, 3760.
Lichtenberg, D. B. (1976). *Phys. Rev. D* **14**, 1412.
Lichtenberg, D. B. (1977). *Phys. Rev. D* **16**, 231.
Lichtenberg, D. B., and Tassie, L. J. (1967). *Phys. Rev.* **155**, 1601.
Lichtenberg, D. B., Tassie, L. J., and Kelemen, P. J. (1968). *Phys. Rev.* **167**, 1535.
Lie, S., and Scheffers, G. (1893). "Vorlesungen Uber Kontinuierliche Gruppen." Teubner, Leipzig.
Lipkin, H. (1965). "Lie Groups for Pedestrians." North-Holland, Amsterdam.
Lipkin, H., and Meshkov, S. (1965). *Phys. Rev. Lett.* **14**, 670.
Lipkin, H., and Scheck, F. (1966). *Phys. Rev. Lett.* **16**, 71.
Lubkin, G. B. (1977). *Phys. Today* **30**, No. 6, 17.
Maki, Z. (1964). *Prog. Theoret. Phys.* **31**, 331.
McGlinn, W. D. (1964). *Phys. Rev. Lett.* **12**, 467.
McNamee, P., and Chilton, F. (1964). *Rev. Mod. Phys.* **36**, 1005.
Marshak, R. E., Riazuddin, and Ryan, C. P. (1969). "Theory of Weak Interactions in Particle Physics." Wiley (Interscience), New York.
Meshkov, S., Levinson, C. A., and Lipkin, H. J. (1963). *Phys. Rev. Lett.* **10**, 631 (reprinted in Gell-Mann and Ne'eman, 1964).
Messiah, A. M. L., and Greenberg, O. W. (1964). *Phys. Rev.* **136**, B248.
Mitra, A. N. (1967). *Ann. Physics* **43**, 126.
Miyamoto, Y. (1966). *Prog. Theoret. Phys.* **35**, 175.
Moffat, J. W. (1965). *Phys. Rev.* **140**, B1681.
Moffat, J. W. (1975). *Phys. Rev. D* **12**, 288.
Morpurgo, G. (1965). *Physics* **2**, 95.
Morpurgo, G. (1968). "14th Intern. Conf. High Energy Physics," Vienna, p. 225. CERN, Geneva.
Ne'eman, Y. (1961). *Nucl. Phys.* **26**, 222 (reprinted in Gell-Mann and Ne'eman 1964).
Ne'eman, Y. (1964). "Proc. Intern. Conf. Nucleon Structure" (R. Hofstadter and L. I. Schiff, eds.), p. 172. Stanford Univ. Press (reprinted in Gell-Mann and Ne'eman, 1964).
Ne'eman, Y. (1967). "Algebraic Theory." Benjamin, New York.
Nishijima, K., and Nakano, T. (1953), *Prog. Theoret. Phys.* **10**, 581.
Oakes, R. J. (1963). *Phys. Rev.* **131**, 2239 (reprinted in Gell-Mann and Ne'eman, 1964).
Okubo, S. (1962). *Prog. Theoret. Phys.* **27**, 949 (reprinted in Gell-Mann and Ne'eman, 1964).
Okubo, S. (1963). *Phys. Lett.* **5**, 165.
Okubo, S. (1975). *Phys. Rev. D* **11**, 3261.
Ono, S. (1974). *Phys. Rev. D* **10**, 3124.
O'Raifeartaigh, L. (1968). *In* "Lectures in Theoretical Physics" (A. O. Barut and W. E. Brittin, eds.), p. 527. Gordan and Breach, New York.
Pais, A. (1966). *Rev. Mod. Phys.* **38**, 215.
Particle Data Group (1969). *Rev. Mod. Phys.* **41**, 109.
Particle Data Group (1976). *Rev. Mod. Phys.* **48**, S1.
Particle Data Group (1977). *Phys. Lett.* **68 B**, 1.
Pati, J. C., and Salam, A. (1973). *Phys. Rev. D* **8**, 1240.
Pati, J. C., and Salam, A. (1974). *Phys. Rev. D* **10**, 275.
Perl, M. L. (1974). "High Energy Hadron Physics." Wiley (Interscience), New York.
Perl, M. L. (1977). "Proc. Fifth International Conf. Experimental Meson Spectroscopy," Northeastern Univ., Boston, to be published.
Peruzzi, I., *et al.* (1976). *Phys. Rev. Lett.* **37**, 569.
Politzer, H. D. (1973). *Phys. Rev. Lett.* **30**, 1346.

Pontriagin, L. S. (1966). "Topological Groups." Gordon and Breach, New York.
Rabl, V., Campbell, G., and Wali, K. C. (1975). *J. Math. Phys.* **16**, 2494.
Racah, G. (1965). "Group Theory and Spectroscopy" (Springer Tracts in Modern Physics, **37**), p. 28. Springer, New York.
Rose, M. E. (1957). "Elementary Theory of Angular Momentum." Wiley, New York.
Rosen, S. P. (1963). *Phys. Rev. Lett.* **11**, 100.
Rosen, S. P. (1965). *Phys. Rev.* **140**, B326.
Rosenzweig, C. (1976). *Phys. Rev. Lett.* **36**, 697.
Roy, P. (1975). "Theory of Lepton–Hadron Processes at High Energies." Clarendon Press, Oxford.
Rubinstein, H. R., Scheck, F., and Socolow, R. H. (1967). *Phys. Rev.* **154**, 1608.
Sakata, S. (1956). *Prog. Theoret. Phys.* **16**, 686.
Sakita, B. (1964). *Phys. Rev.* **136**, B1756.
Sakurai, J. J. (1958). *Nuovo Cimento* **7**, 649.
Salam, A. (1968). *In* "Elementary Particle Theory" (N. Svartholm, ed.), p. 367. Almqvist and Wiksell, Stockholm.
Samios, N. P., Goldberg, M., and Meadows, B. T. (1974). *Rev. Mod. Phys.* **46**, 49.
Schiff, L. I. (1968). "Quantum Mechanics." McGraw-Hill, New York.
Schütz, J. R. (1897). *Gött. Nachr.* p. 110.
Schwinger, J. (1969). *Science* **165**, 757; **166**, 690.
Streater, R. F., and Wightman, A. S. (1964), "TCP, Spin and Statistics and All That." Benjamin, New York.
Sudarshan, E. C. G., and Marshak, R. E. (1958). *Phys. Rev.* **109**, 1860.
Sugawara, H. (1964). *Prog. Theoret. Phys.* **31**, 213.
Tarjanne, P., and Teplitz, V. L. (1963). *Phys. Rev. Lett.* **11**, 447.
Taylor, J. C. (1958). *Phys. Rev.* **110**, 1216.
Taylor, J. C. (1976). "Gauge Theories of Weak Interactions." Cambridge Univ. Press, Cambridge, England.
Van Dam H., and Wigner, E. P. (1965). *Phys. Rev.* **138**, B1576.
Van der Waerden, B. L. (1974). "Group Theory and Quantum Mechanics." Springer, New York.
Weinberg, S. (1967). *Phys. Rev. Lett.* **19**, 1264.
Weinberg, S. (1973). *Phys. Rev. Lett.* **31**, 494.
Weinberg, S. (1974). *Rev. Mod. Phys.* **46**, 255.
Wick, G. C., Wightman, A. S., and Wigner, E. P. (1952). *Phys. Rev.* **88**, 101.
Wigner, E. P. (1937). *Phys. Rev.* **51**, 106.
Wigner, E. P. (1954). *Prog. Theoret. Phys.* **11**, 437.
Wigner, E. P. (1959). "Group Theory." Academic Press, New York.
Wigner, E. P. (1964). *Phys. Today* **17**, No. 3, 34.
Wills, J. G., Lichtenberg, D. B., and Kiehl, J. T. (1977). *Phys. Rev. D* **15**, 3358.
Wolfenstein, L. (1964). *Phys. Rev. Lett.* **13**, 562.
Wolfenstein, L. (1977). *In* "Weak-Interaction Physics—1977" (D. B. Lichtenberg, ed.). American Institute of Physics, New York.
Wu, C. S., Ambler, E., Hayward, R. W., Hoppes, D. D., and Hudson, R. P. (1957). *Phys. Rev.* **105**, 1413.
Wu, T. T., and Yang, C. N. (1964). *Phys. Rev. Lett.* **13**, 380.
Yan, T.-M. (1976). *Ann. Rev. Nucl. Sci.* **26**, 199.
Yang, C. N., and Mills, R. L. (1954). *Phys. Rev.* **96**, 191.
Zweig, G. (1964). CERN preprint 8409/Th. 412, unpublished.
Zweig, G. (1965). *In* "Symmetries in Elementary Particle Physics" (A. Zichichi, ed.). Academic Press, New York.

INDEX

H

I

J

K

L

S

T

U

V

W

X

Y

Z

A B C 8 D 9 E 0 F 1 G 2 H 3 I 4 J 5

Printed in Dunstable, United Kingdom

91343010R00165